Differential Linear Matrix Inequalities

José C. Geromel

Differential Linear Matrix Inequalities

In Sampled-Data Systems Filtering and Control

Springer

José C. Geromel
School of Electrical and Computer
Engineering
University of Campinas – UNICAMP
Campinas, São Paulo, Brazil

ISBN 978-3-031-29753-3 ISBN 978-3-031-29754-0 (eBook)
https://doi.org/10.1007/978-3-031-29754-0

This Springer imprint is published by the registered company Springer Nature Switzerland AG
The registered company address is: Gewerbestrasse 11, 6330 Cham, Switzerland

Preface

This book has been written during the last two years.

On a day in the middle of March 2020, the COVID-19 pandemic arrived in Brazil and suddenly began to spread all over the country. My university, albeit located in the interior of São Paulo state, decided to put all non-essential activities in virtual form. During the one and a half years, I had to work differently than usual, at home, alone, without in-person contact with students and colleagues. During that period, it was necessary to learn how to give classes using digital blackboard, discuss virtually with students, and carry on the research activities.

At the very beginning, I believed that such a situation could not last a long time. However, after one or two months, viewing the increasing number of new infections and deaths, I realized that, contrary to my initial expectations, it would be wise and important to prepare a new way of life at home.

At that moment, I decided to propose to myself a big project—to write a new book. The idea was to *faire le point* about a new research subject that I have contributed a bit and seemed to be at the core of sampled-data control systems analysis and design. I am proud to have accomplished that goal and happy to present the final result. I really hope that it is received with interest by the international research community.

The subject is treated with mathematical rigor and, at the same time, thought to attract potential readers not only among colleagues but also their students. I have tried to keep the reading agreeable and fruitful. To this respect, the book contains, together with the theoretical developments, many solved illustrative examples and the formulation of some *open problems* that could be faced and hopefully solved by interested readers.

I have to confess that the work I have finished has been accomplished with some pain. If I compare with the previous tasks of writing a book that I did four times before—one with colleagues Patrizio Colaneri and Arturo Locatelli from Polytechnic Institute of Milan, Italy, and three with colleagues Alvaro G. B. Palhares, Rubens H. Korogui, and Grace S. Deaecto from University of Campinas, Brazil—this one has been much more difficult and complicated. The reason is simple and clear—this time I have worked alone all the time, without the possibility to share ideas, discuss, and be helped by someone in case of doubts and difficulties. There were no collective moments of joy due to new discoveries and new results.

This is a good and appropriate moment to thank many anonymous people involved in the publication process. Among them, I would like to thank Ms Letícia D. C. Motta for the final reading, for polishing and improving the English language of the manuscript, just before submission to the editor. Finally, I would like to thank the CNPq, National Council for Scientific and Technological Development, and FAPESP, São Paulo Research Foundation, for the financial support of my research activities throughout my professional career.

Campinas, Brazil
November, 2022

José C. Geromel

Brief Scope and Description

This book is entirely devoted to sampled-data control systems analysis and design from a new point of view, which has at its core a mathematical tool named *Differential Linear Matrix Inequality—DLMI*, a natural generalization of *Linear Matrix Inequality—LMI*, that had an important and deep impact on systems and control theory almost thirty years ago. It lasts until now. It is shown that the DLMI is well adapted to deal with the important class of sampled-data control systems in both theoretical and numerical contexts. All design conditions are expressed by convex programming problems, including when robustness against parameter uncertainty is assessed and imposed through state feedback control. Special attention is given to filter, dynamic output feedback, and model predictive control design, as well as nonlinear systems of Lur'e class and Markov jump linear systems.

The subject is treated with mathematical rigor, at the same time, trying to keep the reading agreeable and fruitful for colleagues and students. To this respect, the book contains together with the theoretical developments, many solved illustrative examples and the formulation of some *open problems* that could be faced and hopefully solved by interested readers.

Contents

About the Author

José C. Geromel received the BS and MS degrees in Electrical Engineering from the University of Campinas, UNICAMP, Brazil, in 1975 and 1976, respectively, and the Docteur d'État degree from the University Paul Sabatier, LAAS/CNRS, Toulouse, France, in 1979. He joined the School of Electrical and Computer Engineering at UNICAMP in 1975, where he is a full professor of Control Systems Analysis and Design. In 1987, he held a visiting professor position at Polytechnic Institute of Milan, Milan, Italy. He has been a member of the Editorial Board of the *International Journal of Robust and Nonlinear Control* and Associate Editor of *European Journal of Control and Nonlinear Analysis: Hybrid Systems*. He was awarded in 1994 and 2014 the Zeferino Vaz Award for his teaching and research activities at UNICAMP and, in 2007, the Scopus Award jointly awarded by Elsevier and CAPES, Brazil. Since 1991 he has been a Fellow 1-A of the Brazilian National Council for Scientific and Technological Development, CNPq. Since 1998 he has been a member of the Brazilian Academy of Science. In 1999, he was named Chevalier dans l' Ordre des Palmes Académiques by the French Minister of National Education. In 2010, he received the Docteur Honoris Causa degree from the University Paul Sabatier, Toulouse, France, and was named a member of the Brazilian Order of Scientific Merit by the President of the Federative Republic of Brazil. In 2011, he was named Distinguished Lecturer by the IEEE Control Systems Society. In 2018, he was promoted to Grã-Cruz of the Brazilian Order of Scientific Merit.

He is co-author of the books *Control Theory and Design* with P. Colaneri and A. Locatelli, Academic Press, 1997; Análise Linear de Sistemas Dinâmicos (in Portuguese), with A. G. B. Palhares, Edgard Blucher, 2004; Controle Linear de Sistemas Dinâmicos (in Portuguese), with R. H. Korogui, Edgard Blucher, 2011; and Análise Linear de Sinais (in Portuguese), with G. S. Deaecto, Edgard Blucher, 2019.

Chapter 1
Preliminaries

1.1 Introduction

The aim of this book is to study a new mathematical tool named differential linear matrix inequality (DLMI) which, as the denomination indicates, is a natural generalization of the well known linear matrix inequality (LMI) that has had an important impact in systems and control theory, since the publication of the seminal book [10], almost 30 years ago. With this goal in mind, this book focuses on the following main aspects:

- *The general context on which DLMI can be successfully applied*. For the time being, it can be said that the results presented in this book are restricted to time-invariant continuous-time systems. However, a wider class of linear, Markov jump linear, and nonlinear systems can be treated as well.
- *Numerical determination of a solution whenever it exists*. A simple numerical method of solution is presented and discussed. It is based on the determination of a valid conversion of DLMIs into LMIs, with pre-specified precision.
- *Control and filtering design problems involving sampled-data systems*. As it is clearly demonstrated in the sequel, sampled-data systems control and filtering design problems are well adapted to be handled in this mathematical framework. This is true because DLMIs represent, with particular simplicity, the effect of parameter uncertainty in exponential and similar matrix-valued mappings.

The first and the third items are related in the sense that sampled-data control systems have already been identified as a research area, where DLMIs can be applied with positive impact on both theoretical and numerical viewpoints. The second item is crucial, in order to make DLMIs really amenable to users dealing with control system analysis and design. We hope that this book contributes by bringing to light many other potential applications of DLMIs.

A DLMI has the following form:

J. C. Geromel, *Differential Linear Matrix Inequalities*,
https://doi.org/10.1007/978-3-031-29754-0_1

$$\mathcal{L}(\dot{P}(t), P(t)) < 0, \ t \in [0, h) \tag{1.1}$$

which, in general, is stated together with a boundary condition

$$\mathcal{H}(P_0, P_h) < 0 \tag{1.2}$$

where the elements appearing in both inequalities deserve the following considerations:

- The matrix differential inequality (1.1) must hold in the time interval $t \in [0, h)$ with $h > 0$ being a given scalar. In the context of sampled-data systems, the positive scalar h is denominated *sampling period*.
- The matrix-valued function $P(\cdot)$, of compatible dimensions, is the unknown variable to be determined. Other matrix variables, which are not important at this moment, may appear in inequalities (1.1) and (1.2).
- $\mathcal{L}$ and $\mathcal{H}$ are linear maps. More specifically, (1.2) is an LMI and, if $P = P(t)$ is constant in the entire time interval, then (1.1) reduces to an LMI as well.
- The linear matrix inequality (1.2) represents the boundary condition which, by consequence, depends on the values of $P(t)$ at the extreme points of the time interval under consideration, that is, $P_0 = P(0)$ and $P_h = P(h)$.

Finally, it is evident but it is important to stress that both inequalities are coupled together. It may occur that (1.1) admits a solution which is incompatible with the boundary condition (1.2), in which case the DLMI together with the boundary condition is unfeasible. The numerical method adopted for the solution, must be able to detect this situation and declare that the DLMI one wants to solve does not admit a solution.

1.2 Linear Matrix Inequality

In order to understand the importance of DLMIs, let us first recall the positive impact of LMIs to the effective solution of some control design problems. For a rather complete view on this matter, the reader is requested to see the Bibliography notes of this chapter. Consider a matrix $A \in \mathbb{R}^{n\times n}$ and a symmetric matrix variable $P = P' \in \mathbb{R}^{n\times n}$, such that

$$A'P + PA < 0 \,, \ P > 0 \tag{1.3}$$

which are feasible (with respect to P) if and only if matrix A is Hurwitz stable, that is, all its eigenvalues have negative real part. This is an immediate consequence of the Lyapunov Lemma. Hence, as it is clearly seen, the stability condition (1.3) is expressed by means of two LMIs. There is a strong relationship between the LMIs (1.3) and the Lyapunov equation

$$A'P + PA + Q = 0 \tag{1.4}$$

where $0 < Q = Q' \in \mathbb{R}^{n \times n}$ is some given matrix. It has a unique solution $P > 0$ if and only if matrix A is Hurwitz stable, which means that, under this assumption, both conditions are equivalent and all feasible solutions can be written as

$$P = \int_0^\infty e^{A't} Q e^{At} dt \tag{1.5}$$

In the general case, the feasibility of (1.3) implies that A is Hurwitz stable and the same conclusion can be drawn if, for a given $Q > 0$, a solution P to (1.4) exists and is positive definite. The numerical determination of a solution (if any) to the Lyapunov equation (1.4) is a well established procedure.

Now, let us consider a more demanding problem but in the same stability framework, whose origin is in the area of robust control. We want to know under which conditions a set of matrices of the form $A_\Delta = A + E\Delta C$ for all $||\Delta|| \leq \gamma^{-1}$ is composed by Hurwitz stable matrices, exclusively. This class of parameter uncertainty is called *norm bounded uncertainty*. The factorization

$$\begin{aligned} C'\Delta' E'P + PE\Delta C &= PE\Delta\Delta' E'P + C'C - \big(PE\Delta - C'\big)\big(PE\Delta - C'\big)' \\ &\leq \gamma^{-2} PEE'P + C'C \end{aligned} \tag{1.6}$$

is valid for all feasible perturbations of the class under consideration, which is characterized by $\Delta\Delta' \leq \gamma^{-2} I$ and $P = P'$. This result is important since it allows us to determine an upper bound to the Lyapunov inequality

$$\begin{aligned} A_\Delta' P + PA_\Delta &\leq A'P + PA + \gamma^{-2} PEE'P + C'C \\ &< 0 \end{aligned} \tag{1.7}$$

which, whenever satisfied for some $P = P' > 0$, ensures that A_Δ is Hurwitz stable for all feasible norm bounded uncertainty. Setting $\mu = \gamma^2$ and applying the Schur Complement, the inequalities

$$\begin{bmatrix} A'P + PA & PE & C' \\ \bullet & -\mu I & 0 \\ \bullet & \bullet & -I \end{bmatrix} < 0, \; P > 0 \tag{1.8}$$

are LMIs with respect to the variables (P, μ), which allows the determination of the minimum value of $\mu > 0$, by handling a jointly convex programming problem in only one step. Clearly, this is our interest, since the smallest μ produces the largest uncertainty domain. It is important to mention that, alternatively, under mild assumptions, the same solution can be calculated by searching the minimum $\gamma > 0$, such that the Riccati equation

$$A'P + PA + \gamma^{-2}PEE'P + C'C = 0 \tag{1.9}$$

still admits a positive definite stabilizing solution. Hence, the machinery we dispose of to solve Lyapunov and Riccati equations is good enough to deal with stability and robust stability concepts. In this context, numerical methods to cope with the LMIs (1.3) and (1.8) are not necessary, but, if one wants to go beyond this point, other aspects must be taken into account.

Indeed, let us move our attention to another class of parameter uncertainty called *convex bounded uncertainty*, whose origin is discussed in the Bibliography notes. As before, we want to determine the conditions under which the set of matrices of the form $A_\lambda = \sum_{i\in\mathbb{K}} \lambda_i A_i$, $\forall\lambda \in \Lambda$, is composed by Hurwitz stable matrices exclusively, where Λ is the unity simplex, a set composed by all non-negative vectors $\lambda \in \mathbb{R}^N$ with the sum of components equal to one and matrices $A_i \in \mathbb{R}^{n\times n}$ for all $i \in \mathbb{K} = \{1, \ldots, N\}$ are given. It is clear that A_λ is an element of the convex hull of the set of matrices $\{A_i\}_{i\in\mathbb{K}}$. Linearity indicates that the LMIs

$$A_i'P + PA_i < 0, \ P > 0 \tag{1.10}$$

whenever feasible for all $i \in \mathbb{K}$, imply that

$$A_\lambda'P + PA_\lambda < 0, \ P > 0 \tag{1.11}$$

for all $\lambda \in \Lambda$. This is an interesting and useful result because it determines a set of $N + 1$ LMIs that imposes Hurwitz stability to all matrices in the convex set $\text{co}\{A_i\}_{i\in\mathbb{K}}$. As a result, we can say that this is a genuine use of LMIs, since an alternative condition expressed, for instance, in terms of a Riccati-like equation, in principle, does not appear to exist.

1.3 Differential Linear Matrix Inequality

To make the importance of DLMIs clear, let us consider a problem that frequently arises in state feedback control design of sampled-data systems with a fixed sampling period $h > 0$. The discrete-time evolution of such a system at the sampling times $\{t_k = kh\}_{k\in\mathbb{N}}$ depends on a state feedback gain matrix $L \in \mathbb{R}^{m\times n}$, which makes the closed-loop matrix $A_L = A_d + B_d L$ with

$$A_d = e^{Ah}, \ B_d = \int_0^h e^{At} B dt \tag{1.12}$$

Schur stable, that is, all its eigenvalues have modulus less than one. From the discrete-time version of the Lyapunov Lemma, this property holds if and only if there exists a symmetric matrix variable $S = S' \in \mathbb{R}^{n\times n}$ such that

$$A_L' S A_L - S < 0, \; S > 0 \tag{1.13}$$

Clearly, the existence of a solution in terms of the pair of matrix variables (L, S) with S positive definite depends on the existence of a feedback gain such that A_L is Schur stable. Using LMIs, there is no difficulty to determine (if any) a feasible solution to (1.13). Actually, given the scalar $h > 0$ and matrices (A, B), the pair (A_d, B_d) is calculated from (1.12), and proceeding with the one-to-one change of variables $W = S^{-1} > 0$ and $Y = LS^{-1}$, Schur Complement calculations show that (1.13) is equivalent to

$$\begin{bmatrix} W & WA_d' + Y'B_d' \\ \bullet & W \end{bmatrix} > 0 \tag{1.14}$$

which makes the determination possible, whenever it exists, of a stabilizing feedback gain and the associated Lyapunov matrix by solving a simple LMI. The situation is much more complicated if one wants to take into account parameter uncertainty. This claim is true due to the intricate dependence of matrices A_d and B_d of A and B elements, as illustrated in the next example.

Example 1.1 (Exponential Mapping) *This is a simple example used to illustrate the previous discussion and to put in clear evidence the importance of DLMIs. Consider* $A_\Delta = A + E\Delta C$ *given by*

$$A = \begin{bmatrix} 0 & 1 \\ -1 & -1 \end{bmatrix}, \; E = \begin{bmatrix} 0 \\ 1 \end{bmatrix}, \; C = \begin{bmatrix} 1 & 0 \\ 0 & 1 \end{bmatrix}$$

and $\Delta = [\rho\cos(\theta) \; \rho\sin(\theta)]$ *with* $0 \leq \rho \leq 1$ *and* $0 \leq \theta \leq \pi/2$ *in order to guarantee that* $\Delta\Delta' \leq 1$. *The characteristic equation of matrix* A_Δ *has the form* $s^2 + a_1 s + a_0 = 0$ *with uncertain coefficients* (a_0, a_1) *belonging to the convex circular sector*

$$a_0 \leq 1, \; a_1 \leq 1, \; (a_0 - 1)^2 + (a_1 - 1)^2 \leq 1$$

Adopting $h = 7.5$, *the left hand side of Fig. 1.1 shows the exponential mapping* $e^{A_\Delta h}$ *viewed as a function of* Δ. *For each* Δ *satisfying* $||\Delta|| \leq 1$, *we have determined the characteristic equation of* $e^{A_\Delta h}$ *that has the form* $z^2 + \alpha_1 z + \alpha_0 = 0$. *As expected, the exponential mapping of the convex circular sector becomes a highly non-convex domain, which indicates that it does not preserve convexity.*

The situation is similar if we consider $A \in \mathrm{co}\{A_i\}_{i \in \mathbb{K}}$, *where* $\mathbb{K}$ *is defined by the following* $N = 3$ *extreme matrices:*

$$A_1 = \begin{bmatrix} 0 & 1 \\ 0 & -1 \end{bmatrix}, \; A_2 = \begin{bmatrix} 0 & 1 \\ -1 & 0 \end{bmatrix}, \; A_3 = \begin{bmatrix} 0 & 1 \\ -1 & -1 \end{bmatrix}$$

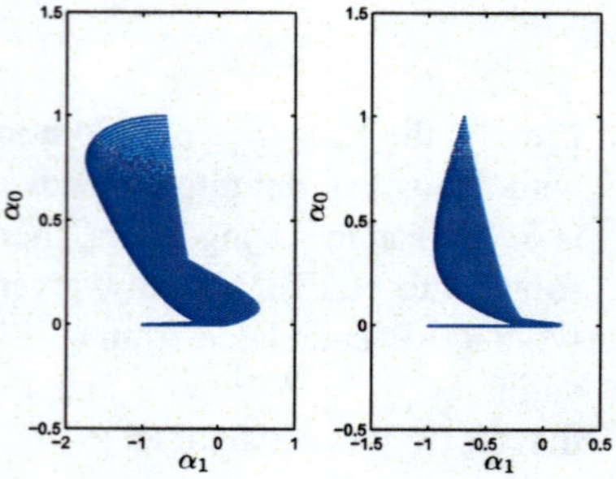

Fig. 1.1 Exponential mapping with convex domain

The characteristic equation of A_λ for $\lambda \in \Lambda$ is of the form $s^2 + a_1 s + a_0 = 0$, where the uncertain coefficients (a_0, a_1) belong to the convex domain

$$a_0 \leq 1, \ a_1 \leq 1, \ a_0 + a_1 \geq 1$$

Once again, adopting $h = 7.5$, for each $\lambda \in \Lambda$, we have determined the characteristic equation $z^2 + \alpha_1 z + \alpha_0 = 0$ of the exponential mapping $e^{A_\lambda h}$. The right hand side of Fig. 1.1 shows the image of this mapping, which exhibits a highly non-convex shape. Based on the results presented so far, the importance of DLMIs becomes evident, mainly in the context of sampled-data systems, to be deeply analyzed in this book. □

In order to overcome this difficulty, let us define the augmented square matrices

$$F = \begin{bmatrix} A & B \\ 0 & 0 \end{bmatrix}, \ H = \begin{bmatrix} I & 0 \\ L & 0 \end{bmatrix} \tag{1.15}$$

of compatible dimensions and state one of the theoretical results that puts in clear evidence the importance of DLMIs in the context of sampled-data control system design. Supposing that the DLMI

$$\dot{P}(t) + F'P(t) + P(t)F < 0, \ t \in [0, h) \tag{1.16}$$

together with the boundary conditions

$$P_h > 0, \ P_h > H'P_0H \tag{1.17}$$

holds for some symmetric matrix-valued function $P(t) : [0, h] \to \mathbb{R}^{(n+m)\times(n+m)}$, then matrix $A_L = A_d + B_d L$ is Schur stable. Actually, as it will be exhaustively studied in the next chapter, any feasible solution of the differential inequality (1.16) satisfies

$$P(t) > e^{F'(h-t)} P_h e^{F(h-t)} \tag{1.18}$$

for all $t \in [0, h)$, whose evaluation at $t = 0$ multiplied to the left by H' and to the right by H, and the boundary conditions (1.17) provide the inequality

$$\begin{aligned} P_h &> H' P_0 H \\ &> H' e^{F'h} P_h e^{Fh} H \end{aligned} \tag{1.19}$$

which from the fact that $P_h > 0$, the discrete-time version of the Lyapunov Lemma allows one to conclude that matrix $e^{Fh} H$ is Schur stable. Finally, simple algebraic manipulations yield

$$\begin{aligned} e^{Fh} H &= \begin{bmatrix} e^{Ah} & \int_0^h e^{At} B dt \\ 0 & I \end{bmatrix} \begin{bmatrix} I & 0 \\ L & 0 \end{bmatrix} \\ &= \begin{bmatrix} A_d + B_d L & 0 \\ L & 0 \end{bmatrix} \end{aligned} \tag{1.20}$$

which shows that matrix $A_L = A_d + B_d L$ is Schur stable as well. This result is somewhat surprising, because the matrices A_d and B_d viewed as functions of matrices A and B are highly nonlinear and non-convex. Even so, the DLMI (1.16) depends linearly on the same pair of matrices, which opens the possibility to use it successfully in robust control design. To this end, consider first the case of norm bounded uncertainty. We want to determine a matrix gain $L \in \mathbb{R}^{m \times n}$, such that the closed-loop matrix $A_L(\Delta) = A_d(\Delta) + B_d(\Delta) L$ remains Schur stable for all uncertainty satisfying the norm bound $||\Delta|| \le \gamma^{-1}$. It is assumed that $A_\Delta = A + E\Delta C$ and $B_\Delta = B + E\Delta D$ which yield $F_\Delta = F + J\Delta G$ with

$$J = \begin{bmatrix} E \\ 0 \end{bmatrix}, \ G = \begin{bmatrix} C & D \end{bmatrix} \tag{1.21}$$

and the pair $A_d(\Delta)$, $B_d(\Delta)$ is given by (1.12) with A and B replaced by A_Δ and B_Δ, respectively. Following the same steps of the previous section, it can be seen that the existence of a feasible solution to the DLMI

$$\begin{bmatrix} \dot{P}(t) + F'P(t) + P(t)F & P(t)J & G' \\ \bullet & -\mu I & 0 \\ \bullet & \bullet & -I \end{bmatrix} < 0, \ t \in [0, h) \tag{1.22}$$

where $\mu = \gamma^2$, together with the boundary conditions (1.17), solves our problem. Actually, the Schur Complement of the two last rows and columns of (1.22) shows that the same matrix-valued function $P(t)$ is also feasible to the Riccati inequality

$$\dot{P}(t) + F'P(t) + P(t)F + \gamma^{-2} P(t) J J' P(t) + G'G < 0, \ t \in [0, h) \tag{1.23}$$

and, by consequence, it is a solution to the DLMI

$$\dot{P}(t) + F_\Delta' P(t) + P(t)F_\Delta < 0, \ t \in [0, h) \tag{1.24}$$

for all Δ such that $||\Delta|| \leq \gamma^{-1}$, from which the conclusion that $A_L(\Delta)$ is Schur stable follows. Two remarks are necessary. The first one concerns the fact that, in general, a constant matrix-valued function $P = P(t)$ for all $t \in [0, h]$ is not feasible, which means that the possibility of working with a time-varying solution is essential to get the stabilizability conditions. On the other hand, the DLMI makes it possible to adopt all well known reasoning to cope with parameter uncertainty in the context of sampled-data control systems.

Now, let us move our attention to the class of convex bounded uncertainty. The parameter uncertainty is such that $(A, B) \in \text{co}\{(A_i, B_i)\}_{i \in \mathbb{K}}$. Our purpose is to determine a matrix gain $L \in \mathbb{R}^{m \times n}$, such that the closed-loop matrix $A_L(\lambda) = A_d(\lambda) + B_d(\lambda)L$ remains Schur stable for all $\lambda \in \Lambda$. Each pair (A_i, B_i) inserted in (1.16) provides the corresponding matrices F_i for each $i \in \mathbb{K}$. The set of DLMIs

$$\dot{P}(t) + F_i' P(t) + P(t)F_i < 0, \ t \in [0, h) \tag{1.25}$$

satisfied for all $i \in \mathbb{K}$ implies, due to linearity, that

$$\dot{P}(t) + F_\lambda' P(t) + P(t)F_\lambda < 0, \ t \in [0, h), \tag{1.26}$$

holds for all $\lambda \in \Lambda$. Together with the boundary conditions (1.17), this means that $A_L(\lambda)$ is Schur stable for all $\lambda \in \Lambda$ where $A_d(\lambda)$ and $B_d(\lambda)$ are given by (1.12) with A and B replaced by A_λ and B_λ, respectively.

In general lines, the same remarks are still valid. In particular, the linearity of the DLMI (1.16) with respect to matrix F and, by consequence, with respect to the pair (A, B) is a major issue that makes it possible to obtain the stabilizability results associated with sampled-data control systems subject to norm and convex bounded parameter uncertainty.

1.3.1 A Fundamental Lemma on Stabilizability

A very important point of concern is the conservativeness of the previous results. The next lemma formally states that, in the context of sampled-data systems free of parameter uncertainty, the conditions we have used are necessary and sufficient for asymptotic stability. This is a central result that has an enormous importance, because it appears, at least indirectly, on the basis of several procedures in sampled-data control system analysis and design.

Lemma 1.1 *There exists a gain matrix $L \in \mathbb{R}^{m \times n}$ such that the closed-loop system matrix $A_L = A_d + B_d L$ is Schur stable if and only if there exists a symmetric* *matrix-valued function* $P(t) : [0, h) \to \mathbb{R}^{(n+m)\times(n+m)}$ *that solves the DLMI* (1.16) *subject to the boundary conditions* (1.17).

Proof For the sufficiency, the existence of a solution to the DLMI (1.16) with final boundary condition $P(h) = P_h > 0$ implies that $P_0 > e^{F'h}P_h e^{Fh}$. Multiplying this inequality to the left by H' and to the right by its transpose and using the boundary condition (1.17), we obtain $P_h > (e^{Fh}H)'P_h(e^{Fh}H)$. Since $P_h > 0$, the conclusion is that the matrix $e^{Fh}H$ is Schur stable which, taking into account (1.20), implies that A_L is also Schur stable.

For the necessity, assuming that A_L is Schur stable for some $L \in \mathbb{R}^{m\times n}$, then, from (1.20), matrix $e^{Fh}H$ is also Schur stable and the same is true for He^{Fh}, because e^{Fh} is not singular and

$$
\begin{aligned}
0 &= \det\left(sI - e^{Fh}H\right) \\
&= \det\left(e^{-Fh}\right)\det\left(sI - e^{Fh}H\right)\det\left(e^{Fh}\right) \\
&= \det\left(sI - He^{Fh}\right)
\end{aligned}
\tag{1.27}
$$

As a consequence of the discrete-time version of the Lyapunov Lemma, there exists $P_0 > 0$ such that $P_0 > (He^{Fh})'P_0(He^{Fh})$ holds. Choosing $P_h > 0$ close enough to $H'P_0H$ such that $P_h > H'P_0H$, the boundary conditions (1.17) are reproduced and the inequality $P_0 > e^{F'h}P_h e^{Fh}$ is also verified. Hence, it remains to prove that there exists a feasible solution to the DLMI (1.16) such that $P(0) = P_0$ and $P(h) = P_h$. To this end, let us now define the following symmetric matrix-valued function $W(t) : [0, h] \to \mathbb{R}^{(n+m)\times(n+m)}$ as being

$$
W(t) = e^{-F't}W_0e^{-Ft} \tag{1.28}
$$

with

$$
W_0 = h^{-1}\left(P_0 - e^{F'h}P_h e^{Fh}\right) > 0 \tag{1.29}
$$

which, by construction, is positive definite in the whole time interval $[0, h]$ and consider the linear differential equation

$$
\dot{P}(t) + F'P(t) + P(t)F = -W(t), \ t \in [0, h) \tag{1.30}
$$

subject to the boundary condition $P(h) = P_h > 0$. Due to linearity, a solution to (1.30) always exists (see Theorem 2.1) and can be expressed as

$$
P(t) = e^{F'(h-t)}P_h e^{F(h-t)} + \int_t^h e^{F'(\xi-t)}W(\xi)e^{F(\xi-t)}d\xi \tag{1.31}
$$

which, evaluated at $t = 0$, yields

$$\begin{aligned} P(0) &= e^{F'h} P_h e^{Fh} + \int_0^h e^{F'\xi} W(\xi) e^{F\xi} d\xi \\ &= e^{F'h} P_h e^{Fh} + h W_0 \\ &= P_0 \end{aligned} \tag{1.32}$$

proving, thus, the necessity. The proof is complete. □

Notice that Lemma 1.1 is valid for sampled-data systems free of parameter uncertainty. As usual, whenever parameter uncertainty is present, the robust stability conditions, including the one presented before, are no more necessary. See the Bibliography notes of Chap. 2, where this limitation is explained by establishing a connection with the well known concept of quadratic stability.

It is clear that whenever the matrix gain $L \in \mathbb{R}^{n \times n}$ is given, then for stabilizability analysis, we have to solve the DLMI (1.16) subject to the boundary conditions (1.17), which are identical to (1.1) and (1.2), respectively. However, for control synthesis, the matrix gain must be involved in the calculations. Fortunately, by doing this, it is possible to show that the boundary conditions (1.17) remain LMIs with respect to all matrix variables involved.

The reader should keep in mind that there are many other aspects to be analyzed as far as DLMIs are concerned. We have given the one, which is, in our opinion, the central motivation that involves conditions for sampled-data systems robust control and filter design. Actually, from this starting point, nonlinear systems of Lur'e type, model predictive control, and Markov jump linear systems can be handled as well. Whenever possible, to increase readability, the theoretical results are illustrated by several numerical examples. In the next section, we provide an outline of each chapter to make clear the content and scope of the book.

1.4 Chapters' Outline

Excluding this chapter, the other ones have a common structure. A brief introduction provides the main points to be address and the most important results that deserve special attention of the reader. In the body of the chapter, citations are not included. The most relevant contributions are clearly indicated at the end, in a section denominated Bibliography notes, where details about existing results in the literature and related information are presented and quoted.

- **Chapter 2: Differential linear matrix inequalities**

This chapter introduces the mathematical structure of DLMIs, how to solve them, including some motivations for the specific and important class of sampled-data robust control, where they can successfully be applied. At the very beginning,

DLMIs and the associated boundary conditions are presented and a class of solutions of interest is defined.

The Lyapunov differential inequality (linear with respect to the unknown variable) is stated and conditions are given to ensure that any feasible solution is positive definite in the time interval under consideration. Moreover, it is shown that this DLMI has a minimal element, which is provided by the solution of the associated Lyapunov differential equation. In the same path, the Riccati differential inequality (quadratic with respect to the unknown variable) is analyzed, taking into account the additional difficulty that stems from the fact that, being nonlinear, a solution may not exist.

In the sequel, a simple and efficient numerical method to deal with DLMIs is presented. It is based on a necessary and sufficient condition that converts DLMIs into a set of equivalent LMIs, within a precision defined by the user. It is applied to the solution of two sampled-data control design problems, whose open-loop system is subject to norm and convex bounded parameter uncertainties. This chapter makes it clear the importance and usefulness of studying DLMIs.

- **Chapter 3: Sampled-data control systems**

This chapter provides the general structure of the sampled-data control system treated throughout the book. In our opinion, it is comprehensive enough because it retains the most important flavors of this class of dynamic linear systems. To increase readability, we repeat here its state space realization

$$\dot{x}(t) = Ax(t) + Bu(t) + Ew_c(t) \tag{1.33}$$

$$y[k] = C_y x[k] + E_y w_d[k] \tag{1.34}$$

$$z(t) = C_z x(t) + D_z u(t) \tag{1.35}$$

which evolves from the initial condition $x(0) = x_0$ at $t_0 = 0$, where $x(\cdot) : \mathbb{R}_+ \to \mathbb{R}^{n_x}$, $u(\cdot) : \mathbb{R}_+ \to \mathbb{R}^{n_u}$, $w_c(\cdot) : \mathbb{R}_+ \to \mathbb{R}^{r_c}$, and $z(\cdot) : \mathbb{R}_+ \to \mathbb{R}^{n_z}$ are the state, the control, the exogenous input, and the controlled output of the continuous-time process, respectively. The measurement $y[\cdot] : \mathbb{N} \to \mathbb{R}^{n_y}$ is available in discrete-time at instants t_k for all $k \in \mathbb{N}$ and is corrupted by the exogenous discrete-time sensor perturbation $w_d[\cdot] : \mathbb{N} \to \mathbb{R}^{r_d}$. Often, but not always, the control signal has the form

$$u(t) = u[k] = v[k] + E_u w_d[k], \ t \in [t_k, t_{k+1}), \forall k \in \mathbb{N} \tag{1.36}$$

in which the sequence of evenly spaced sampling instants $\{t_k\}_{k\in\mathbb{N}}$ is such that $t_0 = 0$, $t_{k+1} - t_k = h > 0, \forall k \in \mathbb{N}$. It is assumed that the discrete-time control signal $v[k]$ to be implemented is corrupted by the exogenous actuator perturbation $w_d[\cdot] : \mathbb{N} \to \mathbb{R}^{r_d}$. In some instances, in particular, for full order dynamic output feedback control design, another possibility is the continuous-time control law

$$u(t) = v(t) + E_u w_c(t), \ t \in \mathbb{R}_+ \tag{1.37}$$

which, as before, uses the information provided by the measurement $y[k]$ available at the sampling instants t_k for all $k \in \mathbb{N}$. It is corrupted by the continuous-time actuator perturbation $w_c(t)$. Next, the stability of LTI continuous- and discrete-time systems is analyzed by means of well known quadratic Lyapunov functions, namely

$$v_c(x(t)) = x(t)'Px(t), \quad v_d(x[k]) = x[k]'Sx[k] \tag{1.38}$$

Moreover, it is established a condition under which the equality $S = P > 0$ holds. The consequences of that to robust stability due to parameter bounded uncertainty are discussed as well.

The performance analysis begins by the introduction of the *equivalent system* that plays an important role from now on. Roughly speaking, a continuous-time LTI system and a discrete-time LTI system are equivalent if, for the same but arbitrary piecewise constant input and the same but arbitrary initial condition, both outputs have identical norms. Based on this concept, the $\mathcal{H}_2$ performance index is defined and calculated. The design of state feedback and dynamic full order output feedback optimal controllers are addressed. The derived problems are convex and are expressed by LMIs. Unfortunately, for the $\mathcal{H}_\infty$ performance index, optimality could not be established through problems expressed by LMIs. This fact increases the importance of DLMIs since, with this new mathematical tool, not only optimality of both performance indices is preserved through convex problems but also robust control to cope with parameter uncertainty can be addressed with no further additional difficulty.

- **Chapter 4: $\mathcal{H}_2$ filtering and control**

This chapter is entirely devoted to introducing the $\mathcal{H}_2$ performance index. From the very beginning, this kind of performance, fully addressed in the precedent chapter, is considered. The performance analysis is done from classical arguments followed by the adoption of a hybrid linear system approach, based on the general state space realization

$$\dot{\psi}(t) = F\psi(t) + J_c w_c(t) \tag{1.39}$$

$$z(t) = G\psi(t) \tag{1.40}$$

$$\psi(t_k) = H\psi(t_k^-) + J_d w_d[k] \tag{1.41}$$

where $t \in [t_k, t_{k+1})$ and $k \in \mathbb{N}$. Between successive jumps that occur evenly at instants $\{t_k\}_{k\in\mathbb{N}}$, the system evolves as an LTI system, whose behavior is well known. In addition, the evaluation of the $\mathcal{H}_2$ performance index is simple to be accomplished by taking into account those properties together with the Hamilton–Jacobi–Bellman inequality and Bellman's Principle of Optimality. The main goal is to express all design conditions through convex programming problems.

The sampled-data state feedback control design is tackled in a general setting. Several aspects and different but equivalent design conditions are presented.

Optimality is characterized and robustness with respect to parameter uncertainty is analyzed. Whenever possible, the equivalent system supports the design of full order output feedback controllers.

A filter of the following class

$$\dot{\hat{x}}(t) = \hat{A}_c(t)\hat{x}(t) \tag{1.42}$$

$$\hat{x}(t_k) = \hat{A}_d\hat{x}(t_k^-) + \hat{B}_d y[k] \tag{1.43}$$

$$\hat{z}_e(t) = \hat{C}_c(t)\hat{x}(t) \tag{1.44}$$

is considered. It is general enough and is inspired by the literature. It provides, in continuous-time, the estimation $z_e(t), t \in \mathbb{R}_+$ from the available discrete-time measurement $y[k], k \in \mathbb{N}$. Optimality and robustness against parameter uncertainty are characterized by means of LMIs and DLMIs. The class of full order output feedback controllers has the following state space realization:

$$\dot{\hat{x}}(t) = \hat{A}_c(t)\hat{x}(t) \tag{1.45}$$

$$\hat{x}(t_k) = \hat{A}_d\hat{x}(t_k^-) + \hat{B}_d y[k] \tag{1.46}$$

$$u(t) = \hat{C}_c(t)\hat{x}(t) + E_u w_c(t) \tag{1.47}$$

As in the context of LTI systems, optimality is characterized through convex problems, but, unfortunately, the same does not hold when facing parameter uncertainty.

- **Chapter 5: $\mathcal{H}_\infty$ filtering and control**

This chapter is entirely devoted to analyze and optimize $\mathcal{H}_\infty$ performance. From the very beginning, it is made clear that the problems to be dealt with cannot be converted to the discrete-time domain exactly, due to the continuous-time nature of the exogenous perturbation $w_c \in \mathcal{L}_2$. On the contrary of what has been done in the case of $\mathcal{H}_2$ performance, the equivalent system still can be calculated, but at the expense of introducing some degree of conservatism, unavoidably. Hence, due to this fact, this chapter is a good example of the usefulness of differential linear matrix inequalities. As before, in the context of state feedback control and linear filtering, parameter uncertainty is naturally handled, with no additional difficulty.

The first and main theorem that provides the theoretical basis for all results presented in the chapter is expressed in terms of a quadratic differential inequality subject to a boundary condition. In the analysis framework, that is, for $\mathcal{H}_\infty$ performance calculation, the aforementioned theoretical result is equivalently expressed through a DLMI subject to an LMI boundary condition. In general lines, the same occurs in state feedback and filter with state space realization (1.42)–(1.44) design cases since joint convexity with respect to all involved matrix variables of the respective design problem has been retained.

The situation is much more difficult for the determination of the stationary, that is, the optimal time-invariant full order dynamic output feedback controller with state space realization

$$\dot{\hat{x}}(t) = \hat{A}_c \hat{x}(t) \tag{1.48}$$

$$\hat{x}(t_k) = \hat{A}_d \hat{x}(t_k^-) + \hat{B}_d y[k] \tag{1.49}$$

$$u(t) = \hat{C}_c \hat{x}(t) + E_u w_c(t) \tag{1.50}$$

The key point is to prove that the scalar-valued function $Z_0(\mu) = \bar{Z} > 0$ where $\bar{Z}$ is the stabilizing solution to the algebraic Riccati equation

$$\begin{aligned} A'\bar{Z} + \bar{Z}A - \big(D_z'C_z + B'\bar{Z}\big)'(D_z'D_z)^{-1}\big(D_z'C_z + B'\bar{Z}\big) + \\ + \mu^{-1}\bar{Z}EE'\bar{Z} + C_z'C_z = 0 \end{aligned} \tag{1.51}$$

evaluated for each possible value of $\mu > 0$ can be replaced, without introducing any kind of conservatism, by an appropriate LMI condition. This has successfully been done preserving, thus, the joint convexity of the control design problem under consideration. This chapter is concluded by the design of a pure discrete-time optimal controller that provides the control signal $u[k]$ based on the output measurement variable $y[k]$, $k \in \mathbb{N}$.

- **Chapter 6: Markov jump linear systems**

In this chapter, Markov jump linear systems are tackled with the main goal of providing sampled-data filter and control design conditions, expressed through DLMIs and LMIs, as a natural generalization of those presented in the previous chapters. At the very beginning, the classical MJLS model is presented and it is given particular attention to the stochastic characteristics of signals belonging to continuous and discrete-time domains and to the Markov chain by introducing the *transition rate matrix* and signal norms. The general mode dependent sampled-data hybrid MJLS is introduced and it is particularized whenever necessary, in order to treat specific problems related to sampled-data state feedback, filtering and dynamic output feedback control design.

The $\mathcal{H}_2$ performance index is defined and it is shown that it can be calculated from a convex programming problem expressed through DLMIs and LMIs. In fact, the feasible solutions to the set of DLMIs are proven to exhibit the same useful positivity property already observed in the context of sampled-data control of LTI systems. It is important to make it clear that it is assumed throughout this chapter that the Markov mode $\theta(t)$ is measured in all time instants $t \in \mathbb{R}_+$, being, thus, part of the filter or of the feedback control law. In this case, they are denominated *mode dependent*. In general, since mode independence can be enforced only at the expense of some conservatism, this aspect is not pursued in this book. The results are similar but not identical to the ones presented in Chaps. 4 and 5 concerning sampled-data LTI systems. Even though the optimal time-invariant filter, state, and output feedback controllers have been calculated by handling convex programming problems. The major difficulty is the existence of coupling terms on the right hand side of the DLMIs

$$\dot{P}_i(t) + F_i' P_i(t) + P_i(t) F_i + G_i' G_i < -\sum_{j \in \mathbb{K}} \lambda_{ij} P_j(t), \ t \in [0, h) \tag{1.52}$$

where $i \in \mathbb{K}$ are the Markov modes. Indeed, if we need to express it in terms of the inverses $Q_i(t) = P_i(t)^{-1}$, $i \in \mathbb{K}$, then the coupling terms become nonlinear and the result can be equivalently rewritten as a set of coupled DLMIs, but with much higher dimension. In some instances, the nonlinear coupling terms prevent the design conditions from being expressed by DLMIs. This is the case of pure discrete-time controllers, whose design, for this reason, remains an open problem deserving further attention.

In the sequel, the $\mathcal{H}_\infty$ performance index is tackled and the same facilities and difficulties are made explicit. The formulas one has to cope with are a bit more complicated, but the main difficulty stems from the presence of the coupling terms. One again, the design conditions that characterize the optimal time-invariant filter, state, and output feedback controllers are expressed through convex programming problems. The important case of pure discrete-time controllers design remains an open problem where it is necessary to find if one exists, of a new and suitable one-to-one change of variables promoting linearity.

- **Chapter 7: Nonlinear system control**

This chapter is entirely devoted to study a class of nonlinear systems known as Lur'e systems. They are characterized as being the feedback connection of an LTI system and a nonlinear function $\phi(\cdot) \in \Phi$ with the components lying in a prescribed sector that defines the set Φ. Hence, it is clear that Lur'e systems have a particular structure that simplifies analysis and control synthesis, but it is general enough to keep intact its nonlinear nature. Many systems in practice are of Lur'e class, which increases the importance of developing control design conditions specially devoted to them. For these reasons, the synthesis of sampled-data control is the main theme of this chapter. The model and the control structure are as follows:

$$\dot{x}(t) = Ax(t) + Bu(t) + Eq(t) \tag{1.53}$$

$$p(t) = C_p x(t) \tag{1.54}$$

$$z(t) = C_z x(t) + D_z u(t) \tag{1.55}$$

$$u(t) = L_x x[k] + L_u u[k-1], \ t \in [t_k, t_{k+1}), \ \forall k \in \mathbb{N} \tag{1.56}$$

where $q(t) = -\phi(p(t))$. We tackle the problem consisting in the determination of the state feedback matrix gain $L = [L_x \ L_u]$ of compatible dimensions such that the closed-loop system (1.53)–(1.56) starting from the initial condition $x(0) = x_0$ solves

$$\inf_{L} \sup_{\phi \in \Phi} \int_0^\infty z(t)'z(t)dt \tag{1.57}$$

It has been possible to prove that the control memory is useless, as far as the optimality of the guaranteed cost (1.57) is concerned. As expected, whenever one deals with nonlinear systems, the design conditions are convex and expressed through DLMIs and LMIs, but they are only sufficient. This implies that only an upper bound, the guaranteed cost, of the objective function of (1.57) is minimized. We are pleased to say, and we hope the reader appreciates it, that the material of this chapter can be viewed as a generalization of the ideas behind the celebrated Popov criterion to the design of sampled-data state feedback control of Lur'e systems. As a by-product, the results reported here are applied to the class of hybrid Lur'e systems as well. Unlike the previous chapters, this one does not treat filter and dynamic output feedback controller design. This is left as a task to be accomplished in a more general and adequate setting, after the stated open problem is adequately addressed and solved.

- **Chapter 8: Model predictive control**

This chapter deals with Model Predictive Control (MPC) in the framework of sampled-data state feedback control design. The very beginning is devoted to providing general information about MPC, including main characteristics, properties, and difficulties to face. Particular attention is given to resolve the issues of optimality, stability, and robustness whenever the plant is subject to parameter uncertainty of norm and convex bounded classes. Perhaps the most challenging issue of the MPC design is the necessity to cope with constraints that force the controlled output to remain in some pre-established convex region.

All efforts go toward the sampled-data control design of the following plant with state space realization:

$$\dot{x}(t) = Ax(t) + Bu(t) + Ew_c(t), \ x(0) = 0 \tag{1.58}$$

$$s[k] = C_s x[k] + D_s u[k] \tag{1.59}$$

$$z(t) = C_z x(t) + D_z u(t) \tag{1.60}$$

$$u(t) = v[k] + E_u w_d[k], \ t \in [t_k, t_{k+1}), \ \forall k \in \mathbb{N} \tag{1.61}$$

where, in principle, the control law with general form

$$v[k] = \varphi(x[k]) \tag{1.62}$$

must be designed such as the closed-loop system is globally asymptotically stable and the worst case (due to the presence of uncertainty) $\mathcal{H}_2$ performance index is optimized. The main feature is how to manipulate the hard constraint $s[k] \in \mathbb{S}, \forall k \in \mathbb{N}$, where $\mathbb{S}$ is a convex set with nonempty interior. In the first step, the guaranteed cost is defined and the optimality conditions are derived for the control design problem

$$\inf_{\varphi(\cdot)} \sup_{(A,B)\in\mathcal{A}} \sum_{i=1}^{r_c+r_d} \int_0^\infty z_i(t)'z_i(t)dt \tag{1.63}$$

where $\mathcal{A}$ is the uncertainty set and the sum is performed in order to catch the effect of impulses injected in each channel of the continuous-time w_c and discrete-time w_d exogenous perturbations. In a second step, the output controlled constrained $s[k] \in \mathbb{S}, \forall k \in \mathbb{N}$, is addressed. At this stage, inevitably, due to the complexity of the problem, some conservatism is included on the design conditions. This aspect is evaluated by solving some academic examples.

At the final section, an MPC strategy which exhibits all properties already mentioned is designed and its possible numerical implementation is discussed. It has the form $u[k] = u_{k|k} = \varphi_{(k)}(x_{k|k})$, where $\varphi_{(k)}(\cdot)$ is a linear function of the measured state at each update time $\{t_k = kh\}_{k\in\mathbb{N}}$. The possibility to implement MPC with an updating strategy multiple of the sampling time is assessed. Examples are solved and compared with other proposals available in the literature.

- **Chapter 9: Numerical experiments**

In this chapter we handle two dynamical systems with practical appeal. The objectives are twofold: first, to design state feedback, dynamic output feedback, and model predictive control by using the design conditions presented in the precedent chapters of this book, and second, to evaluate the computational effort needed to tackle numerically the associated convex programming problems.

The first numerical experiment focuses on the control of an inverted pendulum in order to bring the pendulum to the upright position. The nonlinear model is provided, but the control synthesis is done with the approximated linear model

$$(M_p + m_p)\ddot{\eta} - m_p\ell_p\ddot{\delta} = u \tag{1.64}$$

$$\ell_p\ddot{\delta} - \ddot{\eta} - g\delta = 0 \tag{1.65}$$

where η denotes the position of the cart, δ the angular position of the pendulum, and u the intensity of the horizontal force applied to the cart. This fourth order model does not include any kind of friction, making it much more difficult to control.

The second numerical experiment is derived from the classical multiplier-accelerator aggregated Samuelson model of a simplified economy, that is,

$$C[k] = (1 - a_1)Y[k-1] \tag{1.66}$$

$$I[k] = a_2(Y[k-1] - Y[k-2]) \tag{1.67}$$

$$Y[k] = C[k] + I[k] + G[k] \tag{1.68}$$

where $k \in \mathbb{N}$ and the real parameters a_1 and a_2 have economical denominations. The indicated variables are $C(\cdot)$ the *consumption expenditure*, $Y(\cdot)$ the *national income*, $I(\cdot)$ the *private investment*, and $G(\cdot)$ the *governmental expenditure*. This is a discrete-time model because only discrete-time acquisition data is available for parameter estimation. The interesting feature, reported in [9], is that this discrete-time model reveals a continuous-time Markov jump linear system. See the Bibliography notes of this chapter for details. Hence, a sampled-data control structure appears to be well adapted to deal with this particular class of dynamic systems.

1.5 Notation

The notation used throughout the book is standard. For square matrices, $\mathbf{tr}(\cdot)$ denotes the trace function. For real vectors or matrices, $(')$ refers to their transpose. For symmetric matrices, $(\bullet)$ denotes the symmetric block. The symbols $\mathbb{R}$, $\mathbb{R}_+$, and $\mathbb{N}$ denote the sets of real, non-negative real, and natural (non-negative integer) numbers, respectively. For any symmetric matrix, $X > 0$ $(X \geq 0)$ denotes a positive (semi-)definite matrix. As usual, in continuous-time, the norm of a trajectory $w_c(t) \in \mathbb{R}^{r_c}$ for all $t \in \mathbb{R}_+$ is $||w_c||_2^2 = \int_0^\infty w_c(t)'w_c(t)dt$, while $\mathcal{L}_2$ indicates the set of all trajectories with finite norm. Similarly, in discrete-time, the norm of a trajectory $w_d(t_k) \in \mathbb{R}^{r_d}$ for all $k \in \mathbb{N}$ is $||w_d||_2^2 = \sum_{k=0}^{\infty} w_d(t_k)'w_d(t_k)$ and ℓ_2 indicates the set of all trajectories with finite norm. In the special framework of Markov jump linear systems, these norm definitions are modified accordingly to cope with stochastic processes. To ease the notation, a function or trajectory $f(t)$ evaluated at $t = t_k$ is denoted, in some instances, as $f[k] = f(t_k)$ for all $k \in \mathbb{N}$. Whenever necessary to make explicit the dimensions, the $n \times n$ identity matrix is denoted by I_n, otherwise, it is simply denoted by I. The $\mathcal{Z}$ transform of $g[k]$ defined for all $k \in \mathbb{N}$ is denoted by $G(\zeta)$, whereas the Laplace transform of $g(t)$ defined for all $t \in \mathbb{R}_+$ is denoted by $G(s)$.

1.6 Bibliography Notes

There is no doubt that the book [10], published almost three decades ago, has played a major role in the development of many research areas that had the main goal of applying convex programming tools to control and filtering design. In particular, it brought to light the possibility of solving important design problems whenever they are expressed by means of *Linear Matrix Inequalities (LMIs)*. In this context, the model of parameter uncertainty denominated *convex bounded uncertainty*, introduced in [24], being well adapted to the notion of *quadratic stability* elaborated and proposed in [5], was particularly simple to be handled by LMI solvers. The seminal books [6] and [32] are indicated as precise and valuable sources on matrix analysis. Concerning the Hamilton–Jacobi–Bellman equations that are at the core of several results presented in this book, the reader is requested to see [48].

Chapter 2
Differential Linear Matrix Inequalities

2.1 Introduction

This chapter is entirely devoted to analyzing and making explicit some useful properties of the solutions of several DLMIs that are particularly important in the context of this book. In the general case, a numerical method to solve DLMIs is provided and it is exhaustively applied to the solution of a series of examples in order to have a precise idea of its implementation simplicity, efficiency, and computational burden needed to determine a solution.

Given a scalar $h > 0$, our interest is to solve a DLMI of the form

$$\mathcal{L}(\dot{P}(t), P(t)) < 0, \ t \in [0, h) \tag{2.1}$$

which, in general, is settled together with a boundary condition

$$\mathcal{H}(P_0, P_h) < 0 \tag{2.2}$$

where $\mathcal{L}$ and $\mathcal{H}$ are linear mappings. The main objective is to calculate a solution in terms of the symmetric matrix-valued function $P(t) : [0, h] \rightarrow \mathbb{R}^{n \times n}$ such that the linear inequalities (2.1)–(2.2) hold. The next definition characterizes the solutions of interest.

Definition 2.1 (DLMI Solution) *The symmetric matrix-valued function* $P(t) : [0, h] \rightarrow \mathbb{R}^{n \times n}$ *is a solution of the DLMI (2.1), if* $P(t)$ *is continuous and satisfies (2.1) almost everywhere in the time interval* $[0, h)$.

Notice that Definition 2.1 requires continuity but allows solutions $P(t)$ that fail to be differentiable in some isolated points of the time domain $[0, h)$. In general, a DLMI does not admit a unique solution, accordingly, we have only to search for a generic one that satisfies the requirements expressed in both conditions (2.1)

J. C. Geromel, *Differential Linear Matrix Inequalities*,
https://doi.org/10.1007/978-3-031-29754-0_2

and (2.2). In some instances, a solution is denominated *feasible solution*. In the next section, some DLMIs are defined and solved.

2.2 Lyapunov Differential Inequality

First of all let us consider the Lyapunov differential equation which is one of the key linear equations whose solution is frequently invoked throughout this book. For a given scalar $h > 0$, our interest is to solve

$$\dot{P}(t) + A'P(t) + P(t)A + Q = 0, \ t \in [0, h) \tag{2.3}$$

with given matrices $A \in \mathbb{R}^{n\times n}$ and $Q = Q' \in \mathbb{R}^{n\times n}$. It is subject to an initial $P(0) = P_0$ or to a final $P(h) = P_h$ boundary condition. The symmetric matrix-valued function $P(t) : [0, h] \to \mathbb{R}^{n\times n}$ is the variable to be determined. It is well known that, due to linearity, the solution of (2.3) always exists and has the form

$$P(t) = e^{-A'(t-\tau)}P(\tau)e^{-A(t-\tau)} - \int_{\tau}^{t} e^{-A'(t-\xi)}Qe^{-A(t-\xi)}d\xi \tag{2.4}$$

for any scalars $t \in \mathbb{R}$ and $\tau \in \mathbb{R}$. It is important to mention that this solution becomes unique whenever one of the mentioned boundary conditions is imposed. The next lemma summarizes the first important case of interest.

Lemma 2.1 *The solution of the Lyapunov differential equation* (2.3) *subject to the final boundary condition* $P(h) = P_h$ *is given by*

$$P(t) = e^{A'(h-t)}P_h e^{A(h-t)} + \int_{0}^{h-t} e^{A'\xi}Qe^{A\xi}d\xi \tag{2.5}$$

Moreover, if $Q \geq 0$ *and* $P_h > 0$, *then* $P(t) > 0$, $\forall t \in [0, h]$, *and* $P_0 \geq e^{A'h}P_h e^{Ah}$.

Proof Set $\tau = h > 0$ in (2.4) to obtain

$$\begin{aligned} P(t) &= e^{-A'(t-h)}P_h e^{-A(t-h)} - \int_{h}^{t} e^{-A'(t-\xi)}Qe^{-A(t-\xi)}d\xi \\ &= e^{A'(h-t)}P_h e^{A(h-t)} + \int_{0}^{h-t} e^{A'\xi}Qe^{A\xi}d\xi \\ &\geq e^{A'(h-t)}P_h e^{A(h-t)} \end{aligned} \tag{2.6}$$

for all $t \in [0, h]$. The second equality is exactly (2.5) and the last inequality is due to the fact that $Q \geq 0$. Taking into account that $P_h > 0$, the consequence is that $P(t) > 0$, $\forall t \in [0, h]$. Finally, evaluating the last inequality at $t = 0$, one obtains the desired result, concluding thus the proof. □

It is simple to see that if $Q > 0$, then the inequality depending on both boundary conditions becomes strict, that is,

$$P_0 > e^{A'h} P_h e^{Ah} \tag{2.7}$$

On the other hand, since $P(t)$ is positive definite, it is invertible in the whole time interval $t \in [0, h]$. Consequently, the inverse matrix-valued function $S(t) = P(t)^{-1}$ is well defined, and using its time derivative

$$\dot{S}(t) = -P(t)^{-1}\dot{P}(t)P(t)^{-1} \tag{2.8}$$

in (2.3) multiplied both sides by $P(t)^{-1}$, it is seen that $S(t)$ solves the quadratic differential equation

$$-\dot{S}(t) + AS(t) + S(t)A' + S(t)QS(t) = 0, \ t \in [0, h) \tag{2.9}$$

subject to $S(h) = S_h = P_h^{-1} > 0$. It is important to keep in mind that the result of Lemma 2.1 is no longer valid if we replace the final boundary condition by the initial boundary condition $P(0) = P_0 > 0$. This is true because, in this case, the matrix-valued function $P(t)$ may fail to be positive definite at some time instant $t \in (0, h)$. For completeness, let us consider the Lyapunov differential equation

$$-\dot{P}(t) + AP(t) + P(t)A' + Q = 0, \ t \in [0, h) \tag{2.10}$$

which, at first glance, appears to share the same properties as the previous ones. In fact, this is not exactly the case, as far as the positivity of the solution is concerned. Its general solution is given by

$$P(t) = e^{A(t-\tau)}P(\tau)e^{A'(t-\tau)} + \int_\tau^t e^{A(t-\xi)}Qe^{A'(t-\xi)}d\xi \tag{2.11}$$

for any scalars $t \in \mathbb{R}$ and $\tau \in \mathbb{R}$. Once again, an initial or a final boundary condition ensures that this solution is unique.

Lemma 2.2 *The solution of the Lyapunov differential equation (2.10) subject to the initial boundary condition $P(0) = P_0$ is given by*

$$P(t) = e^{At}P_0e^{A't} + \int_0^t e^{A\xi}Qe^{A'\xi}d\xi \tag{2.12}$$

Moreover, if $Q \geq 0$ and $P_0 > 0$, then $P(t) > 0$, $\forall t \in [0, h]$, and $P_h \geq e^{Ah}P_0e^{A'h}$.

Proof Set $\tau = 0$ in (2.11) to obtain

$$P(t) = e^{At}P_0e^{A't} + \int_0^t e^{A(t-\xi)}Qe^{A'(t-\xi)}d\xi$$

$$= e^{At} P_0 e^{A't} + \int_0^t e^{A\xi} Q e^{A'\xi} d\xi$$

$$\geq e^{At} P_0 e^{A't} \tag{2.13}$$

for all $t \in [0, h]$. The second equality is exactly (2.12) and the last inequality is due to the fact that $Q \geq 0$. Taking into account that $P_0 > 0$, the consequence is that $P(t) > 0$, $\forall t \in [0, h]$. Finally, evaluating the last inequality at $t = h$, one obtains the desired result, concluding thus the proof. □

Since $P(t)$ is positive definite, its inverse $S(t) = P(t)^{-1}$ is well defined and satisfies that quadratic differential equation

$$\dot{S}(t) + A'S(t) + S(t)A + S(t)QS(t) = 0, \ t \in [0, h) \tag{2.14}$$

subject to $S(0) = S_0 = P_0^{-1} > 0$. As expected, the initial condition replaced by the final condition $P(h) = P_h > 0$ in Lemma 2.2 does not necessarily certify that $P(t) > 0$. The role of the boundary conditions is the key factor to impose positive definiteness to the solution. To this end, the correct choice of the initial or the final boundary condition must be made carefully in order to impose to the solution that important and desired property.

One of the most important DLMIs is the Lyapunov differential inequality that can be stated as follows:

$$\dot{P}(t) + A'P(t) + P(t)A + Q < 0, \ t \in [0, h) \tag{2.15}$$

with given matrices $A \in \mathbb{R}^{n\times n}$ and $Q = Q' \in \mathbb{R}^{n\times n}$. This means that any feasible solution satisfies the Lyapunov equation

$$\dot{P}(t) + A'P(t) + P(t)A + Q = -W(t), \ t \in [0, h) \tag{2.16}$$

where $W(t) : [0, h) \to \mathbb{R}^{n\times n}$ is a symmetric and positive definite matrix-valued function, that is, $W(t) > 0$ for all $t \in [0, h)$. The previous results can now be applied with no difficulty to characterize useful properties of the Lyapunov differential inequality (2.15).

Theorem 2.1 *Assume that* $Q \geq 0$. *Any feasible solution* $P(t)$ *to the Lyapunov differential inequality* (2.15), *subject to the final boundary condition* $P(h) = P_h > 0$ *satisfies*

$$P(t) > e^{A'(h-t)} P_h e^{A(h-t)} + \int_0^{h-t} e^{A'\xi} Q e^{A\xi} d\xi \tag{2.17}$$

Moreover, $P(t) > 0$, $\forall t \in [0, h]$.

Proof Similarly to (2.4), the solution of (2.16) can be written as

$$
\begin{aligned}
P(t) &= e^{-A'(t-\tau)}P(\tau)e^{-A(t-\tau)} \\
&\quad - \int_\tau^t e^{-A'(t-\xi)}(Q + W(\xi))e^{-A(t-\xi)}d\xi
\end{aligned}
\tag{2.18}
$$

which, imposing $\tau = h$, becomes

$$
\begin{aligned}
P(t) &= e^{A'(h-t)}P_h e^{A(h-t)} \\
&\quad + \int_t^h e^{A'(\xi-t)}(Q + W(\xi))e^{A(\xi-t)}d\xi \\
&> e^{A'(h-t)}P_h e^{A(h-t)} + \int_0^{h-t} e^{A'\xi}Qe^{A\xi}d\xi
\end{aligned}
\tag{2.19}
$$

and by consequence $P(t) > 0$ for all $t \in [0, h]$, concluding thus the proof. □

Some remarks are in order. Comparing the inequality (2.17) and the solution of the Lyapunov equation (2.3) given by (2.5), it is evident that the unique solution to the Lyapunov equation is a lower bound (in the matrix sense) of any feasible solution to the Lyapunov differential inequality. In other words, we can say that the cone of positive definite solutions of (2.15) admits a minimal element. On the other hand, evaluating (2.17) at $t = 0$, one obtains

$$
\begin{aligned}
P_0 &> e^{A'h}P_h e^{Ah} + \int_0^h e^{A'\xi}Qe^{A\xi}d\xi \\
&\geq e^{A'h}P_h e^{Ah}
\end{aligned}
\tag{2.20}
$$

which reproduces the relationship between the initial and the final boundary conditions already observed in the framework of Lyapunov equations, see (2.7). For the alternative Lyapunov differential inequality

$$
-\dot{P}(t) + AP(t) + P(t)A' + Q < 0,\ t \in [0, h)
\tag{2.21}
$$

similar results are stated in the next theorem.

Theorem 2.2 *Assume that $Q \geq 0$. Any feasible solution $P(t)$ to the Lyapunov differential inequality* (2.21)*, subject to the initial boundary condition $P(0) = P_0 > 0$, satisfies*

$$
P(t) > e^{At}P_0 e^{A't} + \int_0^t e^{A\xi}Qe^{A'\xi}d\xi
\tag{2.22}
$$

Moreover, $P(t) > 0$, $\forall t \in [0, h]$.

Proof Similarly to what have been done with (2.15), we may convert the differential inequality (2.21) into an equivalent differential equation by adding to it the positive definite matrix-valued function $W(t)$. Hence, the solution is written as

$$\begin{aligned} P(t) &= e^{A(t-\tau)}P(\tau)e^{A'(t-\tau)} \\ &\quad + \int_{\tau}^{t} e^{A(t-\xi)}(Q + W(\xi))e^{A'(t-\xi)}d\xi \end{aligned} \tag{2.23}$$

which imposing $\tau = 0$ becomes

$$\begin{aligned} P(t) &= e^{At}P_0e^{A't} \\ &\quad + \int_{0}^{t} e^{A(t-\xi)}(Q + W(\xi))e^{A'(t-\xi)}d\xi \\ &> e^{At}P_0e^{A't} + \int_{0}^{t} e^{A\xi}Qe^{A'\xi}d\xi \end{aligned} \tag{2.24}$$

and consequently $P(t) > 0$ for all $t \in [0, h]$, concluding thus the proof. □

The same remarks made before are still valid. The solution of the corresponding Lyapunov equation is a lower bound to all feasible solutions of the DLMI (2.21). The only difference between this case and the former one is the boundary condition (initial or final) we must impose in order to obtain the reported lower bounds. Once again, setting $t = h$, (2.22) yields

$$\begin{aligned} P_h &> e^{Ah}P_0e^{A'h} + \int_{0}^{h} e^{A'\xi}Qe^{A\xi}d\xi \\ &\geq e^{Ah}P_0e^{A'h} \end{aligned} \tag{2.25}$$

which is the counterpart of (2.20). We have made explicit some properties of two versions of the Lyapunov DLMI. An important problem is how to calculate, among all feasible solutions, the one that optimizes a certain pre-specified objective function. This is addressed and solved in a forthcoming section.

2.3 Riccati Differential Inequality

We now move our attention to another well known differential equation. It is the Riccati differential equation and it has the form

$$\dot{P}(t) + A'P(t) + P(t)A + P(t)RP(t) + Q = 0, \; t \in [0, h) \tag{2.26}$$

with given matrices $A \in \mathbb{R}^{n\times n}$, $Q = Q' \in \mathbb{R}^{n\times n}$, and $R = R' \in \mathbb{R}^{n\times n}$. The analysis of this equation is much more involved. First of all, due to the nonlinear nature of (2.26), it may occur that a solution does not exist in the time interval of interest. Indeed, this claim can be confirmed by a simple example. For instance, the scalar quadratic differential equation $\dot{p}(t) + rp(t)^2 = 0$ can be integrated with no difficulty, yielding $1/p(t) = 1/p(0) + rt, \forall t \geq 0$. For a given initial condition $p(0) = p_0 > 0$, if $r < 0$, then at $t_e = 1/(|r|p_0)$, the solution escapes to infinity meaning that for $h > t_e$ a solution does not exist in the time interval $[0, h]$. On the other hand, for a given initial condition $p(0) = p_0 > 0$, if $r \geq 0$, the solution $p(t) = 1/(1/p_0 + rt)$ obviously exists in any time interval $[0, h]$ with $h > 0$.

Lemma 2.3 *Assume that $Q \geq 0$ and $R \geq 0$. If $P(t)$ is a solution to the Riccati differential equation* (2.26)*, subject to the final boundary condition $P(h) = P_h > 0$, then $P(t) > 0, \ \forall t \in [0, h]$.*

Proof Similarly to (2.4), a solution whenever exists must satisfy

$$\begin{aligned} P(t) &= e^{-A'(t-\tau)}P(\tau)e^{-A(t-\tau)} \\ &\quad - \int_{\tau}^{t} e^{-A'(t-\xi)}\big(Q + P(\xi)RP(\xi)\big)e^{-A(t-\xi)}d\xi \end{aligned} \tag{2.27}$$

which imposing $\tau = h$ becomes

$$\begin{aligned} P(t) &= e^{A'(h-t)}P_h e^{A(h-t)} \\ &\quad + \int_{t}^{h} e^{A'(\xi-t)}\big(Q + P(\xi)RP(\xi)\big)e^{A(\xi-t)}d\xi \end{aligned} \tag{2.28}$$

and consequently under our assumptions $P(t) > 0$ for all $t \in [0, h]$, concluding thus the proof. □

For the purposes of this book, the definiteness of a solution whenever it exists is essential. However, it is important to keep in mind that the existence of a solution is not assured by the result of Lemma 2.3. This fact puts in evidence the need of a numerical procedure able to determine a solution whenever one exists. For completeness, let us consider the Riccati differential equation

$$-\dot{P}(t) + AP(t) + P(t)A' + P(t)RP(t) + Q = 0, \ t \in [0, h) \tag{2.29}$$

where matrices A, Q, and R are the same as in the previous case. Obviously, the remarks concerning the existence of a solution remain valid for this new Riccati equation we want to consider.

Lemma 2.4 *Assume that $Q \geq 0$ and $R \geq 0$. If $P(t)$ is a solution to the Riccati differential equation* (2.29)*, subject to the initial boundary condition $P(0) = P_0 > 0$, then $P(t) > 0, \ \forall t \in [0, h]$.*

Proof Following (2.10), a solution whenever exists must satisfy

$$P(t) = e^{A(t-\tau)}P(\tau)e^{A'(t-\tau)} + \int_{\tau}^{t} e^{A(t-\xi)}\big(Q + P(\xi)RP(\xi)\big)e^{A'(t-\xi)}d\xi \tag{2.30}$$

which imposing $\tau = 0$ becomes

$$P(t) = e^{At}P_0e^{A't} + \int_{0}^{t} e^{A(t-\xi)}\big(Q + P(\xi)RP(\xi)\big)e^{A'(t-\xi)}d\xi \tag{2.31}$$

and consequently under our assumptions $P(t) > 0$ for all $t \in [0, h]$, concluding thus the proof. □

As far as Lyapunov or Riccati equations are concerned there is no incompatibility if both initial $P_0 > 0$ and final $P_h > 0$ are simultaneously imposed. Even though one of them is superfluous, both boundary conditions impose the positivity of the solution to the Riccati differential equations (2.26) and (2.29). Observe that the Lyapunov differential equations always admit a solution, but this is not necessarily true for Riccati differential equations. The next example illustrates some theoretical points raised so far.

Example 2.1 *We have numerically solved, by backward integration, the Riccati differential equation (2.26) with*

$$A = \begin{bmatrix} 0 & 1 \\ -4 & -1 \end{bmatrix}$$

$R = (1/10)I$, $Q = I$, *and the time interval* $[0, 5]$. *Two different final conditions have been considered, namely* $P_h = 0$ *and* $P_h = 2I$. *Figure 2.1 shows the minimum*

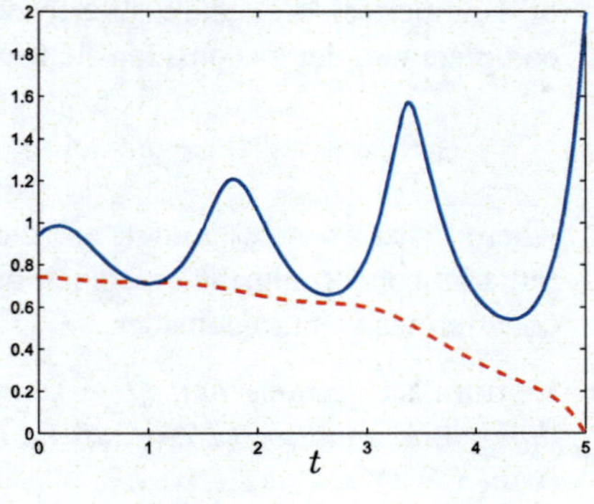

Fig. 2.1 Minimum eigenvalue for two final conditions

eigenvalue of $P(t)$ against time in the interval previously defined. In dashed line, the solution corresponding to $P_h = 0$ shows that $P(t) > 0$ for all $t \in [0, 5)$. In solid line, it is clear that the solution $P(t)$ corresponding to $P_h = 2I$ is also positive definite in the whole time interval $t \in [0, 5]$. Those situations illustrate the result of Lemma 2.3. It is interesting to mention that for $P_h = 2.96I$, the solution ceases to exist (within the adopted precision) in the same time interval. This fact has been verified numerically using Matlab routines. □

The Riccati differential inequality has the form

$$\dot{P}(t) + A'P(t) + P(t)A + P(t)RP(t) + Q < 0, \ t \in [0, h) \tag{2.32}$$

which obviously is nonlinear. However, taking into account the factorization of matrices $R = EE' \geq 0$ and $Q = C'C \geq 0$, the Schur Complement provides the equivalent DLMI formulated in the same time interval, that is,

$$\begin{bmatrix} \dot{P}(t) + A'P(t) + P(t)A & P(t)E & C' \\ \bullet & -I & 0 \\ \bullet & \bullet & -I \end{bmatrix} < 0, \ t \in [0, h) \tag{2.33}$$

This is a key algebraic manipulation that makes it possible the determination of a solution to the nonlinear differential inequality (2.32) by the numerical method provided in a forthcoming section. We proceed by pointing out the main characteristic of the feasible solutions to the DLMI (2.33), equivalent to (2.32).

Theorem 2.3 *Assume that $Q \geq 0$ and $R \geq 0$. Any feasible solution $P(t)$ to the Riccati differential inequality (2.32), subject to the final boundary condition $P(h) = P_h > 0$, is such that $P(t) > P_E(t), \forall t \in [0, h)$, where $P_E(t)$ is a solution to the Riccati differential equation (2.26), subject to the same final boundary condition. Moreover, $P(t) > 0, \ \forall t \in [0, h]$.*

Proof Defining $\Xi(t) = P(t) - P_E(t)$, the factorization

$$P(t)RP(t) - P_E(t)RP_E(t) = P_E(t)R\Xi(t) + \Xi(t)RP_E(t) + \Xi(t)R\Xi(t) \tag{2.34}$$

allows us to express the difference between (2.32) and (2.26) as

$$\dot{\Xi}(t) + A_E(t)'\Xi(t) + \Xi(t)A_E(t) + \Xi(t)R\Xi(t) < 0, \ t \in [0, h) \tag{2.35}$$

where $A_E(t) = A + RP_E(t)$, subject to the final condition $\Xi(h) = \Xi_h = 0$. This is the time-varying version of a Lyapunov differential inequality, and, consequently, any feasible solution satisfies

$$\Xi(t) = \Phi(h, t)'\Xi_h\Phi(h, t) + \int_t^h \Phi(\xi, t)'\big(\Xi(\xi)R\Xi(\xi) + W(\xi)\big)\Phi(\xi, t)d\xi$$

$$= \int_t^h \Phi(\xi, t)' \big(\Xi(\xi) R \Xi(\xi) + W(\xi)\big) \Phi(\xi, t) d\xi \tag{2.36}$$

for some $W(t) > 0, \ \forall t \in [0, h)$, in order to cope with the inequality in (2.35) and where $\Phi(t, \tau)$ is the fundamental matrix associated with $A_E(t)$, that is,

$$\frac{\partial \Phi(t, \tau)}{\partial t} = A_E(t) \Phi(t, \tau), \ \Phi(\tau, \tau) = I \tag{2.37}$$

Hence, from our assumptions, Lemma 2.3 implies that $P_E(t) > 0$ for all $t \in [0, h]$ and (2.36) allows the conclusion that $\Xi(t) = P(t) - P_E(t) > 0$ for all $t \in [0, h)$, concluding thus the proof. □

For this class of Riccati differential inequalities, the same properties of positivity and minimality already established in the context of Lyapunov inequalities hold. Moreover, observe that even though the Riccati differential inequality is time-invariant since matrices A, R, and Q are constant, the proof of Theorem 2.3 needs to handle the time-varying version of a Lyapunov equation. There is no difficulty to state and prove the corresponding result related to the Riccati differential inequality

$$-\dot{P}(t) + AP(t) + P(t)A' + P(t)RP(t) + Q < 0, \ t \in [0, h) \tag{2.38}$$

subject to initial boundary conditions. Even though the proof has the same pattern as the one of the previous theorems, it is given for completeness.

Theorem 2.4 *Assume that $Q \geq 0$ and $R \geq 0$. Any feasible solution $P(t)$ to the Riccati differential inequality (2.38), subject to the initial boundary condition $P(0) = P_0 > 0$, is such that $P(t) > P_E(t), \forall t \in [0, h)$, where $P_E(t)$ is a solution to the Riccati differential equation (2.29), subject to the same initial boundary condition. Moreover, $P(t) > 0, \ \forall t \in [0, h]$.*

Proof Defining $\Xi(t) = P(t) - P_E(t)$, the factorization (2.34) together with the difference between (2.38) and (2.29) yields

$$-\dot{\Xi}(t) + A_E(t)\Xi(t) + \Xi(t)A_E(t)' + \Xi(t)R\Xi(t) < 0, \ t \in [0, h) \tag{2.39}$$

where $A_E(t) = A + RP_E(t)$, subject to the initial condition $\Xi(0) = \Xi_0 = 0$. This is the time-varying version of a Lyapunov differential inequality and, consequently, any feasible solution satisfies

$$\begin{aligned} \Xi(t) &= \Phi(t, 0)\Xi_0\Phi(t, 0)' + \int_0^t \Phi(t, \xi)\big(\Xi(\xi)R\Xi(\xi) + W(\xi)\big)\Phi(t, \xi)' d\xi \\ &= \int_0^t \Phi(t, \xi)\big(\Xi(\xi)R\Xi(\xi) + W(\xi)\big)\Phi(t, \xi)' d\xi \end{aligned} \tag{2.40}$$

for some $W(t) > 0$, $\forall t \in [0, h)$ in order to cope with the inequality in (2.39) and where $\Phi(t, \tau)$ is the fundamental matrix associated with $A_E(t)$ given in (2.37). From our assumptions, Lemma 2.4 implies that $P_E(t) > 0$ for all $t \in [0, h]$ and (2.40) allows the conclusion that $\Xi(t) = P(t) - P_E(t) > 0$ for all $t \in (0, h]$, concluding thus the proof. □

The positivity and minimality of the solution are once again established for this class of Riccati differential inequality. Of course, these properties follow from the adequate choice of the initial boundary condition and are important whenever optimal control design (in some sense) is concerned.

We have presented some key properties exhibited by the feasible solutions to the Lyapunov and Riccati differential inequalities. These are the two most important classes of differential inequalities because they are on the basis of the $\mathcal{H}_2$ and $\mathcal{H}_\infty$ performance indices frequently used in control and filtering design. To close this section, in the next remark, we discuss some minimal properties to these differential inequalities that are useful in the forthcoming chapters of this book.

Remark 2.1 *Let A be Hurwitz stable, $R \geq 0$, and $Q \geq 0$. The set of all feasible solutions $P(t)$ to the Riccati differential inequality (2.32), subject to boundary conditions satisfying $P_h > P_0 > 0$, admits the minimal elements $(\bar{P}_h, \bar{P}_0)$ both arbitrarily close to $\bar{P}$, where $\bar{P} \geq 0$ is the stabilizing solution, that is, $\bar{A} = A + R\bar{P}$ is Hurwitz stable, to the algebraic Riccati equation $A'\bar{P} + \bar{P}A + \bar{P}R\bar{P} + Q = 0$. Defining $\Xi(t) = P(t) - \bar{P}$, it can be seen that it satisfies the quadratic differential inequality $\dot{\Xi}(t) + \bar{A}'\Xi(t) + \Xi(t)\bar{A} + \Xi(t)R\Xi(t) < 0$ for all $t \in [0, h)$ with boundary conditions $\Xi(h) = P_h - \bar{P} > P_0 - \bar{P} = \Xi(0)$. Taking into account that $R \geq 0$, any feasible solution is such that*

$$\Xi(h) > \Xi(0) > e^{\bar{A}'h}\Xi(h)e^{\bar{A}h} > e^{\bar{A}'h}\Xi(0)e^{\bar{A}h}$$

and consequently $\Xi(h) = P_h - \bar{P} > 0$ and $\Xi(0) = P_0 - \bar{P} > 0$ because $\bar{A}$ is Hurwitz stable, confirming thus the claim. Imposing $R = 0$, we conclude that the result is also valid for the differential Lyapunov inequality (2.15). □

The next section provides a numerical method that can be applied to solve DLMIs. The main idea is to convert DLMIs into LMIs without adding any kind of conservatism.

2.4 Numerical Solution

It is very important to keep in mind the context that we have to solve DLMIs. The first point to take into account is that the differential inequality (2.1) and the boundary condition (2.2) must be jointly solved which characterizes a two-point boundary value problem. However, compared with classical versions, the present one is more complicated to handle due to the presence of a differential inequality instead of a differential equation. Of course the second and key point is to design a numerical method that takes advantage of the linearity of maps $\mathcal{L}$ and $\mathcal{H}$.

Let us recall that our purpose is to calculate a symmetric matrix-valued function $P(t) : [0, h) \to \mathbb{R}^{n\times n}$ that fulfills the requirements of Definition 2.1, that is, $P(t)$ is continuous and satisfies (2.1) almost everywhere. To this end, let us introduce the following set of scalar-valued continuous functions $\phi_i(t) : [0, h) \to \mathbb{R}$ for $i \in N_\phi = \{0, 1, \dots, n_\phi\}$ with $n_\phi \geq 1$ and a set of symmetric matrices $X_i \in \mathbb{R}^{n\times n}$ for all $i \in N_\phi$. Hence, we adopt the parametrization

$$P(t) = \sum_{i\in N_\phi} X_i\phi_i(t) \tag{2.41}$$

from which it is clear that, for a given set of functions $\{\phi_i(t)\}_{i\in N_\phi}$, the set of symmetric matrices $\{X_i\}_{i\in N_\phi}$ contains the matrix variables to be determined. There are uncountable possibilities of choosing $\{\phi_i(t)\}_{i\in N_\phi}$, as, for instance, *Piecewise linear*, *Taylor series*, *Fourier series*, and *Splines*, among others. The reader is requested to see the Bibliography notes of this chapter for more information on this subject, including a brief discussion about numerical procedures specially developed to handle DLMIs.

In this book we focus our attention on the set of piecewise linear scalar-valued functions because, from them, one of the simplest methods to cope with DLMIs can be developed. To this end, let us split the time interval $[0, h)$ into n_ϕ subintervals of equal length $\eta = h/n_\phi$ and define the scalar-valued function

$$\phi(t) = \begin{cases} 1 - \frac{|t|}{\eta} & , |t| \leq \eta \\ 0 & , |t| > \eta \end{cases} \tag{2.42}$$

used to construct the set of piecewise functions $\phi_i(t) = \phi(t - i\eta)$ defined in the whole time interval $[0, h)$, for all $i \in N_\phi$. Summing up all contribution, we obtain

$$P(t) = X_i + \left(\frac{X_{i+1} - X_i}{\eta}\right)(t - i\eta) \tag{2.43}$$

valid in the time segment $t \in [i\eta, (i+1)\eta)$ for each $i = 0, 1, \dots, n_\phi - 1$. Continuity is preserved, but differentiability, in general, does not hold at the border points of each time subinterval. The next theorem provides the main result as far as the DLMI numerical solution is concerned.

Theorem 2.5 *Let $h > 0$ and $n_\phi \geq 1$ be given. The piecewise linear matrix-valued function* (2.43) *solves the DLMI* (2.1) *subject to the boundary condition* (2.2) *if and only if the LMIs*

$$\mathcal{L}\left(\frac{X_{i+1} - X_i}{\eta}, X_i\right) < 0 \tag{2.44}$$

$$\mathcal{L}\left(\frac{X_{i+1}-X_i}{\eta}, X_{i+1}\right) < 0 \tag{2.45}$$

$$\mathcal{H}(X_0, X_{n_\phi}) < 0 \tag{2.46}$$

are satisfied for all $i = 0, 1, \ldots, n_\phi - 1$.

Proof Define the time instants $\{t_i = i\eta\}_{i=1}^{n_\phi}$ where, by construction, $t_0 = 0$ and $t_{n_\phi} = h$. Observe that the linear function (2.43) can be rewritten as the convex combination $P(t) = (1-\alpha_i(t))X_i + \alpha_i(t)X_{i+1}$ with

$$\alpha_i(t) = \frac{t-t_i}{t_{i+1}-t_i} \tag{2.47}$$

since $\alpha_i(t)$ covers the entire interval $[0, 1]$ whenever $t \in [t_i, t_{i+1}]$ for each $i = 0, 1, \ldots, n_\phi - 1$. The linear function (2.43) allows us to rewrite the inequalities (2.44) and (2.45) as

$$\mathcal{L}(\dot{P}(t), P(t))\big|_{t=t_i} = \mathcal{L}\left(\frac{X_{i+1}-X_i}{\eta}, X_i\right) < 0 \tag{2.48}$$

$$\mathcal{L}(\dot{P}(t), P(t))\big|_{t=t_{i+1}} = \mathcal{L}\left(\frac{X_{i+1}-X_i}{\eta}, X_{i+1}\right) < 0 \tag{2.49}$$

respectively. Hence, since the conditions (2.44)–(2.45) just impose the feasibility of $P(t)$ at the extreme points of the i-th time interval, necessity follows. Now, taking into account that $\mathcal{L}$ is a linear map, multiplying inequality (2.44) by $1-\alpha_i(t)$, inequality (2.45) by $\alpha_i(t)$, and summing up the results, it follows that $\mathcal{L}(\dot{P}(t), P(t)) < 0$ for all $t \in [t_i, t_{i+1}]$. This shows that $P(t)$ given by the linear function (2.43) is feasible for all $t \in \mathbb{R}$ belonging to the time interval $t \in [t_i, t_{i+1}]$, for each $i = 0, 1, \ldots, n_\phi - 1$, proving thus sufficiency. Finally, keeping in mind that $P_0 = P(t_0) = X_0$ and $P_h = P(t_{n_\phi}) = X_{n_\phi}$, then (2.46) is identical to (2.2), completing thus the proof. □

This is a central result. It indicates that a DLMI is equivalent (in the sense of a necessary and sufficient condition for feasibility) to $2n_\phi$ LMIs. The simplest linear solution

$$P(t) = X_0 + \left(\frac{X_h - X_0}{h}\right)t \tag{2.50}$$

valid in the whole time interval $t \in [0, h)$ can be imposed by means of only two LMIs, namely

$$\mathcal{L}\left(\frac{X_h - X_0}{h}, X_0\right) < 0 \tag{2.51}$$

$$\mathcal{L}\left(\frac{X_h - X_0}{h}, X_h\right) < 0 \tag{2.52}$$

which follows from the adoption of only one ($n_\phi = 1$) time subinterval. On the other extreme, it can be viewed that any feasible continuous, almost everywhere differentiable in the time interval $[0, h)$ solution is reached provided that the number of time subintervals n_ϕ is large enough. Furthermore, this theorem makes it possible the use of the numerical machinery available in the literature to cope with LMIs, to the solution of DLMIs. Within the precision level imposed by the choice of the number of time subintervals n_ϕ, the two-point boundary problem (2.1)–(2.2) is solved in only one shot. The equivalent $2n_\phi + 1$ LMIs (2.44)–(2.46) are jointly handled, which avoids the necessity of an iterative procedure, certainly contributing to increasing numerical efficiency. Of course, the number n_ϕ must be chosen carefully since it represents the trade-off between the precision of the solution and the computational burden involved.

Example 2.2 *With matrices A, R, Q, and $h = 5$ already considered in Example 2.1, we have determined $E = R^{1/2} = (1/\sqrt{10})I$ and $C = Q^{1/2} = I$ and solved*

$$\inf_{P(\cdot)} \{\mathbf{tr}(P_0) : P_0 > 0,\ \rho I > P_h > 0,\ (2.33)\}$$

with $\rho = 1e - 04$. Since $\rho > 0$ is very small, the LMIs $\rho I > P_h > 0$ are a possible way to implement the equality $P_h = 0$, approximately. Applying the result of Theorem 2.5, we have converted this optimization problem into an optimization problem expressed in terms of LMIs only.

Figure 2.2 shows the time evolution of the minimum eigenvalue of the optimal piecewise linear solution $P(t)$ in the time interval $[0, h]$. For $n_\phi = 1$, the problem is infeasible. For $n_\phi = 2$, the solution is not differentiable at $t = 2.5$, but, for $n_\phi = 64$, the corresponding optimal solution seems to be smooth. Table 2.1 shows

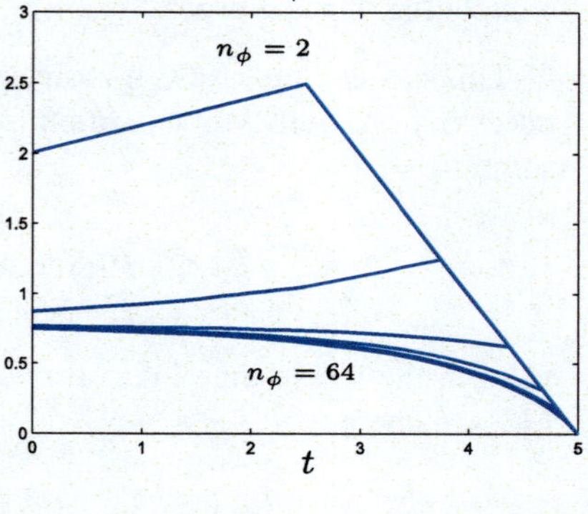

Fig. 2.2 Minimum eigenvalue for several subintervals

Table 2.1 Minimum value of the objective function

n_ϕ	2	4	8	16	32	64
$\mathbf{tr}(P_0)$	10.60	4.71	4.14	4.08	4.05	4.03

the optimal objective function with respect to n_ϕ*. As expected, it decreases as the number of subintervals increases. Figure 2.2 also makes it clear that the solution* $P(t)$ *is always positive definite, due to the fact that* $P_h > 0$*, as the previous theorems indicate. Accordingly, it has been verified that the constraint* $P_0 > 0$ *is superfluous.*

Finally, it is important to establish a post-optimization procedure for the validation of the result presented in Table 2.1, as far as the convergence (within some precision) is concerned. To this end, the result of Theorem 2.3 applies. Indeed, integrating backward the Riccati equation (2.26) from $P(h) = 0$*, it follows that* $\mathbf{tr}(P_0) = 4.00$ *which is very close (see the values presented in Table 2.1) to the minimum value of the objective function of the DLMI optimization problem. Hence, the solution corresponding to* $n_\phi = 64$ *time subintervals is precise enough and provides a near-optimal solution.* □

Example 2.3 *Consider the following matrices:*

$$A = \begin{bmatrix} 0 & 1 & 0 \\ 0 & 0 & 1 \\ 0 & -5 & -9 \end{bmatrix}, \ E = \begin{bmatrix} 0 \\ 0 \\ 1 \end{bmatrix}, \ C' = \begin{bmatrix} 1 \\ 1 \\ 1 \end{bmatrix}, \ L' = \begin{bmatrix} -1 \\ 0 \\ 0 \end{bmatrix}$$

$B = E$*,* $D = 1$*, and* $h = 1.5$ *the sampling period. Our main goal is to determine the maximum norm upper bound* $\mu > 0$ *such that the matrix* $A_L(\Delta) = A_d(\Delta) + B_d(\Delta)L$ *remains Schur stable where*

$$A_d(\Delta) = e^{A_\Delta h}, \ B_d(\Delta) = \int_0^h e^{A_\Delta t} B_\Delta dt$$

and $A_\Delta = A + E\Delta C$ *and* $B_\Delta = B + E\Delta D$ *for all* $||\Delta|| \leq \gamma^{-1} = 1/\sqrt{\mu}$*. This is the robust sampled-data control problem with norm bounded uncertainty, already studied in Chap. 1. A possible solution is accomplished by determining* $\mu > 0$ *and a feasible matrix-valued function* $P(t) : [0, h] \to \mathbb{R}^{4\times 4}$ *to the DLMI*

$$\begin{bmatrix} \dot{P}(t) + F'P(t) + P(t)F & P(t)J & G' \\ \bullet & -\mu I & 0 \\ \bullet & \bullet & -I \end{bmatrix} < 0, \ t \in [0, h) \tag{2.53}$$

which solve the convex programming problem

$$\inf_{P(\cdot),\mu} \{\mu : P_h > 0, \ P_h > H'P_0H, \ (2.53)\}$$

where the augmented matrices are given by

$$F = \begin{bmatrix} A & B \\ 0 & 0 \end{bmatrix}, \ H = \begin{bmatrix} I & 0 \\ L & 0 \end{bmatrix}, \ J = \begin{bmatrix} E \\ 0 \end{bmatrix}, \ G' = \begin{bmatrix} C' \\ D' \end{bmatrix}$$

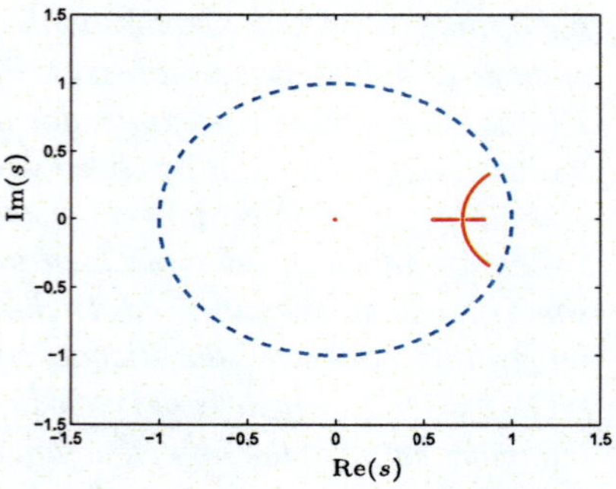

Fig. 2.3 Eigenvalues of the closed-loop uncertain system

The numerical method stated in Theorem 2.5 has been applied with $n_\phi = 64$ *time subintervals and provided the minimum value* $\mu = 0.1948$*, which means that robust stability holds for all* $|\Delta| \leq 2.2657$*. Figure 2.3 shows in dashed line the unitary circle and inside it the eigenvalues of matrix* $A_L(\Delta)$ *for all feasible norm bounded uncertainty. Observe that the eigenvalue locus has points near the unity circle which puts in clear evidence the small conservativeness of the condition as far as robust stability with norm bounded uncertainty is concerned.* □

Example 2.4 *We have used the same matrices given in Example 2.3, but now we consider a convex bounded uncertainty. For comparison, we have considered the following two extreme matrices:*

$$F_1 = F + J\Delta G, \ F_2 = F - J\Delta G$$

the same sampling period $h = 1.5$ *and* $\Delta = 2.2657$*. The goal is to verify if the matrix* $A_L(\lambda) = A_d(\lambda) + B_d(\lambda)L$ *remains Schur stable for all* $\lambda \in \Lambda$*. Based on the results discussed in Chap. 1, a possible answer to this question follows from the determination (if any) of a feasible matrix-valued function* $P(t) : [0, h] \to \mathbb{R}^{4\times 4}$ *to the DLMIs*

$$\begin{bmatrix} \dot{P}(t) + F_i' P(t) + P(t) F_i & G' \\ \bullet & -I \end{bmatrix} < 0, \ t \in [0, h) \tag{2.54}$$

for $i \in \{1, 2\}$ *that solves the convex programming problem*

$$\inf_{P(\cdot)} \left\{ \mathbf{tr}(J' P_h J) : P_h > 0, \ P_h > H' P_0 H, \ (2.54) \right\}$$

Once again, the numerical method stated in Theorem 2.5 has been applied with $n_\phi = 64$ *time subintervals and provided the minimum value* 3.1552 *for the objective function. This means that robust stability holds in the same uncertainty interval already determined in Example 2.3, namely* $|\Delta| \leq 2.2657$*. As before, this puts in* *clear evidence, for this example, the small conservativeness of the condition as far as robust stability with convex bounded uncertainty is concerned.* □

We hope that these examples make clear not only the importance of DLMIs, but the fact that they can be adequately handled by the numerical machinery available in the literature to cope with LMIs. Of course, the price to be paid is the computational burden involved, but, in compensation, important problems in the areas of sampled-data control and filtering design subject to parameter uncertainty become numerically tractable.

2.5 Bibliography Notes

Due to the fact that DLMI is a natural extension of LMI, including for historical reasons, the reader is requested to see [10], the first published book to make it clear the importance of LMIs in the framework of systems and control theory. Almost at the same time, one of the first papers to deal with DLMIs in the theoretical context of nowadays classical $\mathcal{H}_2$ and $\mathcal{H}_\infty$ performance optimization problems has appeared in [46] where, in addition, a one-to-one change of variables used to convert non-convex constraints into LMIs has been proposed. In this timeline, it is important to mention the book [33] where, additionally to many important issues, the common use of mathematical tools, expressed through Lyapunov and Riccati equations, to deal with time-varying systems and sampled-data control systems, has been pointed out.

The first results related to minimal (in matrix sense) and feasible solutions to Lyapunov and Riccati differential equations and differential inequalities have been established in [53] and in the references therein. Furthermore, in this chapter, the importance of DLMIs for robust sampled-data control systems has been illustrated by solving the associated design problem for the class of convex bounded parameter uncertainty where quadratic stability, introduced in [5], is a key concept. Aspects involving robust stability expressed by means of convex conditions are treated with detail in [11].

The continuous piecewise linear solution to DLMIs has been first proposed in [1] in the context of robust control of linear systems via switching. This class of solution and others, together with theoretical results on iterative algorithms and computational burden evaluation, have been given in [28]. Theorem 2.5 is a useful result on the numerical solution of DLMIs that has first appeared in [28]. Recently, DLMIs have been studied in [3] and [31] where more efficient algorithms (when compared to the piecewise linear solution) have been proposed. In general, as expected, the increasing computational efficiency needs more sophisticated numerical procedures and software implementations, see [31] for a quite complete study on this matter.

Chapter 3
Sampled-Data Control Systems

3.1 Introduction

This chapter is devoted to analyzing sampled-data control systems and some classical design strategies that cannot be applied whenever robustness against parameter uncertainty is concerned. This important aspect is left to be treated with details in the forthcoming chapters. The state space realization of the open-loop system is introduced together with the sampled-data control structure. As usual, conditions for sampled-data systems stability are obtained from the adoption of a quadratic Lyapunov function. Together with the so called equivalent discrete-time system, these results constitute the basis for introduction and calculation of performance indices similar to $\mathcal{H}_2$ and $\mathcal{H}_\infty$ transfer function norms that are well established in the context of pure continuous- and discrete-time LTI systems. Several examples, including control design, illustrate the theoretical results. They are solved by well established algorithms based on Lyapunov and Riccati equations as well as LMIs, in order to make a possible comparison clear in terms of numerical efficiency.

3.2 Sampled-Data Systems

The sampled-data system under consideration has the following state space realization:

$$\dot{x}(t) = Ax(t) + Bu(t) + Ew_c(t) \tag{3.1}$$

$$y[k] = C_y x[k] + E_y w_d[k] \tag{3.2}$$

$$z(t) = C_z x(t) + D_z u(t) \tag{3.3}$$

J. C. Geromel, *Differential Linear Matrix Inequalities*,
https://doi.org/10.1007/978-3-031-29754-0_3

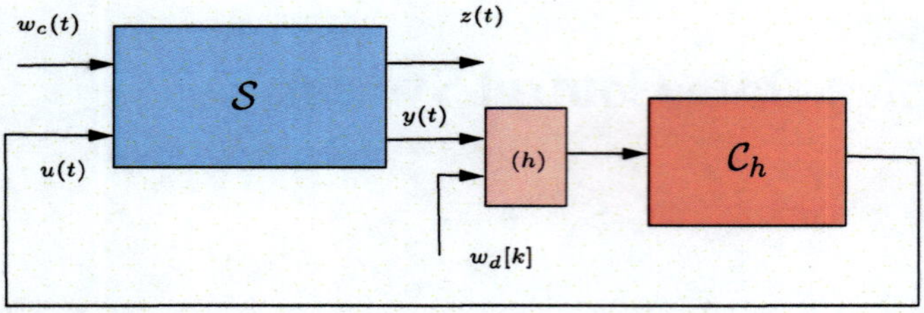

Fig. 3.1 Sampled-data control system structure

and it is assumed that it evolves from an arbitrary initial condition $x(0) = x_0 \in \mathbb{R}^{n_x}$ at $t = t_0 = 0$, where $x(\cdot) : \mathbb{R}_+ \to \mathbb{R}^{n_x}$, $u(\cdot) : \mathbb{R}_+ \to \mathbb{R}^{n_u}$, $w_c(\cdot) : \mathbb{R}_+ \to \mathbb{R}^{r_c}$, and $z(\cdot) : \mathbb{R}_+ \to \mathbb{R}^{n_z}$ are the state, the control, the exogenous perturbation, and the controlled output of the continuous-time process, respectively. The measurement $y[\cdot] : \mathbb{N} \to \mathbb{R}^{n_y}$ is available in discrete-time at instants t_k for all $k \in \mathbb{N}$ and is corrupted by the exogenous discrete-time sensor perturbation $w_d[\cdot] : \mathbb{N} \to \mathbb{R}^{r_d}$. Following the closed-loop system structure depicted in Fig. 3.1, the control signal to be designed may be of two different types, depending on the problem to be tackled, namely:

1. The control signal $u(t) : \mathbb{R}_+ \to \mathbb{R}^{n_u}$ belongs to the continuous-time domain, has the general form

$$u(t) = v(t) + E_u w_c(t), \ t \in \mathbb{R}_+ \tag{3.4}$$

and is supposed to be corrupted (due to the actuator action) by the exogenous continuous-time perturbation $w_c(t)$. The signal $v(\cdot)$ belongs to the continuous-time domain and is synthesized by the controller $\mathcal{C}_h$ from the discrete-time measurements $y[k]$, $k \in \mathbb{N}$.
2. The control signal $u(t) : \mathbb{R}_+ \to \mathbb{R}^{n_u}$ belongs to the continuous-time domain, has the general form

$$u(t) = u[k] = v[k] + E_u w_d[k], \ t \in [t_k, t_{k+1}), \forall k \in \mathbb{N} \tag{3.5}$$

and is supposed to be corrupted (due to the actuator action) by the exogenous discrete-time perturbation $w_d[k]$. The signal $v[\cdot]$ belongs to the discrete-time domain and is synthesized by the controller $\mathcal{C}_h$ from the discrete-time measurements $y[k]$, $k \in \mathbb{N}$. In this case, the control signal is piecewise constant constructed by a non-ideal *zero-order hold* device. For simplicity, we denote $u \in \mathbb{U}$. The sequence of evenly spaced sampling instants $\{t_k\}_{k \in \mathbb{N}}$ is such that $t_0 = 0$, $t_{k+1} - t_k = h > 0$, $\forall k \in \mathbb{N}$. The scalar $h > 0$ is often called *sampling period*.

Figure 3.1 shows the sampled-data control general structure, which deserves the following additional remarks:

- $\mathcal{S}$ denotes a continuous-time LTI system governed by Eqs. (3.1)–(3.3) with inputs $(w_c(t), u(t))$ and outputs $(z(t), y(t))$, for all $t \in \mathbb{R}_+$.
- A non-ideal sampling device with sampling period $h > 0$ provides the measurements $y[k] = y(t_k)$, $\forall k \in \mathbb{N}$ given in (3.2). As indicated, this operation introduces a certain measurement noise modeled as being proportional to the exogenous perturbation w_d.
- The controlled output $z(t)$ given in (3.3) is used for performance evaluation, exclusively. Naturally, the controller is determined such that a certain performance index is optimized.
- $\mathcal{C}_h$ denotes a controller, with input $y[k]$ and output $v(t)$, $t \in \mathbb{R}_+$, or $v[k]$, $k \in \mathbb{N}$, to be designed. It may be of different classes, and among them, the most important are state feedback and full order dynamic output feedback.

The matrices in the model given by the state space equations (3.1)–(3.3) and in the control law (3.4)–(3.5) are real and of compatible dimensions. The initial condition $x_0 \in \mathbb{R}^{n_x}$ may be zero in some instances, whenever specific performance indices are considered. The exogenous perturbations in continuous- and discrete-time are unity impulses or they belong to the sets $\mathcal{L}_2$ or ℓ_2 of norm bounded signals, respectively. Clearly, the fact that the same exogenous signals $w_c(t)$ and $w_d[k]$ appear in (3.1), (3.4) and (3.2), (3.5) simultaneously does not imply a loss of generality, since matrices E, E_y, and E_u can be freely chosen by the designer. In our opinion, this class of sampled-data systems is general enough to our purposes, and for this reason, it is the main object of our attention in the remaining chapters of this book.

3.3 Stability

In this section, we analyze the stability of continuous-time LTI systems and the effect of sampling with period $h > 0$ on stability of sampled-data control systems. From the very beginning, the following definitions are of central importance in the context of LTI systems stability in both continuous-time and discrete-time domains.

Definition 3.1 *A matrix $A \in \mathbb{R}^{n \times n}$ is Hurwitz stable if all its eigenvalues lie in the open left half-plane, the portion of the complex plane composed of all numbers with negative real part.*

Definition 3.2 *A matrix $A \in \mathbb{R}^{n \times n}$ is Schur stable if all its eigenvalues lie in the open unity circle, the portion of the complex plane composed of all numbers with modulus less than one.*

As it is seen in the sequel, these definitions are useful to characterize the stability of continuous-time and discrete-time LTI systems, respectively. Since for these classes of LTI systems and for sampled-data control systems, the various stability concepts are equivalent (as for instance *asymptotic* and *exponential*), only one denomination is adopted. Furthermore, it is interesting to keep in mind that the sampling process together with zero-order hold allows us to change the time domain of some variables from continuous-time to discrete-time.

Consider the following state space realization of an LTI system:

$$\dot{x}(t) = Ax(t), \; x(0) = x_0 \tag{3.6}$$

where $x(\cdot) : \mathbb{R}_+ \to \mathbb{R}^n$ and a quadratic Lyapunov function candidate $v_c(x) = x'Px$ with $0 < P \in \mathbb{R}^{n \times n}$ whose time derivative equals $\dot{v}_c(x) = -x'Qx$ with $0 < Q \in \mathbb{R}^{n \times n}$ provided that $P > 0$ solves the algebraic Lyapunov equation

$$A'P + PA = -Q \tag{3.7}$$

which, under these conditions, admits a symmetric positive definite solution if and only if A is Hurwitz stable. Some remarks are in order. First, for any $h > 0$, since any eigenvalue of e^{Ah} is of the form $z = e^{sh}$, where s is an eigenvalue of A and $|z| = e^{\text{Re}(s)h}$, then the conclusion is that A is Hurwitz stable if and only if e^{Ah} is Schur stable. Second, simple integration in the time interval $[0, h]$ provides

$$\begin{aligned} v_c(x(h)) - v_c(x(0)) &= \int_0^h \dot{v}_c(x(t))dt \\ &= -\int_0^h x(t)'Qx(t)dt \end{aligned} \tag{3.8}$$

which taking into account the solution of (3.6), $x(t) = e^{At}x_0$ valid for an arbitrary initial condition $x_0 \in \mathbb{R}^n$ yields

$$e^{A'h}Pe^{Ah} - P = -\int_0^h e^{A't}Qe^{At}dt \tag{3.9}$$

From the first remark, we can say that there exists a symmetric positive definite solution $P > 0$ to the continuous-time Lyapunov inequality $A'P + PA < 0$ if and only if there exists a symmetric positive definite solution $S > 0$ to the discrete-time Lyapunov inequality $e^{A'h}Se^{Ah} - S < 0$. From the second one, the key fact is that the inequalities $A'P + PA = -Q$ and $e^{A'h}Se^{Ah} - S = -W$ share the same symmetric positive definite solution $P = S > 0$ provided that

$$W = \int_0^h e^{A't}Qe^{At}dt \tag{3.10}$$

There is a limitation that must be understood carefully. If one solves the inequalities $A'P + PA < 0$, $P > 0$, then, from (3.9), it is also true that $e^{A'h}Pe^{Ah} - P < 0$ for every $h > 0$. However, if you solve the inequalities

$e^{A'h}Se^{Ah} - S < 0$, $S > 0$, for a certain $h > 0$, it is not always true that $A'S + SA < 0$. See the next example for a brief discussion on this point.

Example 3.1 *Consider the sampling period $h = 2$ and the following square matrix that defines a second order system:*

$$A = \begin{bmatrix} 0 & 1 \\ -4 & -1 \end{bmatrix}$$

The solution of $A'P + PA = -I$ has been calculated and, as expected, it implies that

$$P = \begin{bmatrix} 2.6250 & 0.1250 \\ 0.1250 & 0.6250 \end{bmatrix} > 0 \Longrightarrow e^{A'h}Pe^{Ah} - P = -\begin{bmatrix} 2.2083 & 0.0786 \\ 0.0786 & 0.5484 \end{bmatrix} < 0$$

On the other hand, we have solved $e^{A'h}Se^{Ah} - S = -I$ and its solution provides

$$S = \begin{bmatrix} 1.4588 & -0.0519 \\ -0.0519 & 1.0680 \end{bmatrix} > 0 \Longrightarrow A'S + SA = -\begin{bmatrix} -0.4155 & 2.7612 \\ 2.7612 & 2.2398 \end{bmatrix} \nless 0$$

Hence, this simple example puts in clear evidence the point raised in our previous discussion. Of course, the solution to the equation $e^{A'h}Se^{Ah} - S = -W$, where $W > 0$ is given in (3.10) with $Q = I$, yields $S = P > 0$. The interpretation is that the discrete-time Lyapunov equation depends on a precise value of $h > 0$, while the continuous-time Lyapunov equation provides a solution valid for every $h > 0$. □

Now, let us move our attention to the system resulting from the sampling of (3.6). Taking into account that $t_0 = 0$ and $t_{k+1} - t_k = h > 0$, $\forall k \in \mathbb{N}$, we have

$$\begin{aligned} x[k+1] &= e^{At_{k+1}}x_0 \\ &= e^{A(t_{k+1}-t_k)}e^{At_k}x_0 \\ &= e^{Ah}x[k] \end{aligned} \tag{3.11}$$

which is the state space realization of the corresponding sampled-data system. As a result, taking again into account the notation $A_d = e^{Ah}$, the sampled-data system can be rewritten as

$$x[k+1] = A_d x[k] \tag{3.12}$$

which evolves from $x[0] = x_0$ for all $k \in \mathbb{N}$. The Lyapunov function candidate $v_d(x) = x'Sx$ with $S > 0$ satisfies the inequality $v_d(x[k+1]) - v_d(x[k]) = -x[k]'Rx[k]$, for all $k \in \mathbb{N}$ if $S > 0$ solves the discrete-time Lyapunov equation

$$A_d'SA_d - S = -R \tag{3.13}$$

which, for $R > 0$, admits a symmetric positive definite solution if and only if A_d is Schur stable. It is important to stress that, alternatively, if we find a solution to $A'P + PA < 0, P > 0$, then (3.13) holds for $S = P > 0$ for some $R > 0$. Actually, if we solve $A'P + PA = -Q < 0, P > 0$, then once again (3.13) holds for $S = P > 0$ whenever $R = W > 0$.

This property allows an interesting interpretation as far as robust stability is concerned. Consider that the continuous-time system (3.6) is subject to norm bounded uncertainty, such that it becomes $\dot{x}(t) = A_\Delta x(t)$, where $A_\Delta = A + E\Delta C$ for all $||\Delta|| \leq \gamma^{-1}$. If A_Δ is Hurwitz stable and there exists $P > 0$ such that $A'_\Delta P + PA_\Delta < 0$, for all $||\Delta|| \leq \gamma^{-1}$, then $A_d(\Delta) = e^{A_\Delta h}$ is Schur stable and $A_d(\Delta)'PA_d(\Delta) - P < 0$ for all $||\Delta|| \leq \gamma^{-1}$. The same reasoning applies to the case of convex bounded uncertainty.

3.4 Performance

Performance indices similar to the well known $\mathcal{H}_2$ and $\mathcal{H}_\infty$ norms that are frequently adopted in practice as valid performance measures are affected by the sampling and hold mechanism. Now, we intend to propose similar performance indices in the framework of sampled-data control systems. Prior to that, a useful result that shows how to convert a sampled-data system to a pure discrete-time one is given. Among others, it is one of the key factors to deal with sampled-data control system analysis and design.

3.4.1 Equivalent System

Sampled-data systems have some particularities that can be exploited in benefit of analysis and control design. Actually, consider a generic sampled-data system with state space realization

$$\dot{x}(t) = Ax(t) + Bu(t) \tag{3.14}$$

$$z(t) = Cx(t) + Du(t) \tag{3.15}$$

evolving from an arbitrary initial condition $x(0) = x_0$ with $u \in \mathbb{U}$, which means that $u(t) = u[k],\ t \in [t_k, t_{k+1})$, for all $k \in \mathbb{N}$. The matrices are of compatible dimensions, that is, $A \in \mathbb{R}^{n \times n}$, $B \in \mathbb{R}^{n \times m}$, $C \in \mathbb{R}^{q \times n}$, and $D \in \mathbb{R}^{q \times m}$. The next definition introduces a pure discrete-time system denominated *equivalent system* which exhibits a very useful and important property.

Definition 3.3 *The discrete-time system with state space realization*

$$x[k+1] = A_d x[k] + B_d u[k] \tag{3.16}$$

$$z_d[k] = C_d x[k] + D_d u[k] \tag{3.17}$$

and arbitrary initial condition $x[0] = x_0$ *is called equivalent to the sampled-data system* (3.14)–(3.15) *if* $||z||_2^2 = ||z_d||_2^2$, *for every feasible input* $u \in \mathbb{U}$.

This definition makes clear by itself the possible existence of an LTI discrete-time system that behaves exactly as the associated continuous-time does, whenever the input is a signal belonging to the set $\mathbb{U}$. Even more surprising is that both are identical as far as the norm of their outputs is concerned. More precisely, from the same but arbitrary initial condition $x_0 \in \mathbb{R}^n$ and the same but arbitrary input sequence $u[k],\ k \in \mathbb{N}$, the equality

$$\int_0^\infty z(t)'z(t)dt = \sum_{k=0}^{\infty} z_d[k]'z_d[k] \tag{3.18}$$

holds. In order to determine the equivalent system, let us introduce once again the augmented matrices

$$F = \begin{bmatrix} A & B \\ 0 & 0 \end{bmatrix},\ G = \begin{bmatrix} C & D \end{bmatrix} \tag{3.19}$$

with dimensions $F \in \mathbb{R}^{(n+m)\times(n+m)}$ and $G \in \mathbb{R}^{q\times(n+m)}$. The next lemma provides the equivalent system satisfying Definition 3.3, by making explicit, as a by-product, a very simple procedure that can be adopted for its numerical calculation.

Lemma 3.1 *Consider that the matrices* $A_d \in \mathbb{R}^{n\times n}$ *and* $B_d \in \mathbb{R}^{n\times m}$ *are determined from the partition of the exponential function evaluated at the sampling period,*

$$e^{Fh} = \begin{bmatrix} A_d & B_d \\ 0 & I \end{bmatrix} \tag{3.20}$$

and the matrices $C_d \in \mathbb{R}^{(n+n)\times n}$ *and* $D_d \in \mathbb{R}^{(n+m)\times m}$ *are determined from the partition of* $G_d = [C_d\ D_d]$ *given by*

$$G_d'G_d = \int_0^h e^{F't}G'Ge^{Ft}dt \geq 0 \tag{3.21}$$

Invoking Definition 3.3, *the discrete-time system* (3.16)–(3.17) *is equivalent to the sampled-data system* (3.14)–(3.15) *with sampling period* $h > 0$.

Proof Taking into account that $u(t) = u[k]$ for all $t \in [t_k, t_{k+1})$, in this time interval, the solution of (3.14) yields

$$x(t) = e^{A(t-t_k)}x(t_k) + \left(\int_0^{t-t_k} e^{A\xi} Bd\xi\right) u[k] \tag{3.22}$$

which when evaluated at $t = t_{k+1}$ together with (3.20), we obtain the first equation (3.16) of the equivalent system. In addition, plugging the solution (3.22) into (3.15), we have

$$\begin{aligned} z(t) &= Ce^{A(t-t_k)}x[k] + \left(C\int_0^{t-t_k} e^{A\xi} Bd\xi + D\right) u[k] \\ &= Ge^{F(t-t_k)} \begin{bmatrix} x[k] \\ u[k] \end{bmatrix} \end{aligned} \tag{3.23}$$

and consequently it is possible to calculate the portion of the output norm corresponding to the k-th time interval, namely

$$\begin{aligned} \int_{t_k}^{t_{k+1}} z(t)'z(t)dt &= \begin{bmatrix} x[k] \\ u[k] \end{bmatrix}' \left(\int_{t_k}^{t_{k+1}} e^{F'(t-t_k)}G'Ge^{F(t-t_k)}dt\right) \begin{bmatrix} x[k] \\ u[k] \end{bmatrix} \\ &= z_d[k]'z_d[k] \end{aligned} \tag{3.24}$$

where the last equality follows from (3.21), together with the second equation (3.17) of the equivalent system. This is the key relationship that allows us to write

$$\begin{aligned} ||z||_2^2 &= \sum_{k=0}^{\infty} \int_{t_k}^{t_{k+1}} z(t)'z(t)dt \\ &= \sum_{k=0}^{\infty} z_d[k]'z_d[k] \\ &= ||z_d||_2^2 \end{aligned} \tag{3.25}$$

completing thus the proof. □

Given a sampled-data control system, there is no major difficulty to calculate the discrete-time equivalent system. It suffices to evaluate the matrix exponential (3.20) and to factorize a symmetric positive semi-definite matrix (3.21), both with dimensions equal to $n + m$. As Lemma 3.1 makes clear, the equivalent system has the same number of state variables n. However, concerning the output, in general, the output dimension of the equivalent system can reach the maximum possible value $n + m$ which is larger than the dimension of the output of the original sampled-data control system $q \leq n$. Notice that the partition of (3.21), as indicated before, provides $C_d \in \mathbb{R}^{(n+m)\times n}$ and $D_d \in \mathbb{R}^{(n+m)\times m}$ and the number of rows of both matrices can be reduced to the rank of $G_d'G_d \in \mathbb{R}^{(n+m)\times(n+m)}$. It is not mandatory, but, if one wants, a linear transformation can be applied to reduce the number of linearly independent output variables of the equivalent system. Moreover, simple algebraic manipulation shows that

$$C_d'C_d = \int_0^h e^{A't}C'Ce^{At}dt \in \mathbb{R}^{n\times n} \tag{3.26}$$

The importance of Lemma 3.1 stems from the fact that the well know $\mathcal{H}_2$ and $\mathcal{H}_\infty$ transfer function norms are expressed in terms of the norms of the system controlled output, respectively.

Example 3.2 *Consider the following matrices that define a sampled-data system:*

$$A = \begin{bmatrix} 0 & 1 & 0 \\ 0 & 0 & 1 \\ -1 & -5 & -9 \end{bmatrix}, \ B = \begin{bmatrix} 0 \\ 0 \\ 1 \end{bmatrix}, \ C' = \begin{bmatrix} 1 \\ 1 \\ 1 \end{bmatrix}, \ D = \begin{bmatrix} 1 \end{bmatrix}$$

with sampling period $h = 1.5$ *and initial condition* $x_0 = [0\ 0\ 1]'$. *From the result of Lemma 3.1, we have determined the equivalent system. It is interesting to verify that the equivalent system has order three but* $C_d \in \mathbb{R}^{4\times 3}$ *and* $D_d \in \mathbb{R}^{4\times 1}$. *Because, in this example,* $\det(G_d'G_d) \approx 0$, *within a good approximation, the dimension of the controlled output* $z_d[k]$ *can be reduced to three by an appropriate change of variables.*

For illustration, Fig. 3.2 shows the partial sums

$$I(t) = \int_0^t z(\tau)'z(\tau)d\tau$$

$$I_d[k] = \sum_{n=0}^{k} z_d[n]'z_d[n]$$

which reproduce the norms $||z||_2^2 = 0.1477$ *and* $||z_d||_2^2 = 0.1477$ *whenever* t *and* k *go to infinity. The function* $I(t)$ *is plotted in solid line and* $I_d[k]$ *in dotted stem*

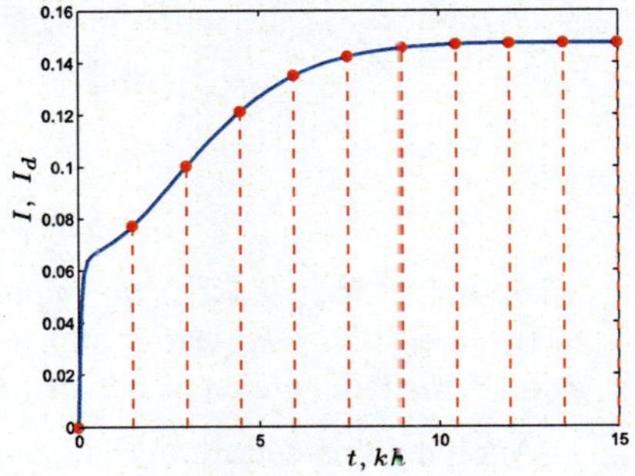

Fig. 3.2 Partial evaluation of norms

format. Both have been obtained with zero input $u[k] = 0$, for all $k \in \mathbb{N}$. The fact that $I(t_k) = I_d[k]$, $k \in \mathbb{N}$, illustrates the result of Lemma 3.1. □

A relevant point is to apply the result of Lemma 3.1 when the sampled-data system is corrupted by continuous-time and discrete-time impulsive perturbations. The main idea is to convert both perturbations to initial conditions and determine appropriate norms of the equivalent system transfer function which is possible, since it is completely defined in the discrete-time domain. As a final remark, it is important to keep in mind that Lemma 3.1 holds for any input $u \in \mathbb{U}$ even though the output norms $||z||_2^2 = ||z_d||_2^2$ are unbounded. These specific considerations and meanings are useful for the algebraic manipulations presented in the next section.

3.4.2 $\mathcal{H}_2$ Performance Analysis and Design

Norms are essential to build up performance indices for control system analysis and design. For LTI systems, they are calculated by manipulating the transfer function or, equivalently, in the specific case of $\mathcal{H}_2$ norm, the impulse response. We now provide these and related concepts in the framework of sampled-data systems.

Firstly, for completeness, let us focus on the $\mathcal{H}_2$ norm of a continuous-time system with the following state space minimal realization:

$$\dot{x}(t) = Ax(t) + Ew_c(t) \tag{3.27}$$

$$z(t) = Cx(t) \tag{3.28}$$

where matrices are of compatible dimensions, that is, $A \in \mathbb{R}^{n\times n}$, $E \in \mathbb{R}^{n\times r}$, and $C \in \mathbb{R}^{q\times n}$. The $\mathcal{H}_2$ norm depends on the impulse response of the system or, equivalently, on a set of initial conditions. Moreover, it is well defined (finite) for LTI systems with strictly proper transfer function only.

Definition 3.4 *The $\mathcal{H}_2$ norm of the transfer function $H_c(s) = C(sI - A)^{-1}E : \mathbb{C} \to \mathbb{C}^{q\times r}$ with A Hurwitz stable is*

$$\begin{aligned} ||H_c(s)||_2^2 &= \sum_{i=1}^{r} ||h_i||_2^2 \\ &= \sum_{i=1}^{r} \int_0^\infty h_i(t)'h_i(t)dt \end{aligned} \tag{3.29}$$

where $h_i(t) : \mathbb{R}_+ \to \mathbb{R}^q$ is the impulse response from the i-th perturbation channel $w_c(t) = e_i\delta(t)$ to the output $z(t)$, with $\delta(t)$ being the continuous-time impulse and $e_i \in \mathbb{R}^r$ the i-th column of the identity matrix. Otherwise, $||H_c(s)||_2^2 = +\infty$.

Applying this definition to (3.27)–(3.28) with zero initial condition $x(0) = 0$, it follows that

$$\begin{aligned} h_i(t) &= C\int_0^t e^{A(t-\tau)} E w_c(\tau) d\tau \\ &= C\int_0^t e^{A(t-\tau)} E e_i \delta(\tau) d\tau \\ &= Ce^{At} E e_i \end{aligned} \tag{3.30}$$

which makes it clear that the impulse response from the i-th perturbation channel of $w_c(t)$ to the output $z(t)$ with zero initial condition $x(0) = 0$ is equal to the response $z(t)$ with zero input $w_c(t) = 0$ and initial condition $x(0) = Ee_i$, for each $i = 1, 2, \ldots, r$. Plugging this result into (3.29), we obtain

$$\begin{aligned} \|H_c(s)\|_2^2 &= \sum_{i=1}^{r} \int_0^\infty e_i' E' e^{A't} C' C e^{At} E e_i dt \\ &= \mathbf{tr}\left(\int_0^\infty E' e^{A't} C' C e^{At} E dt\right) \\ &= \inf_{P>0} \left\{\mathbf{tr}(E'PE) : A'P + PA < -C'C\right\} \end{aligned} \tag{3.31}$$

where the last equality follows from the fact that the Lyapunov inequality admits a minimal matrix element. Hence, the optimal solution is some $P > 0$ arbitrarily close to the one of the algebraic Lyapunov equations $A'P + PA = -C'C$ which, for this reason, can be assumed to be strictly positive definite without loss of generality.

Secondly, let us define the $\mathcal{H}_2$ norm in the framework of discrete-time systems, with state space realization

$$x[k+1] = Ax[k] + Ew_d[k] \tag{3.32}$$

$$z[k] = Cx[k] + Dw_d[k] \tag{3.33}$$

where matrices are real and with the same previous dimensions. The $\mathcal{H}_2$ norm of this discrete-time system is formulated in terms of the transfer function from the perturbation input $w_d[k]$ to the output $z[k]$. Notice that, for discrete-time systems, the norm is also well defined for systems with proper transfer function.

Definition 3.5 *The $\mathcal{H}_2$ norm of the transfer function $H_d(\zeta) = C(\zeta I - A)^{-1}E + D : \mathbb{C} \to \mathbb{C}^{q\times r}$ with A Schur stable is*

$$\begin{aligned} \|H_d(\zeta)\|_2^2 &= \sum_{i=1}^{r} \|h_i\|_2^2 \\ &= \sum_{i=1}^{r}\sum_{k=0}^{\infty} h_i[k]' h_i[k] \end{aligned} \tag{3.34}$$

where $h_i[k] : \mathbb{N} \to \mathbb{R}^q$ is the impulse response from the i-th perturbation channel $w_d[k] = e_i\delta[k]$ to the output $z[k]$, with $\delta[k]$ being the discrete-time impulse and $e_i \in \mathbb{R}^r$ the i-th column of the identity matrix. Otherwise, $||H_d(\zeta)||_2^2 = +\infty$.

Imposing zero initial condition $x[0] = 0$ and setting the input $w_d[k] = e_i\delta[k]$ in (3.32)–(3.33), simple calculations lead to

$$h_i[k] = \begin{cases} De_i & , k = 0 \\ CA^{k-1}Ee_i & , k \geq 1 \end{cases} \tag{3.35}$$

which, when plugging in (3.34), allows the determination of

$$\begin{aligned} ||H_d(\zeta)||_2^2 &= \sum_{i=1}^{r}\left(\sum_{k=1}^{\infty} e_i'E'A'^{k-1}C'CA^{k-1}Ee_i + e_i'D'De_i\right) \\ &= \mathbf{tr}\left(\sum_{k=0}^{\infty} E'A'^kC'CA^kE + D'D\right) \\ &= \inf_{P>0}\left\{\mathbf{tr}(E'PE + D'D) : A'PA - P < -C'C\right\} \end{aligned} \tag{3.36}$$

where the last equality reveals a convex programming problem, whose optimal solution is arbitrarily close to the positive definite solution to the discrete-time algebraic Lyapunov equation. This occurs because it admits a minimal matrix element.

Finally, for future considerations, similarly of what we have done in the continuous-time case, let us convert the effect of the impulse response to an initial condition. To this end, notice that the response of the discrete-time system (3.32)–(3.33) with zero input $w_d[k] = 0$ and initial condition $x[0] = Ee_i$ is $z[k] = CA^kEe_i,\ k \geq 0$. Comparing to (3.35), it follows that $h_i[0] = De_i$ and $h_i[k] = z[k-1],\ \forall k \geq 1$, which means that $h_i[k],\ k \geq 1$, equals the output $z[k]$ delayed by one sampling period, implying that $||h_i||_2^2 = ||z||_2^2 + e_i'D'De_i$, for each $i = 1, 2, \ldots, r$. Whenever $D = 0$, both norms coincide.

Remark 3.1 *There is another equivalent characterization of the $\mathcal{H}_2$ norm of a discrete-time system. Indeed, consider that system (3.32)–(3.33) evolves from $k \in \{-1\}\bigcup\mathbb{N}$ with zero initial condition $x[-1] = 0$ and $w_d[k] = e_i\delta[k+1]$, that is, the discrete-time impulse is applied at time $k = -1$, yielding $w_d[-1] = e_i$ and $w_d[k] = 0,\ \forall k \in \mathbb{N}$. Instead of (3.35), the impulse response becomes*

$$h_i[k] = \begin{cases} De_i & , k = -1 \\ CA^kEe_i & , k \geq 0 \end{cases}$$

which, for $k \geq 0$, is nothing else than the response of the same discrete-time system with zero input $w_d[k] = 0$ and initial condition $x[0] = Ee_i$, for each $i = 1, 2, \ldots, r$, respectively. Hence, $h_i[k] = z[k],\ k \geq 0$, and $||h_i||_2^2 =$

$||z||_2^2 + e_i' D' D e_i$, for each $i = 1, 2, \ldots, r$. As far as the $\mathcal{H}_2$ norm is concerned, both are equivalent. Indeed, taking into account that the constant term $e_i' D' D e_i$ vanishes for strictly proper transfer functions, both norms are equal. □

Let us now move our focus to sampled-data control systems. For the moment, in order to clearly introduce the main idea, let us consider a sampled-data system that operates in open-loop, under the action of continuous-time and discrete-time exogenous perturbations, exclusively. Hence, the model becomes

$$\dot{x}(t) = Ax(t) + Bu(t) + Ew_c(t), \;\; x(0) = 0 \tag{3.37}$$

$$z(t) = C_z x(t) + D_z u(t) \tag{3.38}$$

$$u(t) = w_d[k], \;\; t \in [t_k, t_{k+1}), \;\; \forall k \in \mathbb{N} \tag{3.39}$$

and we are ready to well adapt the previous definitions concerning $\mathcal{H}_2$ norms of continuous-time and discrete-time systems to a model that depends on those two classes of exogenous perturbations, applied sequentially.

Definition 3.6 *The $\mathcal{H}_2$ performance index associated with the sampled-data system (3.37)–(3.39) is given by*

$$\varrho_2^2 = \sum_{i=1}^{r_c + r_d} ||z_i||_2^2 \tag{3.40}$$

where $z(t) = z_i(t)$, $1 \leq i \leq r_c$, is the output due to a continuous-time impulse $\delta(t)$ in the i-th channel of the input $w_c(t)$ and $z(t) = z_i(t)$, $r_c + 1 \leq i \leq r_c + r_d$, is the output due to a discrete-time impulse $\delta[k]$ in the i-th channel of the input $w_d[k]$.

The determination of this performance index needs the following two successive steps to be performed:

- **Continuous-time impulsive input**—The continuous-time impulse responses $1 \leq i \leq r_c$ are obtained by setting $w_d[k] = 0$ for all $k \in \mathbb{N}$. Doing this, the sampled-data system (3.37)–(3.39) reduces to

$$\dot{x}(t) = Ax(t) + Ew_c(t) \tag{3.41}$$

$$z(t) = C_z x(t) \tag{3.42}$$

and consequently we obtain

$$\begin{aligned} \sum_{i=1}^{r_c} ||z_i||_2^2 &= ||H_c(s)||_2^2 \\ &= ||C_z(sI - A)^{-1}E||_2^2 \end{aligned} \tag{3.43}$$

whenever matrix A is Hurwitz stable.

- **Discrete-time impulsive input**—The discrete-time impulse responses $r_c + 1 \leq i \leq r_c + r_d$ are obtained by setting $w_c(t) = 0$ for all $t \in \mathbb{R}_+$. Doing this, the equivalent system associated with (3.37)–(3.39) has the form

$$x[k+1] = A_d x[k] + B_d w_d[k] \tag{3.44}$$

$$z[k] = C_{zd} x[k] + D_{zd} w_d[k] \tag{3.45}$$

which yields

$$\begin{aligned}\sum_{i=r_c+1}^{r_c+r_d} ||z_i||_2^2 &= ||H_d(\zeta)||_2^2 \\ &= ||C_{zd}(\zeta I - A_d)^{-1} B_d + D_{zd}||_2^2\end{aligned} \tag{3.46}$$

whenever matrix A_d is Schur stable.

The $\mathcal{H}_2$ performance index stated in Definition 3.6 follows immediately from these calculations by simply adding the contribution of each exogenous perturbation, that is,

$$\varrho_2^2 = ||H_c(s)||_2^2 + ||H_d(\zeta)||_2^2 \tag{3.47}$$

As expected, this performance index depends on two types of norms, due to the continuous-time and discrete-time nature of the inputs $w_c(t)$ and $w_d[k]$, respectively. However, remembering that the solution $P > 0$ of $A'P + PA = -C_z'C_z$ can alternatively be determined from $A_d'PA_d - P = -C_{zd}'C_{zd}$, then it follows that

$$\begin{aligned}||C_z(sI - A)^{-1}E||_2^2 &= \mathbf{tr}(E'PE) \\ &= ||C_{zd}(\zeta I - A_d)^{-1}E||_2^2\end{aligned} \tag{3.48}$$

which makes it clear that the impulse response of a continuous-time system is exactly converted to the discrete-time impulse response of its equivalent system. Hence, we are able to express the $\mathcal{H}_2$ performance index in the discrete-time domain, exclusively, that is,

$$\varrho_2^2 = ||H_{sd}(\zeta)||_2^2 \tag{3.49}$$

where $H_{sd}(\zeta) : \mathbb{C} \to \mathbb{C}^{n_z \times (r_c + r_d)}$ is the augmented transfer function

$$H_{sd}(\zeta) = C_{zd}\,(\zeta I - A_d)^{-1}\begin{bmatrix} E & B_d \end{bmatrix} + \begin{bmatrix} 0 & D_{zd} \end{bmatrix} \tag{3.50}$$

which can be interpreted as the one that retains the effects of all external perturbations acting on the sampled-data system. The issue of main importance is that, due to Lemma 3.1, it is completely formulated in the discrete-time domain. It is clear that the $\mathcal{H}_2$ performance index is finite whenever matrix A_d is Schur stable. Consequently, as we have previously done, the LMI

$$A_d' P A_d - P < -C_{zd}' C_{zd} \tag{3.51}$$

defines a convex set which, whenever nonempty, admits a minimal matrix element and consequently

$$\varrho_2^2 = \inf_{P>0} \{\mathbf{tr}(E'PE) + \mathbf{tr}(B_d' P B_d) : (3.51)\} + \mathbf{tr}(D_{zd}' D_{zd}) \tag{3.52}$$

Since the minimal matrix element is the positive definite solution to the Lyapunov equation, the value of the norm ϱ_2^2 can be directly calculated from it. In general, this requires a smaller computational effort.

Remark 3.2 *From the previous results, in particular the equality (3.48), it is interesting to put in evidence that the output $z(t) = C_z x(t)$ of the continuous-time system $\dot{x}(t) = Ax(t) + Ew_c(t)$ with zero initial condition $x(0) = 0$ and impulsive input $w_c(t) = w\delta(t)$ for some $w \in \mathbb{R}^{r_c}$ and the output $z_d[k] = C_{zd}x[k]$ of the discrete-time system $x[k+1] = A_d x[k] + Ew_c[k]$ with zero initial condition $x[0] = 0$ and impulsive input $w_c[k] = w\delta[k]$ are not identical, but they have the same norm, that is, $||z||_2^2 = ||z_d||_2^2$ for all $w \in \mathbb{R}^{r_c}$. This is a mere consequence of Lemma 3.1 that can also be obtained from the simple observation that the Lyapunov equations $A'P + PA = -C_z'C_z$ and $A_d' P A_d - P = -C_{zd}' C_{zd}$ admit the same solution $P > 0$ provided that $A_d = e^{Ah}$ is Schur stable and*

$$C_{zd}' C_{zd} = \int_0^h e^{A't} C_z' C_z e^{At} dt$$

which naturally implies that $||z||_2^2 = w'E'PEw = ||z_d||_2^2$. We want to put in clear evidence the notation we are introducing here. As far as we are concerned with impulsive inputs, the effect of the continuous-time input $w_c(t)$ in the equivalent system is denoted $w_c[k]$ according to the fact that it acts in a pure discrete-time system. Hence, the $\mathcal{H}_2$ performance index is determined from the equivalent discrete-time system

$$\begin{aligned} x[k+1] &= A_d x[k] + B_d u[k] + Ew_c[k], \ x[0] = 0 \\ y[k] &= C_y x[k] + E_y w_d[k] \\ z[k] &= C_{zd} x[k] + D_{zd} u[k] \end{aligned}$$

with a discrete-time impulse $\delta[k]$ in each channel of $w_c[k]$ and $w_d[k]$, respectively.
□

To put these results in a clearer perspective, let us consider the sampled-data control system (3.1)–(3.3) with $C_y = I$, which means that the whole state variable is available for feedback. The goal is to determine the matrix gain $L \in \mathbb{R}^{n_u \times n_x}$ such that with the state feedback control $v[k] = Ly[k]$, $k \in \mathbb{N}$, the closed-loop sampled-data control system is asymptotically stable and the $\mathcal{H}_2$ performance index is minimized. At this first moment, we also assume that the system under consideration is perturbation free, that is, $E_y = 0$ and $E_u = 0$. Hence, from (3.2) together with (3.5), we have $u[k] = Lx[k]$. Taking into account (3.52) and Remark 3.2, this optimal control problem is formulated as

$$\inf_{P>0,L} \{\mathbf{tr}(E'PE) : A_L'PA_L - P < -C_L'C_L\} \tag{3.53}$$

where $C_L = C_{zd} + D_{zd}L$ and $A_L = A_d + B_dL$ are the closed-loop system matrices. There are two possible ways to solve it. The first one follows from the fact that the optimality condition reveals that the optimal matrix gain must satisfy

$$L = -R_d^{-1}\left(D_{zd}'C_{zd} + B_d'PA_d\right) \tag{3.54}$$

where $R_d = D_{zd}'D_{zd} + B_d'PB_d$ and $P > 0$ is the positive definite stabilizing solution to the discrete-time algebraic Riccati equation

$$P = A_d'PA_d - \left(D_{zd}'C_{zd} + B_d'PA_d\right)' R_d^{-1}\left(D_{zd}'C_{zd} + B_d'PA_d\right) + C_{zd}'C_{zd} \tag{3.55}$$

which can be proven to be unique and such that A_L is Schur stable. Efficient numerical methods to solve this nonlinear equation are available in the literature. Another possibility is to express the matrix inequality in (3.53) as

$$\begin{bmatrix} P & A_L' & C_L' \\ \bullet & P^{-1} & 0 \\ \bullet & \bullet & I \end{bmatrix} > 0 \tag{3.56}$$

which, when multiplied both sides by $\mathrm{diag}(P^{-1}, I, I)$ and defining the one-to-one change of variables $P^{-1} = Q > 0$ and $Y = LP^{-1}$, yields the LMI

$$\begin{bmatrix} Q & QA_d' + Y'B_d' & QC_{zd}' + Y'D_{zd}' \\ \bullet & Q & 0 \\ \bullet & \bullet & I \end{bmatrix} > 0 \tag{3.57}$$

which provides the optimal matrix gain from the inverse transformation $L = YQ^{-1}$, where the indicated matrix variables solve the jointly convex optimization problem

$$\inf_{Q>0,Y} \left\{\mathbf{tr}\left(E'Q^{-1}E\right) : (3.57)\right\} \tag{3.58}$$

As we have just formulated, the $\mathcal{H}_2$ optimal sampled-data control problem is solved in the discrete-time domain. This is an immediate consequence of the equivalent system provided by Lemma 3.1, see also Remark 3.2. However, it must be clear that, in this way, one is not able to cope with robustness against parameter uncertainty. The next example illustrates this theoretical result.

Example 3.3 *This example is inspired by Example 3.2. Consider a sampled-data system with state space realization matrices*

$$A = \begin{bmatrix} 0 & 1 & 0 \\ 0 & 0 & 1 \\ 0 & -5 & -9 \end{bmatrix}, \; B = \begin{bmatrix} 0 \\ 0 \\ 1 \end{bmatrix}, \; C_z' = \begin{bmatrix} 1 \\ 1 \\ 1 \end{bmatrix}, \; D_z = \begin{bmatrix} 1 \end{bmatrix}$$

with sampling period $h = 1.5$ *and* $E = [\, 0\ 0\ 1]'$. *From the result of Lemma 3.1, we have determined the equivalent system and we have solved numerically the algebraic Riccati equation (3.55) to get the optimal gain matrix*

$$L = \begin{bmatrix} -0.8590 & -0.8120 & -0.1401 \end{bmatrix}$$

and the associated minimum cost $\varrho_2^2 = 0.0404$. *The eigenvalues of the closed-loop system matrix* A_L *are* $\{0.6216 \pm j0.0883, -0.0074\}$, *which confirms that it is Schur stable. As already mentioned, this example shows that the sampled-data state feedback control design in the context of* $\mathcal{H}_2$ *performance introduced here reduces to the well known control design procedure of a pure discrete-time LTI system that makes possible the use of well known algebraic Riccati equation or LMI solvers. This is possible thanks to the calculation of the equivalent system as characterized in Definition 3.3.* □

It can be noticed that the algebraic Riccati equation (3.55) does not depend on matrix E appearing in the cost of problem (3.58) which is a consequence of the existence of a minimal matrix element of the Lyapunov inequality. Now, if we drop the assumption $E_u = 0$, then $u[k] = Lx[k] + E_u w_d[k]$ in which case, the problem to be solved turns out to be

$$\inf_{P>0,L} \{\mathbf{tr}(E'PE) + \mathbf{tr}(E_u' B_d' P B_d E_u) : A_L' P A_L - P < -C_L' C_L\} \tag{3.59}$$

whose optimal solution, due to the reasons raised before, is given by (3.54) and (3.55). Of course the minimum value of ϱ_2^2 must be adjusted by adding to it the constant $\mathbf{tr}(E_u' D_{zd}' D_{zd} E_u)$ similar to what we have done in (3.52). Finally, if we also drop the assumption $E_y = 0$, then $u[k] = Lx[k] + (E_u + LE_y)w_d[k]$ and the problem becomes extremely hard to solve. Unfortunately, until the present moment, we have to recognize that this control design problem cannot be approached neither by the algebraic Riccati equation solution nor by LMI solvers.

In this framework, it is natural to go further by investigating the design of a full order dynamic output feedback controller C_h with state space realization

$$\hat{x}[k+1] = \hat{A}\hat{x}[k] + \hat{B}y[k] \tag{3.60}$$

$$v[k] = \hat{C}\hat{x}[k] + \hat{D}y[k] \tag{3.61}$$

with initial condition $\hat{x}[0] = 0$. The key assumption is that $\hat{x}[\cdot] : \mathbb{N} \to \mathbb{R}^{n_x}$, which indicates that the number of state variables of the sampled-data system is equal to the number of the controller state variables. For this reason, it is denominated *full order* dynamic output feedback controller. According to Fig. 3.1, considering the equivalent system given in Remark 3.2 together with the controller (3.60)–(3.61) and the control action (3.5), the closed-loop system has the state space realization

$$\varphi[k+1] = A_{cl}\varphi[k] + E_{cl}w[k] \tag{3.62}$$

$$z[k] = C_{cl}\varphi[k] + D_{cl}w[k] \tag{3.63}$$

where $\varphi[k]' = [x[k]' \ \hat{x}[k]'] \in \mathbb{R}^{2n_x}$, $w[k]' = [w_c[k]' \ w_d[k]'] \in \mathbb{R}^{r_c+r_d}$, and the indicated matrices of compatible dimensions are

$$A_{cl} = \begin{bmatrix} A_d + B_d\hat{D}C_y & B_d\hat{C} \\ \hat{B}C_y & \hat{A} \end{bmatrix}, \ E_{cl} = \begin{bmatrix} E & B_dE_u + B_d\hat{D}E_y \\ 0 & \hat{B}E_y \end{bmatrix} \tag{3.64}$$

and

$$C_{cl} = \begin{bmatrix} C_{zd} + D_{zd}\hat{D}C_y & D_{zd}\hat{C} \end{bmatrix}, \ D_{cl} = \begin{bmatrix} 0 & D_{zd}E_u + D_{zd}\hat{D}E_y \end{bmatrix} \tag{3.65}$$

Our goal now is to determine the optimal controller $\mathcal{C}_h$ that minimizes the $\mathcal{H}_2$ performance index associated with the discrete-time system (3.62)–(3.63). To accomplish this task, we have to solve

$$\inf_{P>0,\mathcal{C}_h} \{\mathbf{tr}(E_{cl}'PE_{cl} + D_{cl}'D_{cl}) : A_{cl}'PA_{cl} - P < -C_{cl}'C_{cl}\} \tag{3.66}$$

which is a jointly non-convex programming problem with respect to the matrices $P > 0$ and $(\hat{A}, \hat{B}, \hat{C}, \hat{D})$ that define the full order output feedback controller $\mathcal{C}_h$. Fortunately, there exists a one-to-one change of variables that converts this control design problem into a convex one expressed through LMIs. We proceed in this vein because the same algebraic manipulations will be useful in all forthcoming chapters. The Bibliography notes of this chapter give more information about this procedure, including the historical origin of the mentioned one-to-one change of variables, which is the key issue to express the jointly nonlinear constraints of (3.66) as equivalent LMIs.

From the very beginning, let us introduce the following symmetric positive definite matrices partitioned in four square matrix blocks with dimensions $n_x \times n_x$, namely

$$P = \begin{bmatrix} X & V \\ V' & \hat{X} \end{bmatrix}, \; P^{-1} = \begin{bmatrix} Y & U \\ U' & \hat{Y} \end{bmatrix} \tag{3.67}$$

and we observe that matrices V and U, if necessary after arbitrary small perturbations, can be assumed to be non-singular, with no loss of generality. Moreover, defining the full rank square matrix

$$\Gamma = \begin{bmatrix} I & Y \\ 0 & U' \end{bmatrix} \tag{3.68}$$

it is easily seen that

$$\Gamma' P \Gamma = \begin{bmatrix} X & I \\ I & Y \end{bmatrix} \tag{3.69}$$

making it clear that the equivalent constraints $P > 0 \leftrightarrow \Gamma' P \Gamma > 0$ are LMIs. With similar algebraic manipulations that involve only matrix multiplications, several relationships among matrices of the closed-loop system state space realization (3.62)–(3.63) are given in the next remark.

Remark 3.3 *The key issue to solve problem (3.66) is the change of variables*

$$(M, K, L, D) \iff (\hat{A}, \hat{B}, \hat{C}, \hat{D})$$

defined as

$$\begin{bmatrix} M - X A_d Y & K \\ L & D \end{bmatrix} = \begin{bmatrix} V & X B_d \\ 0 & I \end{bmatrix} \begin{bmatrix} \hat{A} & \hat{B} \\ \hat{C} & \hat{D} \end{bmatrix} \begin{bmatrix} U' & 0 \\ C_y Y & I \end{bmatrix}$$

which is a one-to-one transformation since matrices V and U are invertible. This transformation is of particular importance because it allows us to express the following matrices:

$$\Gamma' P A_{cl} \Gamma = \begin{bmatrix} X A_d + K C_y & M \\ A_d + B_d D C_y & A_d Y + B_d L \end{bmatrix}$$

$$C_{cl} \Gamma = \begin{bmatrix} C_{zd} + D_{zd} D C_y & C_{zd} Y + D_{zd} L \end{bmatrix}$$

$$\Gamma' P E_{cl} = \begin{bmatrix} X E & X B_d E_u + K E_y \\ E & B_d E_u + B_d D E_y \end{bmatrix}$$

$$D_{cl} = \begin{bmatrix} 0 & D_{zd}E_u + D_{zd}DE_y \end{bmatrix}$$

linearly *with respect to the set of the new matrix variables* (M, K, L, D). *In this case, with some abuse of language, we can say that the matrices have been* linearized. *Moreover, adopting it, the control design problem (3.66) becomes convex with linear objective function and LMI constraints.* □

This remark appears very simple, but the reader must appreciate an important historical algebraic manipulation whose proposal is appropriately quoted in the Bibliography notes. The same idea can be used to convert a huge number of control design problems to convex ones, many of them going beyond the subject treated in this book. Furthermore, it is important to mention that in the forthcoming chapters it will be generalized to cope with DLMIs instead of LMIs.

We are now in a position to express the control design problem (3.66) in terms of LMIs. The proof takes into account the fact that the matrix variables transformation introduced in Remark 3.3 is one-to-one.

Theorem 3.1 *Consider the following LMIs expressed in terms of the matrix variables* M, K, L, *and* D *and the symmetric matrix variables* W_c, W_d, X, *and* Y *of compatible dimensions, namely*

$$\begin{bmatrix} X & I & A_d'X + C_y'K' & A_d' + C_y'D'B_d' & C_{zd}' + C_y'D'D_{zd}' \\ \bullet & Y & M' & YA_d' + L'B_d' & YC_{zd}' + L'D_{zd}' \\ \bullet & \bullet & X & I & 0 \\ \bullet & \bullet & \bullet & Y & 0 \\ \bullet & \bullet & \bullet & \bullet & I \end{bmatrix} > 0 \tag{3.70}$$

$$\begin{bmatrix} W_c & E'X & E' \\ \bullet & X & I \\ \bullet & \bullet & Y \end{bmatrix} > 0 \tag{3.71}$$

$$\begin{bmatrix} W_d & E_u'B_d'X + E_y'K' & E_u'B_d' + E_y'D'B_d' & E_u'D_{zd}' + E_y'D'D_{zd}' \\ \bullet & X & I & 0 \\ \bullet & \bullet & Y & 0 \\ \bullet & \bullet & \bullet & I \end{bmatrix} > 0 \tag{3.72}$$

The optimal control design problem (3.66) is equivalent to the convex programming problem

$$\inf_{M,K,L,D,W_c,W_d,X,Y} \{\mathbf{tr}(W_c) + \mathbf{tr}(W_d) : (3.70)\text{-}(3.72)\} \tag{3.73}$$

Proof Firstly, let us rewrite the constraint of problem (3.66) as

$$\begin{bmatrix} P & A_{cl}'P & C_{cl}' \\ \bullet & P & 0 \\ \bullet & \bullet & I \end{bmatrix} > 0 \tag{3.74}$$

which, multiplying it to the left by diag(Γ', Γ', I) and to the right by the transpose, using the matrix calculations and the one-to-one change of variables given in Remark 3.3 the inequality (3.70) follows. Secondly, the objective function of problem (3.66) can be rewritten as

$$\mathbf{tr}(E_{cl}'PE_{cl} + D_{cl}'D_{cl}) = \mathbf{tr}(E_{cl}'P\Gamma(\Gamma'P\Gamma)^{-1}\Gamma'PE_{cl} + D_{cl}'D_{cl}) \tag{3.75}$$

which taking again into account Remark 3.3, it can be split into two terms bounded above by matrices W_c and W_d satisfying the LMIs (3.71) and (3.72), respectively. The proof is complete. □

The existence of a solution to the problem (3.73) depends on the existence of a controller C_h that makes the matrix A_{cl} Schur stable. Under this mild condition, once the optimal solution is found, it remains to apply the inverse transformation to get matrices $(\hat{A}, \hat{B}, \hat{C}, \hat{D})$ of the controller state space realization (3.60)–(3.61). From Remark 3.3, to build the inverse transformation, one needs the matrix variables determined from the optimal solution of (3.73) and matrices V and U. They are determined by simply imposing that the partitioned matrices in (3.67) must be such that $PP^{-1} = I$. To this end, let us observe that any feasible solution to the LMIs (3.70)–(3.72) satisfies

$$\begin{bmatrix} X & I \\ I & Y \end{bmatrix} > 0 \iff X > Y^{-1} > 0 \tag{3.76}$$

which, by (3.69), implies that $P > 0$. Given X, Y satisfying (3.76) and U nonsingular the partitioned equation $PP^{-1} = I$ provides

$$XY + VU' = I \rightarrow V = (I - XY)U'^{-1} \tag{3.77}$$

$$XU + V\hat{Y} = 0 \rightarrow \hat{Y} = U'\left(Y - X^{-1}\right)^{-1} U \tag{3.78}$$

$$V'Y + \hat{X}U' = 0 \rightarrow \hat{X} = U^{-1}Y\left(X - Y^{-1}\right)YU'^{-1} \tag{3.79}$$

and, more importantly, if we plug these solutions into the forth equation, we obtain

$$\begin{aligned} V'U + \hat{X}\hat{Y} &= U^{-1}\left((I - YX) + (YX - I)YX(YX - I)^{-1}\right)U \\ &= U^{-1}(I - YX)\big((YX - I) - YX\big)(YX - I)^{-1}U \\ &= I \end{aligned} \tag{3.80}$$

which means that the non-singular matrix U can be arbitrarily chosen and, at the same time, it can keep the equation $PP^{-1} = I$ satisfied. A possible choice is

$$U = Y > 0 \rightarrow V = Y^{-1} - X < 0 \tag{3.81}$$

in which case, the one-to-one change of variables introduced in Remark 3.3 becomes completely well defined and operational. It can be verified that the same reasoning holds if instead of U, a matrix V non-singular is arbitrarily chosen.

It is imperative to compare the state feedback controller and the dynamic output feedback controller whenever the more general sampled-data control system model is adopted. As we have seen, the state feedback sampled-data control design problem is not convex and, by consequence, it is very difficult to solve by Riccati equation or LMI solvers. On the contrary, the full dynamic output feedback sampled-data control design problem is convex and it can be solved with no difficulty by any LMI solver. This is possible because the problem (3.73) has a linear objective function and constraints expressed by LMIs. For this reason, even when the whole state can be measured (which indicates that $C_y = I$), the design of a full dynamic output feedback controller is more indicated since the conditions for existence and optimality are globally solved. The next example illustrates this important aspect of sampled-data control system design.

Example 3.4 *The sampled-data control system is defined by matrices A, B, E, C_z, and D_z given in Example 3.3 and $h = 1.5$. The solution of the problem (3.73) has been implemented using Matlab routines. The following cases have been solved:*

(a) The whole state is available for feedback ($C_y = I$) and the control is immune to external perturbations, that is, $E_u = 0$ and $E_y = 0$. The optimal solution has a minimum cost equal to $\varrho_2^2 = 0.0404$, identical to the one already calculated in Example 3.3.

(b) The whole state is available for feedback ($C_y = I$) but with control and measurement perturbations

$$E_y = \begin{bmatrix} 1 & 0 \\ 1 & 0 \\ 1 & 0 \end{bmatrix}, \quad E_u = \begin{bmatrix} 0 & 1 \end{bmatrix}$$

Now the cost raises to $\varrho_2^2 = 2.1667$. It is important to emphasize that, in this case, the state feedback control design problem is not convex and is very hard to solve. However, the full order dynamic output feedback control design problem is convex and its global optimal solution is determined with no major difficulty.

(c) We have considered that only the first state variable is available for feedback with control and measurement perturbations

$$C_y = \begin{bmatrix} 1 & 0 & 0 \end{bmatrix}, \quad E_y = \begin{bmatrix} 1 & 0 \end{bmatrix}, \quad E_u = \begin{bmatrix} 0 & 1 \end{bmatrix}$$

The optimal solution of problem (3.73) provides the minimum cost $\rho_2^2 = 2.6182$ *and the optimal controller (3.60)–(3.61) with transfer function*

$$\mathcal{C}_h(\zeta) = \frac{-0.2401\zeta^3 + 0.08362\zeta^2 - 0.001399\zeta - 0.0004028}{\zeta^3 - 0.9682\zeta^2 + 0.2824\zeta + 0.006377}$$

Observe that this controller of full order (three) operates under continuous-time and discrete-time exogenous impulsive perturbations. Closed-loop stability and optimal $\mathcal{H}_2$ *performance are granted.*

This example puts in evidence an important fact. The optimal $\mathcal{H}_2$ *performance controller for the most general sampled-data control system can be determined through the solution of a jointly convex programming problem expressed by LMIs.* □

Example 3.5 *The sampled-data control system is defined by* $h = 1.5$ *and the matrices of the state space realization given in Example 3.4, item (c) but*

$$A = \begin{bmatrix} 0 & 1 & 0 \\ 0 & 0 & 1 \\ 0 & 5 & -9 \end{bmatrix}$$

with eigenvalues $\{0.0000, 0.5249, -9.5249\}$, *which indicates that the open-loop system is clearly unstable. The minimum cost is* $\varrho_2^2 = 240.7332$ *and the associated optimal controller transfer function is given by*

$$\mathcal{C}_h(\zeta) = \frac{-4.36\zeta^3 + 4.299\zeta^2 - 0.01205\zeta - 0.0001042}{\zeta^3 + 0.05453\zeta^2 + 0.08504\zeta + 0.0004305}$$

It has been verified that the matrix A_{cl} *is Schur stable. Figure 3.3 shows the time simulation of the closed-loop system evolving after application, simultaneously, at* $t_0 = 0$, *of one continuous-time impulse in* $w_c(t)$ *and two discrete-time impulses, one in each channel of* $w_d[k]$. *It can be viewed that the sampled-data controller is very effective even though the sampling period* $h = 1.5$ *is relatively large.* □

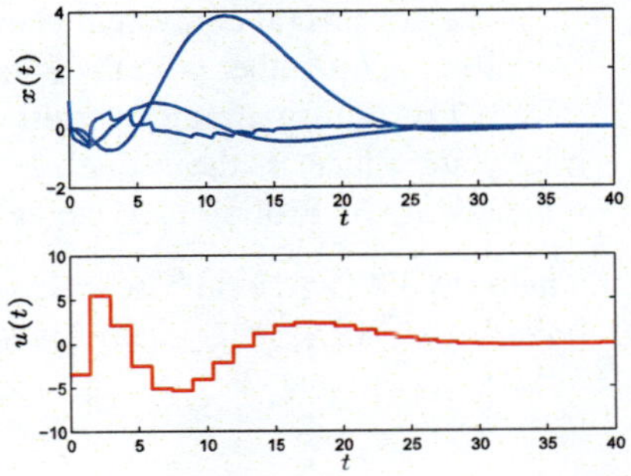

Fig. 3.3 State and control trajectories

This section ends by putting again in evidence the importance of the full order dynamic output feedback controller design procedure introduced in this book. A general sampled-data control system model has been considered keeping the optimal control design problem convex with linear objective function and constraints expressed by LMIs. This has been possible due to the adoption of a well known one-to-one variable transformation that, in addition, is useful to deal with DLMIs in the next sections.

3.4.3 $\mathcal{H}_\infty$ Performance Analysis and Design

The notion of $\mathcal{H}_\infty$ norm in the context of LTI systems is essential whenever one needs to cope with model or, more specifically, parameter uncertainty. Hence, it can be said that robustness and the $\mathcal{H}_\infty$ norm are strongly related in the sense that the study of the former is based on the properties of the latter. In this section, the calculation of $\mathcal{H}_\infty$ norm is done by using Riccati equations and LMI solvers.

Firstly, let us consider the following continuous-time system state space realization:

$$\dot{x}(t) = Ax(t) + Ew_c(t) \tag{3.82}$$

$$z(t) = Cx(t) + Dw_c(t) \tag{3.83}$$

where $A \in \mathbb{R}^{n\times n}$, $E \in \mathbb{R}^{n\times r}$, $C \in \mathbb{R}^{q\times n}$, and $D \in \mathbb{R}^{q\times r}$. It is assumed that the system evolves from zero initial condition $x(0) = x_0 = 0$ and $w_c \in \mathcal{L}_2$, which means that the continuous-time exogenous signal is norm bounded.

Definition 3.7 *The $\mathcal{H}_\infty$ norm of the transfer function $H_c(s) = C(sI - A)^{-1}E + D : \mathbb{C} \to \mathbb{C}^{q\times r}$ with A Hurwitz stable is*

$$||H_c(s)||_\infty^2 = \sup_{0\neq w_c\in\mathcal{L}_2} \frac{||z||_2^2}{||w_c||_2^2} \tag{3.84}$$

where $z(t) : \mathbb{R}_+ \to \mathbb{R}^q$ is the response of the continuous-time system (3.82)–(3.83), with $x(0) = 0$, to the input $w_c(t) : \mathbb{R}_+ \to \mathbb{R}^r$. Otherwise, $||H_c(s)||_\infty^2 = +\infty$.

There are several equivalent characterizations of the $\mathcal{H}_\infty$ norm that follow from Definition 3.7 together with the Parseval Theorem and Kalman–Yakubovich–Popov (KYP) Lemma, two major results in the framework of systems analysis that we briefly introduce in the sequel. For more details and proofs, see the references indicated in the Bibliography notes of this chapter.

Theorem 3.2 (Parseval Theorem) *Consider that $z(t) : \mathbb{R}_+ \to \mathbb{R}^q$ is a norm bounded signal $z \in \mathcal{L}_2$ with Laplace transform $\hat{z}(s)$, and the following equality holds:*

$$\int_0^\infty z(t)'z(t)dt = \frac{1}{\pi}\int_0^\infty \hat{z}(-j\omega)'\hat{z}(j\omega)d\omega \tag{3.85}$$

The main aspect to highlight is that, for norm bounded signals, the Fourier transform of $z(t)$ equals the Laplace transform assessed at $s = j\omega$. On the other hand, taking into account that $\hat{z}(s) = H_c(s)\hat{w}_c(s)$, then $z \in \mathcal{L}_2$ whenever $w_c \in \mathcal{L}_2$ because A is Hurwitz stable. As a consequence, the inequality

$$\begin{aligned}\int_0^\infty z(t)'z(t)dt &= \frac{1}{\pi}\int_0^\infty \hat{w}_c(-j\omega)'H_c(-j\omega)'H_c(j\omega)\hat{w}_c(j\omega)d\omega \\ &< \frac{\gamma^2}{\pi}\int_0^\infty \hat{w}_c(-j\omega)'\hat{w}_c(j\omega)d\omega \\ &= \gamma^2\int_0^\infty w_c(t)'w_c(t)dt \end{aligned} \tag{3.86}$$

holds for all $w_c \in \mathcal{L}_2$ provided that $H_c(-j\omega)'H_c(j\omega) < \gamma^2 I$ for all $\omega \in \mathbb{R}_+$. This yields the equivalent formulation

$$\begin{aligned}||H_c(s)||_\infty^2 &= \inf_{\mu>0}\left\{\mu : H_c(-j\omega)'H_c(j\omega) < \mu I,\ \forall\omega \geq 0\right\} \\ &= \sup_{\omega\geq 0}\sigma_{max}^2\left(H_c(j\omega)\right)\end{aligned} \tag{3.87}$$

where $\mu = \gamma^2$, and $\sigma_{max}(\cdot)$ denotes the maximum singular value. The numerical determination of the sup operator indicated in the right hand side of (3.87) is simple and well established in the literature. A possibility is to draw the so-called *singular value diagram*. From these calculations, it is seen that $||H_c(s)||_\infty^2 < \gamma^2$, but, formally, it remains to show that the minimum feasible value of $\mu = \gamma^2$ provides the square of the $\mathcal{H}_\infty$ norm. To demonstrate that this fact actually occurs, we need the following simplified version of the KYP Lemma, also referred to as *Bounded-Real Lemma*.

Lemma 3.2 (Bounded-Real Lemma) *Assume that A is a Hurwitz stable matrix. The following statements are equivalent:*

(i) $H_c(-j\omega)'H_c(j\omega) < \mu I,\ \forall\omega \in \mathbb{R}_+$.
(ii) *There exists a symmetric positive definite matrix $P > 0$ of compatible dimensions such that the LMI*

$$\begin{bmatrix} A'P + PA & PE & C' \\ \bullet & -\mu I & D' \\ \bullet & \bullet & -I \end{bmatrix} < 0 \tag{3.88}$$

is feasible.

This lemma provides a simple way to calculate the maximum singular value in the right hand side of (3.87) by solving the convex programming problem

$$\sup_{\omega \geq 0} \sigma_{max}^2\big(H_c(j\omega)\big) = \inf_{\mu, P>0} \{\mu : (3.88)\} \tag{3.89}$$

which is an important alternative as far as control synthesis is concerned. The second statement of this lemma has an additional theoretical implication. Actually, performing the Schur Complement to the last row and column of the LMI (3.88) and multiplying the result to the left by the arbitrary vector $[x' \; w_c'] \neq 0$ and to the right by its transpose, we obtain

$$(Cx + Dw_c)'(Cx + Dw_c) - \mu w_c' w_c + 2x' P(Ax + Ew_c) < 0 \tag{3.90}$$

which by the adoption of the cost-to-go quadratic function $V_c(x) = x' P x$ can be rewritten in the final form

$$\sup_{w_c} \big\{ z'z - \mu w_c' w_c + \nabla V_c(x)'(Ax + Ew_c) \big\} < 0 \tag{3.91}$$

which is nothing else than the celebrated Hamilton–Jacobi–Bellman inequality associated with the optimal control problem

$$\sup_{w_c} \left\{ \int_0^\infty \big(z(t)'z(t) - \mu w_c(t)' w_c(t)\big)\, dt \; : \; (3.82)\text{-}(3.83) \right\} \tag{3.92}$$

Let $\mu > 0$ be given. The LMI (3.88) being infeasible indicates that $\mu > 0$ is so small that the condition *i)* of Lemma 3.2 is violated. On the contrary, if the LMI (3.88) is feasible, then (3.91) holds and consequently problem (3.92) admits a solution whose objective function satisfies the constraint

$$\int_0^\infty \big(z(t)'z(t) - \mu w_c(t)' w_c(t)\big)\, dt < V_c(x_0) - \lim_{t \to \infty} V_c(x(t)) \tag{3.93}$$

for all $0 \neq w_c \in \mathcal{L}_2$. Since the fact that $x_0 = 0$ and A Hurwitz stable makes the right hand side of (3.93) equal to zero, we can conclude that

$$\frac{||z||_2^2}{||w_c||_2^2} < \mu, \; \forall w_c \neq 0 \in \mathcal{L}_2 \tag{3.94}$$

yielding, together with Definition 3.7 and Eq. (3.89), the expected result

$$||H_c(s)||_\infty^2 = \sup_{\omega \geq 0} \sigma_{max}^2\big(H_c(j\omega)\big) = \inf_{\mu, P>0} \{\mu : (3.88)\} \tag{3.95}$$

There is a major information somehow hidden in these calculations that we want to bring to light: the cost-to-go function associated with the $\mathcal{H}_\infty$ problem (3.92) is a quadratic function of the form $V_c(x) = x' P x$. In other words, it can be said that the quadratic function $V_c(x)$ solves the optimality conditions of (3.92), expressed by the Hamilton–Jacobi–Bellman equation because the inequality (3.91) characterizes all feasible solutions.

There is no major difficulty to solve the problem indicated in (3.91). Actually, taking into account (3.83), the solution with respect to w_c exists if and only if $R_c = \mu I - D'D > 0$, and under this condition, it is given by

$$w_c = R_c^{-1}(D'C + E'P)x \tag{3.96}$$

in which case the inequality (3.91), valid for all $0 \neq x$, reduces to

$$A'P + PA + (D'C + E'P)'R_c^{-1}(D'C + E'P) + C'C < 0 \tag{3.97}$$

which, as expected, the Schur Complement reproduces the LMI (3.88). The Hamilton–Jacobi–Bellman equation is nothing else than (3.97) with equality, meaning that the optimal solution is on the border of that matrix inequality. The next example illustrates the theoretical results presented so far.

Example 3.6 *Consider the continuous-time system (3.82)–(3.83) with the following state space matrix data:*

$$A = \begin{bmatrix} 0 & 1 & 0 \\ 0 & 0 & 1 \\ -1 & -1 & -2 \end{bmatrix},\ E = \begin{bmatrix} 1\,0 \\ 1\,0 \\ 1\,0 \end{bmatrix},\ C = \begin{bmatrix} 1\,0\,0 \\ 0\,1\,0 \end{bmatrix},\ D = \begin{bmatrix} 1\,0 \\ 0\,1 \end{bmatrix}$$

where A is Hurwitz stable. Figure 3.4 shows the singular values of the transfer function $H_c(j\omega)$ against $\omega > 0$ in logarithmic scale. The maximum singular value is shown in solid line which provides $\sup_\omega \sigma_{max}(H_c(j\omega)) \approx 14.44$ indicated by the dotted line. Hence, from (3.95), we obtain the norm $||H_c(s)||_\infty^2 \approx 208.61$. For completeness, we have solved the Riccati equation associated with the inequality (3.97), and we have verified that a stabilizing positive definite solution exists for all $\mu \geq 208.62$, approximately. Finally, we have solved the convex programming problem appearing on the right hand side of (3.95) that yielded the optimal value

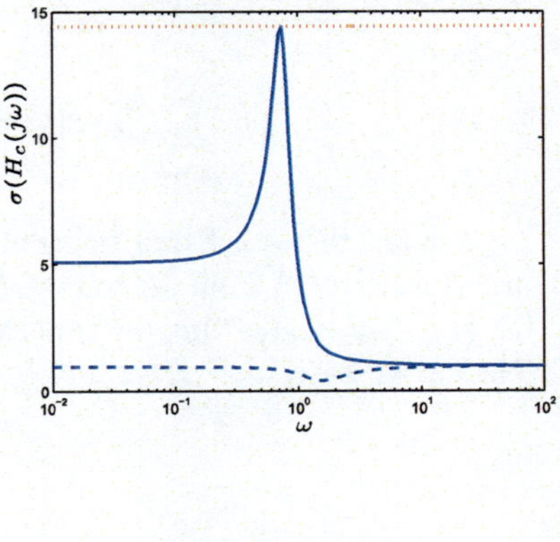

Fig. 3.4 Singular values

$$\inf_{\mu, P>0} \{\mu : (3.88)\} \approx 208.63$$

which once again coincides, within the adopted precision, with the $\mathcal{H}_\infty$ norm of the transfer function under consideration. It is interesting to mention that the matrix solution of the Riccati equation and the one of the convex programming problems may be numerically different, although both are associated with the same value (within the adopted precision) of the minimum squared norm level μ. □

We now move our attention to the following discrete-time system with state space realization:

$$x[k+1] = Ax[k] + Ew_d[k] \tag{3.98}$$

$$z[k] = Cx[k] + Dw_d[k] \tag{3.99}$$

where, as before, $A \in \mathbb{R}^{n\times n}$, $E \in \mathbb{R}^{n\times r}$, $C \in \mathbb{R}^{q\times n}$, and $D \in \mathbb{R}^{q\times r}$. In this context, it is assumed that the system evolves from zero initial condition $x[0] = x_0 = 0$ and $w_d \in \ell_2$ constrains the discrete-time exogenous signal to be norm bounded.

Definition 3.8 *The $\mathcal{H}_\infty$ norm of the transfer function $H_d(\zeta) = C(\zeta I - A)^{-1}E + D : \mathbb{C} \to \mathbb{C}^{q\times r}$ with A Schur stable is*

$$||H_d(\zeta)||_\infty^2 = \sup_{0\neq w_d \in \ell_2} \frac{||z||_2^2}{||w_d||_2^2} \tag{3.100}$$

where $z[k] : \mathbb{N} \to \mathbb{R}^q$ is the response of system (3.98)– (3.99), with $x[0] = 0$, to the input $w_d[k] : \mathbb{N} \to \mathbb{R}^r$. Otherwise, $||H_d(\zeta)||_\infty^2 = +\infty$.

The same path taken in the study of continuous-time systems can be followed once again for dealing with discrete-time systems. In the discrete-time domain, the versions of the Parseval Theorem and KYP Lemma, respectively, are at the core of the algebraic manipulations needed. As expected, with these mathematical tools, there is no major difficulty to obtain the desired results.

Theorem 3.3 (Parseval Theorem) *Consider that $z[k] : \mathbb{N} \to \mathbb{R}^q$ is a norm bounded signal $z \in \ell_2$ with $\mathcal{Z}$ transform $\hat{z}(\zeta)$, and the following equality holds:*

$$\sum_{k=0}^{\infty} z[k]'z[k] = \frac{1}{\pi}\int_0^{\pi} \hat{z}(e^{-j\omega})'\hat{z}(e^{j\omega})d\omega \tag{3.101}$$

The fact that A is Schur stable allows the immediate determination of the Fourier transform $\hat{z}(e^{j\omega})$ from the transfer function of the system under consideration, that is, $\hat{z}(e^{j\omega}) = H_d(e^{j\omega})\hat{w}_d(e^{j\omega})$ because, in this case, $z \in \ell_2$ is implied by $w_d \in \ell_2$. Hence, we have

$$\sum_{k=0}^{\infty} z[k]'z[k] = \frac{1}{\pi}\int_0^{\pi} \hat{w}_d(e^{-j\omega})'H_d(e^{-j\omega})'H_d(e^{j\omega})\hat{w}_d(e^{j\omega})d\omega$$
$$< \frac{\gamma^2}{\pi}\int_0^{\pi} \hat{w}_d(e^{-j\omega})'\hat{w}_d(e^{j\omega})d\omega$$
$$= \gamma^2 \sum_{k=0}^{\infty} w_d[k]'w_d[k] \tag{3.102}$$

which holds for all $w_d \in \ell_2$ provided that $H_d(e^{-j\omega})'H_d(e^{j\omega}) < \gamma^2 I$ for all $0 \leq \omega \leq \pi$, and consequently these calculations yield

$$||H_d(\zeta)||_\infty^2 = \inf_{\mu>0} \left\{\mu : H_d(e^{-j\omega})'H_d(e^{j\omega}) < \mu I,\ 0 \leq \omega \leq \pi\right\}$$
$$= \sup_{0\leq\omega\leq\pi} \sigma_{max}^2\big(H_d(e^{j\omega})\big) \tag{3.103}$$

where $\mu = \gamma^2$, and $\sigma_{max}(\cdot)$ denotes the maximum singular value. As in the previous case of continuous-time systems, the drawing of the *singular value diagram* can be used in order to determine the value of the sup operator indicated on the right hand side of (3.103). However, to keep the mathematical rigor, it remains to show that the minimum value of $\mu = \gamma^2$ equals the square of the $\mathcal{H}_\infty$ norm. This is accomplished by using the following simplified discrete-time version of the KYP Lemma denominated *Bounded-Real Lemma*.

Lemma 3.3 (Bounded-Real Lemma) *Assume that A is a Schur stable matrix. The following statements are equivalent:*

(i) $H_d(e^{-j\omega})'H_d(e^{j\omega}) < \mu I$, *for all* ω *such that* $0 \leq \omega \leq \pi$.
(ii) *There exists a symmetric positive definite matrix* $P > 0$ *of compatible dimensions such that the LMI*

$$\begin{bmatrix} A'PA - P & A'PE & C' \\ \bullet & E'PE - \mu I & D' \\ \bullet & \bullet & -I \end{bmatrix} < 0 \tag{3.104}$$

is feasible.

The maximum singular value on the right hand side of (3.103) is assessed by solving the convex programming problem

$$\sup_{0\leq\omega\leq 0} \sigma_{max}^2\big(H_d(e^{j\omega})\big) = \inf_{\mu,P>0} \{\mu : (3.104)\} \tag{3.105}$$

which is an important alternative as far as control synthesis is concerned. The second statement of this lemma has an additional theoretical implication. Actually, performing the Schur Complement to the last row and column of the LMI (3.104) and multiplying the result to the left by the arbitrary vector $[x' \ w_d'] \neq 0$ and to the right by its transpose, we conclude that

$$\sup_{w_d} \left\{ z'z - \mu w_d' w_d + V_d(Ax + Ew_d) \right\} - V_d(x) < 0 \tag{3.106}$$

where $V_d(x) = x'Px$ is a quadratic cost-to-go function. Hence, (3.106) is immediately recognized as the discrete-time version of the celebrated Hamilton–Jacobi–Bellman inequality associated with the optimal control problem

$$\sup_{w_d} \left\{ \sum_{k=0}^{\infty} \left(z[k]'z[k] - \mu w_d[k]' w_d[k] \right) \, : \, (3.98)\text{-}(3.99) \right\} \tag{3.107}$$

The LMI (3.104) is infeasible whenever $\mu > 0$ violates the condition (*i*) of Lemma 3.3. However, if it is feasible, then (3.106) holds and consequently problem (3.107) admits a solution whose objective function satisfies

$$\sum_{k=0}^{\infty} \left(z[k]'z[k] - \mu w_d[k]' w_d[k] \right) dt < V_d(x_0) - \lim_{k \to \infty} V_d(x[k]) \tag{3.108}$$

for all $0 \neq w_d \in \ell_2$. Since the fact that $x_0 = 0$ and A Schur stable makes the right hand side of (3.108) equal to zero, we conclude that

$$\frac{||z||_2^2}{||w_d||_2^2} < \mu, \ \forall w_d \neq 0 \in \ell_2 \tag{3.109}$$

yielding, together with Definition 3.8 and Eq. (3.105), the expected result

$$||H_d(\zeta)||_\infty^2 = \sup_{0 \le \omega \le \pi} \sigma_{max}^2 \left(H_d(e^{j\omega}) \right) = \inf_{\mu, P > 0} \{ \mu \, : \, (3.104) \} \tag{3.110}$$

Based on these results, we can affirm that the cost-to-go function associated with the $\mathcal{H}_\infty$ problem (3.107) is a quadratic function of the form $V_d(x) = x'Px$. Such quadratic function can be enforced, with no loss of generality, to be a solution to the optimality conditions of the optimal control problem (3.107), expressed by the associated Hamilton–Jacobi–Bellman equation. Keeping in mind the output equation (3.99), the solution with respect to w_d exists if and only if $R_d = \mu I - D'D - E'PE > 0$, and under this condition, it is given by

$$w_c = R_d^{-1}(D'C + E'PA)x \tag{3.111}$$

in which case the inequality (3.106), valid for all $0 \neq x$, reduces to

$$A'PA - P + (D'C + E'PA)'R_d^{-1}(D'C + E'PA) + C'C < 0 \qquad (3.112)$$

which, as expected, the Schur Complement reproduces the LMI (3.104). The Hamilton–Jacobi–Bellman equation is nothing else than (3.112) with equality meaning that the optimal solution is on the border of that matrix inequality. The next example illustrates these results.

Example 3.7 *Let us consider the discrete-time system (3.98)–(3.99) with*

$$A = \begin{bmatrix} 0 & 1 & 0 \\ 0 & 0 & 1 \\ 0 & -0.8 & 0 \end{bmatrix}, \; E = \begin{bmatrix} 1 & 0 \\ 1 & 0 \\ 1 & 0 \end{bmatrix}, \; C = \begin{bmatrix} 1 & 0 & 0 \\ 0 & 1 & 0 \end{bmatrix}, \; D = \begin{bmatrix} 1 & 0 \\ 0 & 1 \end{bmatrix}$$

where A is Schur stable. The singular value diagram gives $\sup_\omega \sigma_{max}(H_c(j\omega)) \approx 9.10$ which yields the squared norm $||H_d(\zeta)||_\infty^2 \approx 82.88$. For completeness, we have solved the Riccati equation associated with the inequality (3.112) for $\mu = 82.89$ and we have found the positive definite stabilizing solution

$$P_{ric} = \begin{bmatrix} 1.0159 & -0.1074 & 0.1334 \\ -0.1074 & 9.0381 & -0.5593 \\ 0.1334 & -0.5593 & 10.1549 \end{bmatrix}$$

For comparison, we have solved the convex programming problem appearing in the right hand side of (3.105), which provided the minimum $\mathcal{H}_\infty$ level $\mu \approx 82.89$, associated with the optimal positive definite matrix variable

$$P = \begin{bmatrix} 13.3658 & -2.7380 & 12.6487 \\ -2.7380 & 9.6607 & -3.2328 \\ 12.6487 & -3.2328 & 22.9149 \end{bmatrix}$$

It is interesting and important to compare P_{ric} and P. They are quite different, but they are associated with practically the same value of $\mu = ||H_d(\zeta)||_\infty^2$. Moreover, the left hand side of (3.112) evaluated for $P = P_{ric}$ produces the null matrix (within a precision of 10^{-13}), while evaluated at $P > 0$ produces a non-null definite negative matrix with eigenvalues $\approx \{-25.5315, 0, 0\}$. This example illustrates the fact that, within some precision, there exist many different near optimal matrix solutions associated with the same minimum value of μ. □

From the theoretical basis acquired by the study of continuous- and discrete-time LTI systems, let us now analyze the $\mathcal{H}_\infty$ performance of sampled-data systems. To this end and to make the reading easier, let us quote again the open-loop model

$$\dot{x}(t) = Ax(t) + Bu(t) + Ew_c(t), \; x(0) = 0 \qquad (3.113)$$

$$z(t) = C_z x(t) + D_z u(t) \qquad (3.114)$$

$$u(t) = w_d[k], \; t \in [t_k, t_{k+1}), \; \forall k \in \mathbb{N} \tag{3.115}$$

and state the definition that characterizes our main concern in this section.

Definition 3.9 *The $\mathcal{H}_\infty$ performance index associated with the sampled-data system (3.113)–(3.115) is given by*

$$\varrho_\infty^2 = \sup_{0 \neq (w_c, w_d) \in \mathcal{L}_2 \times \ell_2} \frac{||z||_2^2}{||w_c||_2^2 + ||w_d||_2^2} \tag{3.116}$$

where $z(t) : \mathbb{R}_+ \to \mathbb{R}^q$ is its response to the continuous-time $w_c(t) : \mathbb{R}_+ \to \mathbb{R}^{r_c}$ and to the discrete-time $w_d[k] : \mathbb{N} \to \mathbb{R}^{r_d}$ exogenous perturbations, assuming that they are not null, simultaneously.

It is important to recognize the difficulty one has to calculate the sup operator appearing on the right hand side of (3.116). It stems from the fact that we have to manipulate signals in continuous- and discrete-time domains jointly. In this framework, on the contrary of what we have done in the study of $\mathcal{H}_2$ performance, now, the equivalent system does not provide the exact value of the index ϱ_∞, but it merely yields a lower bound to it. This is stated in the next lemma.

Lemma 3.4 *Let $h > 0$ be given and assume that A is Schur stable. The following lower bound $||H_{lb}(\zeta)||_\infty^2 \leq \varrho_\infty^2$ where*

$$H_{lb}(\zeta) = C_{zd} \, (\zeta I - A_d)^{-1} \left[\tfrac{E_d}{\sqrt{h}} \; B_d \right] + \left[\tfrac{D_d}{\sqrt{h}} \; D_{zd} \right] \tag{3.117}$$

holds.

Proof As already mentioned, a lower bound can be determined by further constraining the continuous-time signal $w_c(t)$ to be piecewise constant, that is, $w_c \in \mathcal{L}_2 \cap \mathbb{U}$ and taking into account that

$$\int_0^\infty w_c(t)' w_c(t) dt = h \sum_{k=0}^\infty w_c[k]' w_c[k] \tag{3.118}$$

which indicates that the square norm $||w_c||_2^2$ of a continuous-time piecewise constant signal in $\mathcal{L}_2$ is equal to the square norm $||\sqrt{h} w_c||_2^2$ of its samples $w_c[k] = w_c(t_k)$ for all $k \in \mathbb{N}$. In this particular situation, the equivalent system is determined and provides

$$\hat{z}(e^{j\omega h}) = H_{lb}(e^{j\omega h}) \begin{bmatrix} \sqrt{h} \hat{w}_c(e^{j\omega h}) \\ \hat{w}_d(e^{j\omega h}) \end{bmatrix} \tag{3.119}$$

which is valid for all $(w_c, w_d) \in \ell_2 \times \ell_2$. Hence, Definition 3.9 yields

$$\varrho_\infty^2 \geq \sup_{0 \neq (w_c, w_d) \in \mathcal{L}_2 \cap \mathbb{U} \times \ell_2} \frac{||z||_2^2}{||w_c||_2^2 + ||w_d||_2^2}$$

$$= \sup_{0 \neq (w_c, w_d) \in \ell_2 \times \ell_2} \frac{||z||_2^2}{||\sqrt{h} w_c||_2^2 + ||w_d||_2^2}$$

$$= ||H_{lb}(\zeta)||_\infty^2 \tag{3.120}$$

where the last equality is due to (3.119). The proof is complete. □

At first glance, this result is important because the transfer function $H_{lb}(\zeta)$ represents a pure discrete-time system and, consequently, there is no difficulty to use it for sampled-data control design. However, mainly in the context of $\mathcal{H}_\infty$ performance, the minimization of a lower bound is not a guarantee that the true value of ϱ_∞ is actually reduced. Indeed, in general, control design based on lower bounds is not suited and should be avoided. Even though, restricted to the analysis context, the next lemma provides an upper bound to the $\mathcal{H}_\infty$ performance index.

Lemma 3.5 *Let $h > 0$ be given and assume that A is Hurwitz stable. The following upper bound $\varrho_\infty^2 \leq ||H_{ub}(s)||_\infty^2$ where*

$$H_{ub}(s) = C_z (sI - A)^{-1} \left[E \ \sqrt{h} B \right] + \left[0 \ \sqrt{h} D_z \right] \tag{3.121}$$

holds.

Proof The reasoning is exactly the inverse of that applied in the proof of the previous lemma. The upper bound follows from the weakening of the signal w_d to be in $\mathcal{L}_2$. To this end, with $w_d \in \ell_2$, let us construct $w_d \in \mathcal{L}_2 \cap \mathbb{U}$ with $w_d(t_k) = w_d[k]$ for all $k \in \mathbb{N}$ and pay attention to the fact that for this signal we have

$$\int_0^\infty w_d(t)' w_d(t) dt = h \sum_{k=0}^\infty w_d[k]' w_d[k] \tag{3.122}$$

which shows that the square norm $||w_d||_2^2$ of a discrete-time signal in ℓ_2 is equal to the square norm $||w_d/(\sqrt{h})||_2^2$ of a continuous-time signal in $\mathcal{L}_2 \cap \mathbb{U}$. As a consequence, this class of signals yields

$$\hat{z}(j\omega) = H_{ub}(j\omega) \begin{bmatrix} \hat{w}_c(j\omega) \\ \hat{w}_d(j\omega)/\sqrt{h} \end{bmatrix} \tag{3.123}$$

which, remaining valid for all $(w_c, w_d) \in \mathcal{L}_2 \times \mathcal{L}_2$, together with Definition 3.9, enables us to conclude that

$$\varrho_\infty^2 = \sup_{0 \neq (w_c, w_d) \in \mathcal{L}_2 \times \mathcal{L}_2 \cap \mathbb{U}} \frac{||z||_2^2}{||w_c||_2^2 + ||w_d/\sqrt{h}||_2^2}$$

$$\leq \sup_{0\neq(w_c,w_d)\in\mathcal{L}_2\times\mathcal{L}_2} \frac{||z||_2^2}{||w_c||_2^2 + ||w_d/\sqrt{h}||_2^2}$$
$$= ||H_{ub}(s)||_\infty^2 \quad (3.124)$$

where the last equality is due to (3.123). The proof is complete. □

It is important to keep in mind that these bounds concern the open-loop version of the sampled-data control system and that, unfortunately, they cannot be generalized to cope with the closed-loop version of the same system. To make this statement clearer, suppose that we have designed a control law $u[k] = Lx[k],\ k \in \mathbb{N}$, that minimizes the lower bound of the $\mathcal{H}_\infty$ performance defined by Lemma 3.4. This is clearly possible and simple to do because we have to design a linear control law to a pure discrete-time LTI system. Next, we want to determine the upper bound associated with the closed-loop system, but for that we need to calculate the continuous-time transfer function $H_{ub}(s)$, which is an extremely hard or even impossible task. This is because any control law depending on some state variable naturally introduces a discrete-time component in the state space realization of the closed-loop system.

Example 3.8 *Let us determine, from the results of the previous lemmas, the lower and the upper bounds for a fifth order sampled-data system with two inputs and two outputs. All flavors in terms of continuous- and discrete-time perturbations are present. Hence, we believe that this system is representative enough to illustrate the results we have just presented. It is given by the state space realization (3.113)–(3.115) with the following matrices:*

$$A = \begin{bmatrix} 0 & 1 & 0 & 0 & 0 \\ 0 & 0 & 1 & 0 & 0 \\ 0 & 0 & 0 & 1 & 0 \\ 0 & 0 & 0 & 0 & 1 \\ -3 & -5 & -8 & -5 & -3 \end{bmatrix},\ B = \begin{bmatrix} 1\,0 \\ 1\,0 \\ 1\,0 \\ 1\,0 \\ 1\,0 \end{bmatrix},\ E = \begin{bmatrix} 0\,1 \\ 0\,1 \\ 0\,1 \\ 0\,1 \\ 0\,1 \end{bmatrix}$$

and

$$C_z = \begin{bmatrix} 1\,0\,0\,0\,0 \\ 0\,1\,0\,0\,0 \end{bmatrix},\ D_z = \begin{bmatrix} 1\,0 \\ 0\,1 \end{bmatrix}$$

Following the proof of Lemma 3.4, let us consider the open-loop sampled-data system with the input $w_c \in \mathcal{L}_2 \cap \mathbb{U}$, that is,

$$\dot{x}(t) = Ax(t) + Bu(t) + Ew_c(t),\ x(0) = 0$$
$$z(t) = C_z x(t) + D_z u(t)$$
$$u(t) = w_d[k],\ t \in [t_k, t_{k+1}),\ \forall k \in \mathbb{N}$$

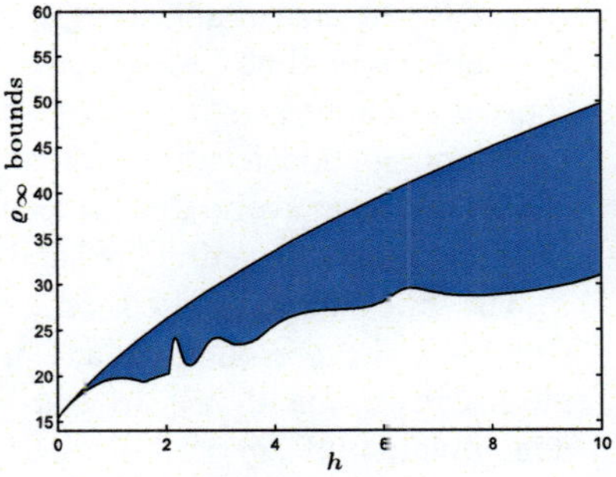

Fig. 3.5 $\mathcal{H}_\infty$ gap as a function of h

$$w_c(t) = w_c[k], \ t \in [t_k, t_{k+1}), \ \forall k \in \mathbb{N}$$

and determine the associated equivalent system by adopting the replacements $A \leftarrow A$, $B \leftarrow [B \ E]$, $C \leftarrow C_z$, *and* $D \leftarrow [D_z \ 0]$, *which yield, after partitioning, the desired matrices* A_d, $[B_d \ E_d]$, C_{zd}, *and* $[D_{zd} \ D_d]$, *respectively. Finally, according to (3.117), the norm compensation* $1/\sqrt{h}$ *on matrices* E_d *ad* D_d *must be included.*

Figure 3.5 shows the existing gap between the lower and upper bounds of ϱ_∞ *as a function of the sampling period in the time interval* $h \in (0, 10]$. *As expected, for small values of the sampling period, the bounds virtually coincide. In this region, the continuous-time, as well as the discrete-time, model represents the sampled-data system within a reasonable precision. Unfortunately, this conclusion does not remain true as the sampling period increases. In general, this behavior often occurs.* □

It is clear that the mathematical tools that we have presented so far are not sufficient to approach successfully optimal $\mathcal{H}_\infty$ performance design in the context of sampled-data control systems. For this reason, the importance of DLMIs becomes evident since, with them, this class of optimal control design problems is adequately and completely solved in both theoretical and numerical frameworks. In addition, analysis and design of robust control for sampled-data systems subject to convex bounded parameter uncertainty are naturally included because DLMIs rely linearly on the system data. The development of several topics based on DLMIs is our main goal to be pursued in the forthcoming chapters of this book.

3.5 Bibliography Notes

This chapter is not a survey. Important aspects of sampled-data control system analysis and design have been tackled having in mind the developments we want to present in the next chapters. Two basic and celebrated books on sampled-data systems are [13] and [33] where the reader can find an expressive amount of material dealing with analysis and control design involving stability, $\mathcal{H}_2$ and $\mathcal{H}_\infty$

norms based on the solutions of Lyapunov and Riccati equations, see also [34]. For continuous-time LTI systems, the same concepts are treated in many different sources, including in the book [14] and the references therein. For a historical view on sampling and Riccati equation, the reader is requested to see the references [51] and [53], respectively. Recent results involving switched and sampled-data control systems can be viewed in [16] and [31].

The presentation based on *equivalent system* is new and, in our opinion, simplifies the presentation and makes the results easier to understand. Similar ideas can be found in [36] and [41]. Moreover, the reader perhaps noticed our denomination of $\mathcal{H}_2$ and $\mathcal{H}_\infty$ performance indices instead of norms. In particular, the $\mathcal{H}_2$ performance index does not coincide with the $\mathcal{H}_2$ norms generally adopted, as for instance, in the previous cited books and in the references [4, 12], and [36]. The reason is quite simple: our definition of $\mathcal{H}_2$ performance is valid and arises more naturally as a generalization of the $\mathcal{H}_2$ norm of LTI systems.

The presentation of $\mathcal{H}_\infty$ analysis of LTI systems is based on two celebrated results, namely the Parseval Theorem and the Kalman–Yakubovich–Popov Lemma. Due to their theoretical importance, the literature has plenty of papers and books dealing with them, formulated in both continuous- and discrete-time domains. The first topic can be found in reference [39] formulated in the context of Fourier transform and series. For the second topic, we recommend the paper [45] and the references therein, where the reader can find its detailed proof together with historical information about it and related results. Specifically, the Bounded-Real Lemma is tackled in the book [10] with particular attention to its LMI representation.

Finally, we would like to give special credit to the reference [46] where, as far as we know, an outstanding result has been presented for the first time. It is used until now, in many different circumstances, in the context of dynamic full order controllers design, with the main proposal of reducing nonlinear matrix inequalities to LMIs. In the forthcoming chapters of this book, a new one-to-one matrix transformation is derived, but it is based on the result of that quoted seminal reference.

Chapter 4
$\mathcal{H}_2$ Filtering and Control

4.1 Introduction

In this chapter, the core of our concerns is the $\mathcal{H}_2$ performance optimization of sampled-data systems. Albeit we have already studied in the previous chapters some aspects involving this important performance index, now our goal is to tackle some other aspects related to norm and convex bounded parameter uncertainty. To this end, in a first step of analysis, we need to express the objective function and the optimal design problems through DLMIs. State feedback and full order dynamic feedback controllers and linear filters are designed such that a guaranteed $\mathcal{H}_2$ performance level and robustness against parameter uncertainty are preserved. Convexity is the central property to be pursued. Several illustrative examples are presented and discussed.

4.2 $\mathcal{H}_2$ Performance Analysis

Let us consider the open-loop sampled-data system already introduced in Chap. 3 that has the form

$$\dot{x}(t) = Ax(t) + Bu(t) + Ew_c(t), \; x(0) = 0 \tag{4.1}$$

$$z(t) = C_z x(t) + D_z u(t) \tag{4.2}$$

$$u(t) = w_d[k], \; t \in [t_k, t_{k+1}), \; \forall k \in \mathbb{N} \tag{4.3}$$

where it is appropriate to remember that $x(\cdot) : \mathbb{R}_+ \to \mathbb{R}^{n_x}$, $u(\cdot) : \mathbb{R}_+ \to \mathbb{R}^{n_u}$, $w_c(\cdot) : \mathbb{R}_+ \to \mathbb{R}^{r_c}$, and $z(\cdot) : \mathbb{R}_+ \to \mathbb{R}^{n_z}$ are the state, the control, the exogenous input, and the controlled output of the continuous-time process, respectively. In open-loop operation, the sampled-data control belongs to the time-invariant set $\mathbb{U}$

J. C. Geromel, *Differential Linear Matrix Inequalities*,
https://doi.org/10.1007/978-3-031-29754-0_4

composed of all piecewise constant trajectories fully defined by the exogenous discrete-time sensor perturbation $w_d[\cdot] : \mathbb{N} \to \mathbb{R}^{r_d}$. To improve readability, we repeat here the definition of $\mathcal{H}_2$ performance often discussed in the previous chapter.

Definition 4.1 The $\mathcal{H}_2$ performance index associated with the sampled-data system (4.1)–(4.3) is given by

$$\varrho_2^2 = \sum_{i=1}^{r_c+r_d} ||z_i||_2^2 \tag{4.4}$$

where $z(t) = z_i(t),\ 1 \le i \le r_c$ is the output due to a continuous-time impulse $\delta(t)$ in the i-th channel of the input $w_c(t)$ and $z(t) = z_i(t), r_c + 1 \le i \le r_c + r_d$ is the output due to a discrete-time impulse $\delta[k]$ in the i-th channel of the input $w_d[k]$.

Making reference to the results of the previous chapter, in that opportunity, we have established that $\varrho_2^2 = ||H_{sd}(\zeta)||_2^2$, where

$$H_{sd}(\zeta) = C_{zd}(\zeta I - A_d)^{-1} \left[E \ B_d \right] + \left[0 \ D_{zd} \right] \tag{4.5}$$

is an augmented transfer function with the matrix A_d being Schur stable. Under this assumption, there exists a symmetric matrix $X_0 > 0$ satisfying the discrete-time Lyapunov inequality

$$A_d' X_0 A_d - X_0 < -C_{zd}' C_{zd} \tag{4.6}$$

such that the performance index under consideration is calculated by solving the following convex programming problem:

$$\varrho_2^2 = \inf_{X_0 > 0} \left\{ \mathbf{tr} \left(\begin{bmatrix} E' \\ B_d' \end{bmatrix} X_0 \begin{bmatrix} E' \\ B_d' \end{bmatrix}' \right) + \mathbf{tr} \left(\begin{bmatrix} 0 \\ D_{zd}' \end{bmatrix} \begin{bmatrix} 0 \\ D_{zd}' \end{bmatrix}' \right) : (4.6) \right\} \tag{4.7}$$

Now, alternatively, we want to go further by calculating the value of this performance index with the help of DLMIs with the advantages made explicit in due time. To this end, to tackle the evaluation of the $\mathcal{H}_2$ performance index by a convex problem expressed by DLMIs, we need to include the control variable in a sort of augmented state variable in order to take into account the existence of a sample and hold device in the control channel. Hence, inspired by the equivalent system, we consider the augmented matrices

$$F = \begin{bmatrix} A & B \\ 0 & 0 \end{bmatrix},\ G = \left[C_z \ D_z \right],\ J_c = \begin{bmatrix} E \\ 0 \end{bmatrix},\ J_d = \begin{bmatrix} 0 \\ I \end{bmatrix} \tag{4.8}$$

all with compatible dimensions. Moreover, the symmetric matrix variable $P_0 \in \mathbb{R}^{(n_x+n_u)\times(n_x+n_u)}$ is partitioned keeping intact the dimensions of the state and control variables, respectively, as follows:

$$P_0 = \begin{bmatrix} X_0 & V_0 \\ V_0' & \hat{X}_0 \end{bmatrix} \tag{4.9}$$

where the block matrices are such that $X_0 \in \mathbb{R}^{n_x \times n_x}$, $V_0 \in \mathbb{R}^{n_x \times n_u}$, and $\hat{X}_0 \in \mathbb{R}^{n_u \times n_u}$. In many instances, it satisfies the constraint $P_0 > 0$. The proof of the main result of this section depends on a procedure that is generally denominated *variable elimination*, in the context of LMIs, that can be applied in the context of DLMIs as well. This is deeply discussed in the next remark, and for more information about that procedure, the reader is requested to see the Bibliography notes of this chapter.

Remark 4.1 Consider the following LMI composed of four blocks:

$$Q = \begin{bmatrix} Q_{11} & V \\ V' & Q_{22} \end{bmatrix} > 0$$

with compatible dimensions. There exists a matrix V such that it is feasible if and only if $Q_{11} > 0$ and $Q_{22} > 0$. In the affirmative case, a possible solution is $V = 0$. This statement is quite obvious, and it puts in evidence that $V = 0$ is the less constrained feasible solution because, in this case, necessity meets sufficiency to assure that $Q > 0$. Let us now consider a nine-block LMI

$$Q = \begin{bmatrix} Q_{11} & V & Q_{13} \\ \bullet & Q_{22} & Q_{23} \\ \bullet & \bullet & Q_{33} \end{bmatrix} > 0$$

which, by the calculation of the Schur Complement corresponding to the third row and column, is equivalent to

$$Q_{33} > 0, \quad \begin{bmatrix} Q_{11} - Q_{13}Q_{33}^{-1}Q_{13}' & V - Q_{13}Q_{33}^{-1}Q_{23}' \\ \bullet & Q_{22} - Q_{23}Q_{33}^{-1}Q_{23}' \end{bmatrix} > 0$$

The result of four blocks can be applied, and we conclude that there exists V such that the nine-block LMI is feasible if and only if

$$\begin{bmatrix} Q_{11} & Q_{13} \\ \bullet & Q_{33} \end{bmatrix} > 0, \quad \begin{bmatrix} Q_{22} & Q_{23} \\ \bullet & Q_{33} \end{bmatrix} > 0$$

and the less constrained solution follows from the choice $V = Q_{13}Q_{33}^{-1}Q_{23}'$. Observe that in both cases, we have obtained the two LMIs that preserve feasibility by simply eliminating the row and the column where the matrix variable V stands. Hence, in the general case (an LMI with an arbitrary number of matrix blocks), a matrix variable V in the (i, j) block exists such that $Q > 0$, if and only if the LMIs

$\bar{Q}_{ii} > 0$ and $\bar{Q}_{jj} > 0$ are feasible. They are obtained by eliminating from $Q > 0$ the i-th row and column and the j-th row and column, respectively. In the general case, the analytic determination of the less constrained feasible solution may not be a simple task. □

We now come back to the calculation of the $\mathcal{H}_2$ performance index proposed in Definition 4.1 or by solving a convex programming problem equivalent to (4.7), which is the same. This is stated and proven in the next theorem.

Theorem 4.1 *Let $h > 0$ be given and consider the block-diagonal matrix $H = \text{diag}(I, 0) \in \mathbb{R}^{(n_x+n_u)\times(n_x+n_u)}$. The open-loop sampled-data system (4.1)–(4.3) is globally asymptotically stable with ϱ_2^2 performance index if and only if the DLMI*

$$\dot{P}(t) + F'P(t) + P(t)F + G'G < 0, \ t \in [0, h) \tag{4.10}$$

subject to the LMI boundary conditions

$$P_h > 0, \ P_h > H'P_0H \tag{4.11}$$

is feasible. In the affirmative case, the equality

$$\varrho_2^2 = \inf_{P(\cdot)} \{\mathbf{tr}(J_c'P_hJ_c) + \mathbf{tr}(J_d'P_0J_d) : (4.10)\text{–}(4.11)\} \tag{4.12}$$

holds.

Proof It is done in two steps. In the first step, it is proven that if a feasible solution exists, then the sampled-data system is globally asymptotically stable and it produces an upper bound to the $\mathcal{H}_2$ performance index. In a second step, assuming global stability, it is shown how to construct a feasible solution that reproduces the performance index exactly.

Take a generic feasible solution to the DLMI (4.10) that satisfies the boundary conditions (4.11). Theorem 2.1 establishes that

$$P_0 > e^{F'h}P_he^{Fh} + \int_0^h e^{F't}G'Ge^{Ft}dt \tag{4.13}$$

which clearly implies that $P_0 > 0$ for any feasible solution subject to a positive definite final boundary condition, namely $P_h > 0$. Hence, taking into account (4.11), and remembering our previous block partition $G_d = [C_{zd} \ D_{zd}]$ of the matrix factorization

$$\int_0^h e^{F't}G'Ge^{Ft}dt = G_d'G_d \tag{4.14}$$

the inequality (4.13) provides

$$
\begin{aligned}
P_0 &> e^{F'h} H' P_0 H e^{Fh} + G_d' G_d \\
&= \begin{bmatrix} A_d' \\ B_d' \end{bmatrix} X_0 \begin{bmatrix} A_d & B_d \end{bmatrix} + \begin{bmatrix} C_{zd}' \\ D_{zd}' \end{bmatrix} \begin{bmatrix} C_{zd} & D_{zd} \end{bmatrix}
\end{aligned} \tag{4.15}
$$

which, by Schur complement calculations involving each element of the sum, can be rewritten in the equivalent form

$$
\begin{bmatrix} X_0 & V_0 & A_d' & C_{zd}' \\ \bullet & \hat{X}_0 & B_d' & D_{zd}' \\ \bullet & \bullet & X_0^{-1} & 0 \\ \bullet & \bullet & \bullet & I \end{bmatrix} > 0 \tag{4.16}
$$

The matrix variable V_0 can be eliminated. Indeed, the fact that (4.16) holds, the result reported in Remark 4.1 allows us to say that with respect to the remaining variables, this inequality is equivalent to the LMIs

$$
\begin{bmatrix} X_0 & A_d' & C_{zd}' \\ \bullet & X_0^{-1} & 0 \\ \bullet & \bullet & I \end{bmatrix} > 0, \tag{4.17}
$$

and

$$
\begin{bmatrix} \hat{X}_0 & B_d' & D_{zd}' \\ \bullet & X_0^{-1} & 0 \\ \bullet & \bullet & I \end{bmatrix} > 0 \tag{4.18}
$$

that follow from the elimination of the second row and column, and the first row and column of the LMI (4.16), respectively. Using again Schur complements, the first one is equivalent to $A_d' X_0 A_d - X_0 < -C_{zd}' C_{zd}$, which, due to the fact that $X_0 > 0$, requires A_d Schur stable and, by consequence, the global stability of the open-loop sampled-data system. Moreover, the second LMI becomes $\hat{X}_0 > B_d' X_0 B_d + D_{zd}' D_{zd}$. In addition, we also have

$$
\begin{aligned}
\mathbf{tr}(J_c' P_h J_c) + \mathbf{tr}(J_d' P_0 J_d) &> \mathbf{tr}(J_c' H' P_0 H J_c) + \mathbf{tr}(J_d' P_0 J_d) \\
&= \mathbf{tr}(E' X_0 E) + \mathbf{tr}(\hat{X}_0) \\
&> \mathbf{tr}(E' X_0 E) + \mathbf{tr}(B_d' X_0 B_d + D_{zd}' D_{zd}) \\
&= \mathbf{tr}\left(\begin{bmatrix} E' \\ B_d' \end{bmatrix} X_0 \begin{bmatrix} E' \\ B_d' \end{bmatrix}' \right) + \mathbf{tr}\left(\begin{bmatrix} 0 \\ D_{zd}' \end{bmatrix} \begin{bmatrix} 0 \\ D_{zd}' \end{bmatrix}' \right) \\
&\geq ||H_{sd}(\zeta)||_2^2
\end{aligned} \tag{4.19}
$$

where the last inequality is an immediate consequence of (4.7). We proceed to the second step by constructing a feasible solution such that (4.19) holds arbitrarily close to equality. Assuming that $A_d = e^{Ah}$ is Schur stable, we can say that there exists $X_0 > 0$ satisfying $A_d' X_0 A_d - X_0 < -C_{zd}' C_{zd}$ solution to (4.7), and by consequence, we can set $\hat{X}_0 > B_d' X_0 B_d + D_{zd}' D_{zd}$ arbitrarily close to equality as well. Hence, these matrices yield $\mathbf{tr}(J_c' P_h J_c) + \mathbf{tr}(J_d' P_0 J_d) > ||H_{sd}(\zeta)||_2^2$ but, by construction, arbitrarily close to equality. Calculating the Schur complement of those two matrix inequalities, it is readily seen that (4.17) and (4.18) are, respectively, satisfied. The elimination procedure depicted in Remark 4.1 assures the existence of a matrix V_0 such that (4.16) holds, whose Schur complement calculation with respect to the two rows and columns provides inequality (4.15). Taking $P_h > H' P_0 H$ arbitrarily close to equality, the boundary conditions (4.11) are satisfied and the same is true for inequality (4.13), which involves the initial $P_0 > 0$ and final $P_h > 0$ boundary conditions that we have just proposed. Now, let us introduce the well-defined matrix-valued function $W(t) = e^{-F't} W_0 e^{-Ft}$ with

$$W_0 = h^{-1} \left(P_0 - e^{F'h} P_h e^{Fh} - \int_0^h e^{F't} G' G e^{Ft} dt \right) > 0 \tag{4.20}$$

in the whole time interval $[0, h]$. The linear differential equation

$$\dot{P}(t) + F' P(t) + P(t) F + G' G = -W(t), \; t \in [0, h) \tag{4.21}$$

subject to the boundary condition $P(h) = P_h > 0$ always admits a solution that, as we know, can be expressed as

$$P(t) = e^{F'(h-t)} P_h e^{F(h-t)} + \int_t^h e^{F'(\xi - t)} (G' G + W(\xi)) e^{F(\xi - t)} d\xi \tag{4.22}$$

which, evaluated at $t = 0$, yields

$$\begin{aligned} P(0) &= e^{F'h} P_h e^{Fh} + \int_0^h e^{F'\xi} (G' G + W(\xi)) e^{F\xi} d\xi \\ &= e^{F'h} P_h e^{Fh} + \int_0^h e^{F'\xi} G' G e^{F\xi} d\xi + h W_0 \\ &= P_0 \end{aligned} \tag{4.23}$$

By construction, the matrix-valued function $W(t)$ is positive definite for all $t \in [0, h]$, allowing the conclusion that the matrix-valued function $P(t)$, the solution of the linear equation (4.21), satisfies the DLMI (4.10), subject to the boundary conditions (4.11), completing thus the proof. □

Notice that the statement of Theorem 4.1 does not explicitly require that $P(t) > 0$, for all $t \in [0, h]$. However, this is implicitly fulfilled because, as we know, any feasible solution to the DLMI (4.10) subject to a positive definite final boundary condition $P(h) = P_h > 0$ is positive definite in the entire time interval under consideration. Hence, in this and similar cases, no kind of conservatism is included if, for convenience, we constrain the search to positive definite feasible solutions only.

The result of Theorem 4.1 is important in the framework of sampled-data control systems. Indeed, it is a tipping point for the treatment of robust sampled-data control systems analysis and design. It stems from the fact that the DLMI (4.10) depends linearly on matrix F and the same is true for the objective function of problem (4.12) because adding the new matrix variables $W_c > 0$, $W_d > 0$ and the LMIs

$$\begin{bmatrix} W_c & J_c' P_h \\ \bullet & P_h \end{bmatrix} > 0, \quad \begin{bmatrix} W_d & J_d' P_0 \\ \bullet & P_0 \end{bmatrix} > 0 \tag{4.24}$$

the Schur complement with respect to the last rows and columns shows that the problem (4.12) is clearly equivalent to

$$\varrho_2^2 = \inf_{P(\cdot), W_c, W_d} \{\mathbf{tr}(W_c) + \mathbf{tr}(W_d) : (4.10)\text{–}(4.11), (4.24)\} \tag{4.25}$$

where it is to be noticed that we have used the positivity of the matrix variables $P_0 > 0$ and $P_h > 0$ to get the LMIs in (4.24). Observe that the DLMI depends only on the matrices of the original sampled-data system (4.1)–(4.3) and not, as usually happens, on the matrices $(A_d, B_d, C_{zd}, D_{zd})$ that depend on a nonlinear (exponential) matrix mapping. This occurs because any feasible solution of the DLMI depends, implicitly, on the matrix exponential mapping, as it has been shown by the results presented in the previous chapters.

To put this scenario in a clearer perspective, consider that the matrices (A, B, C_z, D_z) of the open-loop sampled-data system (4.1)–(4.3) are uncertain but belong to a known convex bounded set

$$(A, B, C_z, D_z) \in \text{co}\{(A_i, B_i, C_{zi}, D_{zi})\}_{i \in \mathbb{K}} \tag{4.26}$$

defined by the convex combination (convex hull) of the $i \in \mathbb{K} = \{1, \ldots, N\}$ extreme matrices with appropriate dimensions. That is, each feasible quadruple of matrices is given by

$$(A_\lambda, B_\lambda, C_{z\lambda}, D_{z\lambda}) = \sum_{i \in \mathbb{K}} \lambda_i (A_i, B_i, C_{zi}, D_{zi}) \tag{4.27}$$

where $\lambda \in \mathbb{R}^N$ is some element of the unity simplex Λ. Clearly, now, the transfer function $H_{sd}(\zeta, \lambda)$ depends on $\lambda \in \Lambda$, and our interest is to determine an upper bound, denominated *guaranteed $\mathcal{H}_2$ performance index*, such that

$$\sup_{\lambda \in \Lambda} ||H_{sd}(\zeta, \lambda)||_2^2 \leq \varrho_{2rob}^2 \tag{4.28}$$

which characterizes the worst parameter uncertainty as far as the impact of robustness on the sampled-data system performance is concerned.

Corollary 4.1 *Let $h > 0$ be given, and consider the block-diagonal matrix $H = \text{diag}(I, 0) \in \mathbb{R}^{(n_x+n_u)\times(n_x+n_u)}$. The uncertain open-loop sampled-data system (4.1)–(4.3) is globally asymptotically stable for all $\lambda \in \Lambda$ with ϱ_{2rob}^2 guaranteed performance index if the set of DLMIs*

$$\begin{bmatrix} \dot{P}(t) + F_i' P(t) + P(t) F_i & G_i' \\ \bullet & -I \end{bmatrix} < 0, \; i \in \mathbb{K}, \; t \in [0, h) \tag{4.29}$$

subject to the LMI boundary conditions

$$P_h > 0, \; P_h > H' P_0 H \tag{4.30}$$

is feasible. In the affirmative case, the equality

$$\varrho_{2rob}^2 = \inf_{P(\cdot)} \{\mathbf{tr}(J_c' P_h J_c) + \mathbf{tr}(J_d' P_0 J_d) : (4.29)\text{–}(4.30)\} \tag{4.31}$$

holds.

Proof Take $\lambda \in \Lambda$ arbitrary. Multiply each DLMI in (4.29) by $\lambda_i \geq 0$, $i \in \mathbb{K}$, add the result, and perform the Schur complement with respect to the last row and column, to obtain

$$\dot{P}(t) + F_\lambda' P(t) + P(t) F_\lambda + G_\lambda' G_\lambda < 0, \; t \in [0, h) \tag{4.32}$$

From Theorem 4.1, we conclude that for ϱ_{2rob}^2 given in (4.31), the upper bound $||H_{sd}(\zeta, \lambda)||_2^2 \leq \varrho_{2rob}^2$ holds for all $\lambda \in \Lambda$, completing the proof. □

Albeit Theorem 4.1 provides the exact value of the $\mathcal{H}_2$ performance index, the same cannot be said of Corollary 4.1. Actually, it is clear that it only provides an upper bound of the same performance index, when the state space model of the open-loop sampled-data system under consideration is corrupted by convex bounded parameter uncertainty. This is an immense step forward, viewed the intricate dependence of the transfer function $H_{sd}(\zeta, \lambda)$ with respect to $\lambda \in \Lambda$. The next example illustrates and discusses the quality of the proposed performance index upper bound.

Example 4.1 Consider the open-loop sampled-data system (4.1)–(4.3) with sampling period $h = 1.5$ and matrices of the state space realization given by

$$A = \begin{bmatrix} 0 & 1 & 0 \\ 0 & 0 & 1 \\ -1 & -5 & -9 \end{bmatrix}, \; B = E = \begin{bmatrix} 0 \\ 0 \\ 1 \end{bmatrix}, \; C_z' = \begin{bmatrix} 1 \\ 1 \\ 1 \end{bmatrix}, \; D_z = \begin{bmatrix} 1 \end{bmatrix}$$

We have determined the squared norm $||H_{sd}(\zeta)||_2^2 = 2.3455$ and the optimal solution of (4.12) with the time interval $[0, h]$ divided in n_ϕ subintervals. Table 4.1 shows that ϱ_2^2 tends asymptotically $||H_{sd}(\zeta)||_2^2$ as n_ϕ increases. It is interesting to point out that, in this case, with only one subinterval, problem (4.12) provides a value of the $\mathcal{H}_2$ performance index within 25% of precision, and for 32 subintervals, the precision becomes less than 1%. As expected, the determination of the true value of the $\mathcal{H}_2$ performance index by the result of Theorem 4.1 is confirmed.

Now, suppose that matrices A and B are given by

$$A = \begin{bmatrix} 0 & 1 & 0 \\ 0 & 0 & 1 \\ -1 & -5 & a \end{bmatrix}, \; B = \begin{bmatrix} 0 \\ 0 \\ b \end{bmatrix}$$

where the parameters (a, b) are uncertain. It is supposed that they can assume any value in the line segment defined by the vertices $(-9, 1)$ and $(-6, 2)$. Table 4.1 also shows the value of ϱ_{2rob}^2 provided by the optimal solution of problem (4.31) with respect to the number of time subintervals n_ϕ. For comparison purpose, we have determined by brute force that $\sup_{\lambda \in \Lambda} ||H_{sd}(\zeta, \lambda)||_2^2 = 3.81$, which is an acceptable upper bound if we take into account the extent of the parameter uncertainty acting on matrices A and B, simultaneously. As clearly indicated in Corollary 4.1, only an upper bound corresponding to the worst case $\mathcal{H}_2$ performance is determined. □

4.2.1 Hybrid Linear Systems

Hybrid systems are largely used to deal with sampled-data control systems design. In this chapter, we have just provided an alternative way to determine the $\mathcal{H}_2$ performance index working directly with the classical concepts derived from the equivalent system. We now tackle the same problem by adopting the classical ideas stemming from the area of hybrid systems. To this end, let us associate to the open-loop sampled-data system (4.1)–(4.3) the hybrid linear system

$$\dot{\psi}(t) = F\psi(t) + J_c w_c(t) \tag{4.33}$$

Table 4.1 Nominal and robust $\mathcal{H}_2$ performance

n_ϕ	1	2	4	8	16	32	64
ϱ_2^2	2.91	2.67	2.48	2.41	2.38	2.36	2.35
ϱ_{2rob}^2	6.39	5.72	5.22	5.02	4.93	4.89	4.86

$$z(t) = G\psi(t) \tag{4.34}$$

$$\psi(t_k) = H\psi(t_k^-) + J_d w_d[k] \tag{4.35}$$

where $t \in [t_k, t_{k+1})$ and $k \in \mathbb{N}$. Several aspects making clear this association must be analyzed carefully.

First, the state variable of the hybrid system is composed of two components, namely $\psi(t) = [\psi_1(t)' \; \psi_2(t)']' \in \mathbb{R}^{n_x+n_u}$. Hence, taking into account the matrices given in (4.8) and in Theorem 4.1, the equations (4.33)–(4.35) are split into the following continuous-time linear equations:

$$\dot{\psi}_1(t) = A\psi_1(t) + B\psi_2(t) + Ew_c(t) \tag{4.36}$$

$$\dot{\psi}_2(t) = 0 \tag{4.37}$$

$$z(t) = C_z\psi_1(t) + D_z\psi_2(t) \tag{4.38}$$

and the jump equations

$$\psi_1(t_k) = \psi_1(t_k^-) \tag{4.39}$$

$$\psi_2(t_k) = w_d[k] \tag{4.40}$$

Equations (4.37) and (4.40) imply that $\psi_2(t) = \psi_2(t_k) = w_d[k]$ in the whole k-th time interval $t \in [t_k, t_{k+1})$. In addition, the Eq. (4.39) imposes the continuity of $\psi_1(t)$ at the jump instant t_k. These observations all together allow us to identify the two components of the state variable of the hybrid system, namely $\psi_1(t) = x(t)$ and $\psi_2(t) = u(t)$, where $x(t)$ is the state variable and $u(t)$ is the control variable (with a zero-order hold) of the open-loop sampled-data system. The initial condition of the hybrid system $\psi(0)$ viewed as a result of the continuous-time impulses $\delta(t)$ in each channel of the perturbation $w_c(t)$ and the discrete-time impulses $\delta[k]$ in each channel of the perturbation $w_d[k]$ will be analyzed in detail afterward. For that, the main goal is to reproduce the $\mathcal{H}_2$ performance index already established in Theorem 4.1, but in the context of hybrid linear systems. However, we go beyond this goal since the next theorem states a general result valid for any hybrid linear system with state space representation of the form (4.33)–(4.35).

Theorem 4.2 *Let $h > 0$ be given. The hybrid linear system defined by* (4.33)–(4.35) *is globally asymptotically stable with ϱ_2^2 performance index if and only if the DLMI* (4.10) *subject to the LMI boundary conditions* (4.11) *is feasible. In the affirmative case, ϱ_2^2 is given by* (4.12).

Proof Initially, for the sufficiency, consider that the hybrid linear system (4.33)–(4.35) evolves from a given initial condition $\psi(0) \neq 0$, and it is free of exogenous perturbations, that is, $w_c = 0$ and $w_d = 0$. The cost function denoted $V_c(\psi, t)$ is well defined in the time interval $t \in [0, h)$. Since the DLMI admits a solution, the Hamilton–Jacobi–Bellman inequality

$$\frac{\partial V_c'}{\partial \psi} F\psi + \psi' G' G \psi + \frac{\partial V_c}{\partial t} < 0 \tag{4.41}$$

holds for the quadratic cost function $V_c(\psi, t) = \psi' P(t)\psi$ for all $t \in [0, h)$, which, by simple integration, yields

$$V_c(\psi(t_k), t_k) > \int_{t_k}^{t_{k+1}} z(t)' z(t) dt + V_c(\psi(t_{k+1}^-), t_{k+1}) \tag{4.42}$$

Now, taking into account that $P_h > 0$ implies $P_0 > 0$, define the positive definite quadratic function $v_d(\psi) = \psi' P_0 \psi$. Due to the time-invariant nature of the hybrid system, the function $V_c(\psi, t) = \psi' P(t - t_k)\psi$ remains feasible in any subsequent time interval, $k \in \mathbb{N}$, provided that the same boundary conditions (4.11) are imposed. Hence, by construction, we have $V_c(\psi(t_k), t_k) = v_d(\psi(t_k))$, and using the jump equation $\psi(t_{k+1}) = H\psi(t_{k+1}^-)$ together with the boundary conditions, we also have

$$\begin{aligned} V_c(\psi(t_{k+1}^-), t_{k+1}) &= \psi(t_{k+1}^-)' P_h \psi(t_{k+1}^-) \\ &> \psi(t_{k+1}^-)' H' P_0 H \psi(t_{k+1}^-) \\ &= \psi(t_{k+1})' P_0 \psi(t_{k+1}) \\ &= v_d(\psi(t_{k+1})) \end{aligned} \tag{4.43}$$

The last two inequalities imply that there exists $\varepsilon > 0$ sufficiently small such that $v_d(\psi(t_{k+1})) \le (1-\varepsilon) v_d(\psi(t_k))$ assuring that the sequence $v_d(\psi(t_k))$ converges to zero as $k \in \mathbb{N}$ goes to infinity, that is, the hybrid system is globally asymptotically stable. On the other hand, the telescoping sum yields

$$\begin{aligned} \int_0^\infty z(t)' z(t) dt &= \sum_{k=0}^{\infty} \int_{t_k}^{t_{k+1}} z(t)' z(t) dt \\ &< \sum_{k=0}^{\infty} \Big(v_d(\psi(t_k)) - v_d(\psi(t_{k+1})) \Big) \\ &= v_d(\psi(0)) - \lim_{k \to \infty} v_d(\psi(t_k)) \\ &= v_d(\psi(0)) \end{aligned} \tag{4.44}$$

This inequality, rewritten as $||z||_2^2 < v_d(\psi(0))$, provides an upper bound to the squared norm of the controlled output $z(t)$ whenever the hybrid system is free of exogenous perturbations and evolves from the initial condition $\psi(0)$. Hence, plugging in the i-th channel of w_c a continuous-time impulse, occurring at t_0^-, that is, just before the first jump, the jump equation $\psi(t_0) = H\psi(t_0^-)$ with

$\psi(t_0^-) = J_{ci}$, where J_{ci} is the i-th column of matrix J_c, yields the initial condition $\psi(0) = H\psi(t_0^-) = HJ_{ci}$, and consequently, we have

$$\begin{aligned}
\sum_{i=1}^{r_c} ||z_i||_2^2 &< \sum_{i=1}^{r_c} v_d(HJ_{ci}) \\
&= \sum_{i=1}^{r_c} J_{ci}'H'P_0HJ_{ci} \\
&\leq \sum_{i=1}^{r_c} J_{ci}'P_hJ_{ci} \\
&= \mathbf{tr}(J_c'P_hJ_c)
\end{aligned} \tag{4.45}$$

where the second inequality is due to the boundary conditions (4.11). On the other hand, plugging in the i-th channel of w_d a discrete-time impulse, occurring at $k = 0$, the jump equation yields the initial condition $\psi(0) = J_{di}$, where J_{di} is the i-th column of matrix J_d, so as

$$\begin{aligned}
\sum_{i=1}^{r_d} ||z_i||_2^2 &< \sum_{i=1}^{r_d} v_d(J_{di}) \\
&= \sum_{i=1}^{r_d} J_{di}'P_0J_{di} \\
&= \mathbf{tr}(J_d'P_0J_d)
\end{aligned} \tag{4.46}$$

Finally, adding (4.45) and (4.46), we conclude that any feasible solution to the DLMI subject to the boundary conditions provides an upper bound to the $\mathcal{H}_2$ performance index.

Necessity follows from Bellman's principle of optimality applied to the hybrid linear system, free of exogenous perturbations. Let us consider that the hybrid linear system is globally asymptotically stable and define the positive cost-to-go function

$$V_c(\xi, t_k) = \int_{t_k}^{\infty} \psi(t)'G'G\psi(t)dt \tag{4.47}$$

where for $t \geq t_k$, the trajectory $\psi(t)$ is provided by the linear, time-invariant differential equation $\dot{\psi}(t) = F\psi(t)$, subject to the initial condition $\psi(t_k) = \xi$ and the jump equation $\psi(t_{k+1}) = H\psi(t_{k+1}^-)$. Assuming that the $\mathcal{H}_2$ performance index exists and is finite, any upper bound to it is provided by the stationary version of the cost-to-go function that satisfies

$$v_d(\psi(t_k)) > \int_{t_k}^{t_{k+1}} \psi(t)'G'G\psi(t)dt + v_d(\psi(t_{k+1})) \tag{4.48}$$

for all $k \in \mathbb{N}$. The smallest upper bound is reached whenever (4.48) is arbitrarily close to equality. In this particular situation, $v_d(\psi(t_k))$ is positive definite and vanishes as $k \in \mathbb{N}$ goes to infinity indicating that the hybrid linear system is globally asymptotically stable indeed. In addition, we can also conclude that $v_d(\xi) = \xi'S\xi$ for some $S > 0$, that is, the stationary cost-to-go function can be assumed to be quadratic, with no loss of generality. To prove this claim, by induction, assume that this actually occurs at sampling time t_{k+1}, and due to the fact that $\psi(t_{k+1}) = H\psi(t_{k+1}^-) = He^{Fh}\psi(t_k)$, then (4.48) yields

$$\begin{aligned} v_d(\psi(t_k)) &> \int_{t_k}^{t_{k+1}} \psi(t_k)'e^{F'(t-t_k)}G'Ge^{F(t-t_k)}\psi(t_k)dt + v_d\left(He^{Fh}\psi(t_k)\right) \\ &= \psi(t_k)'\left(\int_0^h e^{F'\tau}G'Ge^{F\tau}dt + e^{F'h}H'SHe^{Fh}\right)\psi(t_k) \end{aligned} \tag{4.49}$$

which shows that the same remains true in the sampling time t_k, since the smallest upper bound is a quadratic function as well. Finally, if we set the pair of positive definite matrices (P_0, P_h) arbitrarily close to the pair $(S, H'SH)$ and split the whole time interval $t_k \mapsto t_{k+1}$ in two successive events, namely $t_k \mapsto t_{k+1}^-$ and $t_{k+1}^- \mapsto t_{k+1}$, inequality (4.48) implies that there exists a solution $P(t)$, defined in the time interval $[0, h)$, to the DLMI (4.10) subject to the LMI boundary conditions (4.11). Moreover, evolving from $t_0 = 0$ we obtain the exact evaluation $||z||_2^2 = v_d(\psi(0)) = \psi(0)'P_0\psi(0)$ yielding thus the exact value of ϱ_2^2. The proof is concluded. □

The determination of the $\mathcal{H}_2$ performance index associated with the open-loop sampled-data system from the adoption of a hybrid linear model is relevant in the sense that it provides a simple alternative route that can be used to prove the validity of the results obtained so far in the context of classical continuous- and discrete-time linear models. While our attention is devoted to performance evaluation of linear sampled-data systems exclusively, there is no major difference in adopting one or another view. However, in some instances, when dealing with other classes of sampled-data systems, and control design problems, there are clear advantages to expressing them as hybrid systems. The generalization discussed in the next remark supports this claim.

Remark 4.2 The hybrid system approach can be applied to prove that Theorem 4.1 remains valid if matrices (F, G) are time-varying but fully defined in the time interval $[0, h]$. This means that $F(t) = F(t - t_k)$ and $G(t) = G(t - t_k)$ for all $k \in \mathbb{N}$. Under this assumption, it can be seen that the sufficiency part of Theorem 4.2 remains entirely valid. For the necessity, the key point is to show that the stationary cost-to-go function $v_d(\xi)$ is quadratic. This follows from the fact that in the time-varying case under consideration $\psi(t) = \Phi(t, t_k)\psi(t_k)$, where $\Phi(t, t_k)$ is the state transition matrix associated with $\dot{\psi}(t) = F(t)\psi(t)$, satisfying $\Phi(t_k, t_k) = I$, for all

$k \in \mathbb{N}$. The linear dependence of the solution $\psi(t)$ with respect to $\psi(t_k)$ in the time interval $t \in [t_k, t_{k+1})$ and the fact that $\Phi(t, t_k) = \Phi(t - t_k, 0)$ prove, by induction, that this claim remains true as well. □

Many other similar situations will be promptly identified in the forthcoming sections and chapters. That is, for each specific design problem, the hybrid system model is settled and the conditions of Theorem 4.1 are imposed and solved. Last but not least, it is the fact that the previous theorem has been proven without taking into account that $H = \text{diag}(I, 0)$. That is, it remains valid in the general framework of hybrid linear systems of the form (4.33)–(4.35).

4.3 State Feedback Design

Let us consider a sampled-data control system with the following state space realization:

$$\dot{x}(t) = Ax(t) + Bu(t) + Ew_c(t), \; x(0) = 0 \tag{4.50}$$

$$z(t) = C_z x(t) + D_z u(t) \tag{4.51}$$

$$u(t) = Lx[k] + (E_u + LE_y)w_d[k], \; t \in [t_k, t_{k+1}), \; \forall k \in \mathbb{N} \tag{4.52}$$

derived from our general model (3.1)–(3.3) with $C_y = I$, since, by assumption, the whole state variable is available for feedback. Looking at the hybrid linear system model, matrices F, G, and J_c remain the same as before, see (4.8), but matrices H and J_d change to

$$H = \begin{bmatrix} I & 0 \\ L & 0 \end{bmatrix}, \; J_d = \begin{bmatrix} 0 \\ E_u + LE_y \end{bmatrix} \tag{4.53}$$

with appropriate dimensions. Of course the difficulty we have to face is to include the matrix $L \in \mathbb{R}^{n_u \times n_x}$ in the set of variables to be determined. We consider two cases, when $E_y = 0$ or $E_y \neq 0$. It is important to keep in mind that in the first case, treated by the next two theorems and a corollary, matrix J_d does not depend on the gain matrix L, but in the second case it does.

Theorem 4.3 *Let $h > 0$ be given. Consider that together with the DLMI*

$$\begin{bmatrix} -\dot{Q}(t) + Q(t)F' + FQ(t) & Q(t)G' \\ \bullet & -I \end{bmatrix} < 0, \; t \in [0, h) \tag{4.54}$$

subject to the LMI boundary conditions

$$\begin{bmatrix} Q_h & Q_h \begin{bmatrix} I \\ 0 \end{bmatrix} \\ \bullet & W \end{bmatrix} > 0, \quad \begin{bmatrix} W & \begin{bmatrix} W & K' \end{bmatrix} \\ \bullet & Q_0 \end{bmatrix} > 0 \tag{4.55}$$

the optimal solution of the convex programming problem

$$\varrho_2^2 = \inf_{Q(\cdot),W,K} \left\{ \mathbf{tr}(J_c' Q_h^{-1} J_c) + \mathbf{tr}(J_d' Q_0^{-1} J_d) : (4.54)\text{–}(4.55) \right\} \tag{4.56}$$

provides the matrix gain $L = KW^{-1}$*. Then, the closed-loop sampled-data control system (4.50)–(4.52) is globally asymptotically stable and operates with minimum* ϱ_2^2 *performance index.*

Proof The proof follows from a series of nested Schur complement calculations. The boundary conditions (4.55) are equivalent to

$$Q_h^{-1} > \begin{bmatrix} I \\ 0 \end{bmatrix} W^{-1} \begin{bmatrix} I & 0 \end{bmatrix}, \quad W^{-1} > \begin{bmatrix} I & L' \end{bmatrix} Q_0^{-1} \begin{bmatrix} I \\ L \end{bmatrix} \tag{4.57}$$

$W > 0$, $Q_0 > 0$, and $Q_h > 0$, respectively. Moreover, both together are equivalent to $Q_h^{-1} > H' Q_0^{-1} H'$. It has been established that any feasible solution to the DLMI (4.54) subject to positive definite boundary conditions, namely $Q_0 > 0$ and $Q_h > 0$, is such that $Q(t) > 0$ for all $t \in [0, h]$. Hence, multiplying both sides of the DLMI (4.54) by $\mathrm{diag}(Q(t)^{-1}, I)$ and using the fact that the time derivative of $P(t) = Q(t)^{-1}$ is equal to $\dot{P}(t) = -Q(t)^{-1}\dot{Q}(t)Q(t)^{-1}$, then $P(t)$ solves the DLMI (4.10) subject to the boundary conditions (4.11) with $P_h = Q_h^{-1} > 0$ and $P_0 = Q_0^{-1} > 0$. Finally, problem (4.56) is exactly (4.12) but including the matrix gain L in the set of matrix variables. The proof is concluded by direct application of Theorem 4.1. □

The control design problem we are dealing with has a particularity that must be underlined, that is, the matrix gain $L \in \mathbb{R}^{n_u \times n_x}$ appears only once in the boundary conditions (4.11). Actually, albeit the DLMI does not depend on it directly, it is responsible for modifying the jump condition and, by consequence, the trajectory of the closed-loop sampled-data control system in order to reach optimality.

Let us analyze the boundary conditions $P_h > H' P_0 H$, where H is given in (4.53), $P_h > 0$, and $P_0 > 0$ is partitioned into four blocks as indicated in (4.9). The following factorization holds

$$\begin{aligned} \begin{bmatrix} I \\ L \end{bmatrix}' P_0 \begin{bmatrix} I \\ L \end{bmatrix} &= \begin{bmatrix} I \\ L \end{bmatrix}' \begin{bmatrix} X_0 & V_0 \\ V_0' & \hat{X}_0 \end{bmatrix} \begin{bmatrix} I \\ L \end{bmatrix} \\ &= (X_0 - V_0 \hat{X}_0^{-1} V_0') + \left(L + \hat{X}_0^{-1} V_0' \right)' \hat{X}_0 \left(L + \hat{X}_0^{-1} V_0' \right) \end{aligned} \tag{4.58}$$

and, for this reason, it can be verified by simple algebraic manipulations that

$$\begin{bmatrix} I \\ L \end{bmatrix}' P_0 \begin{bmatrix} I \\ L \end{bmatrix} \geq X_0 - V_0 \hat{X}_0^{-1} V_0', \ \forall L \in \mathbb{R}^{n_u \times n_x} \tag{4.59}$$

which means that the minimal element (in matrix sense) is reached if and only if $L = -\hat{X}_0^{-1} V_0'$. In our present context, this simple result is of particular importance. Indeed, let us first define the matrix

$$I_x = \begin{bmatrix} I \\ 0 \end{bmatrix} \in \mathbb{R}^{(n_x+n_u)\times n_x} \tag{4.60}$$

which is often used from now on, and notice that the previous results lead to

$$\begin{aligned} H' P_0 H &= I_x \begin{bmatrix} I \\ L \end{bmatrix}' P_0 \begin{bmatrix} I \\ L \end{bmatrix} I_x' \\ &\geq I_x \big(X_0 - V_0 \hat{X}_0^{-1} V_0' \big) I_x', \ \forall L \in \mathbb{R}^{n_u \times n_x} \end{aligned} \tag{4.61}$$

and the equality holds if and only if $L = -\hat{X}_0^{-1} V_0'$. This is a key inequality that allows us to introduce and prove an important simplification on the control design conditions provided in Theorem 4.3.

Theorem 4.4 *Let $h > 0$ be given. Consider that together with the DLMI*

$$\begin{bmatrix} -\dot{Q}(t) + Q(t)F' + FQ(t) & Q(t)G' \\ \bullet & -I \end{bmatrix} < 0, \ t \in [0, h) \tag{4.62}$$

subject to the LMI boundary conditions

$$Q_0 > 0, \ I_x' \big(Q_0 - Q_h \big) I_x > 0 \tag{4.63}$$

the optimal solution of the convex programming problem

$$\varrho_2^2 = \inf_{Q(\cdot)} \Big\{ \mathbf{tr}(J_c' Q_h^{-1} J_c) + \mathbf{tr}(J_d' Q_0^{-1} J_d) : (4.62)\text{–}(4.63) \Big\} \tag{4.64}$$

provides the matrix gain $L = -\hat{X}_0^{-1} V_0'$ constructed with the partitions indicated in (4.9) of matrix $P_0 = Q_0^{-1}$. Then, the closed-loop sampled-data control system (4.50)–(4.52) is globally asymptotically stable and operates with minimum ϱ_2^2 performance index.

Proof The fact that $Q_0 > 0$ implies that any feasible solution to the DLMI (4.62) is positive definite for all $t \in [0, h]$, in particular $Q_h > 0$. From Theorem 4.1, the boundary conditions to be satisfied are $P_h > 0$ and $P_h > H' P_0 H'$. Since the gain matrix L acts only on matrix H, then the optimal choice of this matrix variable is such that $H' P_0 H'$ coincides with the minimal element, making the matrix constraint $P_h > H'P_0H'$ the less restrictive possible. From the previous algebraic manipulations, this leads to $L = -\hat{X}_0^{-1}V_0'$, in which case the boundary condition becomes

$$\begin{aligned} P_h &> H'P_0H \\ &= I_x(X_0 - V_0\hat{X}_0^{-1}V_0')I_x' \\ &= I_x(I_x'P_0^{-1}I_x)^{-1}I_x' \end{aligned} \tag{4.65}$$

where the last equality is due to the matrix inversion formula that states that the inverse of the first block-diagonal element of P_0^{-1} equals $X_0 - V_0\hat{X}_0^{-1}V_0'$. Hence, the Schur complement of (4.65) yields the equivalent condition

$$\begin{bmatrix} Q_h^{-1} & I_x \\ I_x' & I_x'Q_0I_x \end{bmatrix} > 0 \tag{4.66}$$

whose Schur complement with respect to the first row and column reproduces the inequality (4.63), completing thus the proof. □

This is an interesting result. First of all, Theorems 4.3 and 4.4 provide the optimal state feedback sampled-data control corresponding to the $\mathcal{H}_2$ performance index. In this sense, they are obviously equivalent. However, comparing both results, the last one is preferable because the problem to be solved involves a smaller number of matrix variables (W and K have been eliminated) and simpler boundary conditions. In Theorem 4.4, the optimal matrix gain is constructed, after optimization, by the partition of the boundary optimal matrix $P_0 = Q_0^{-1}$ shown in (4.9).

As we have already established, the optimal state feedback matrix gain L can be alternatively calculated by converting the sampled-data system to its equivalent discrete-time version. As far as numerical efficiency is concerned, this procedure must be adopted, but it cannot be applied whenever the open-loop system is subject to parameter uncertainty. Fortunately, this is the major importance of the last two theorems, and the robust state feedback control can be designed by the next corollary of Theorem 4.4.

Corollary 4.2 *Let $h > 0$ be given. Consider that together with the set of DLMIs*

$$\begin{bmatrix} -\dot{Q}(t) + Q(t)F_i' + F_iQ(t) & Q(t)G_i' \\ \bullet & -I \end{bmatrix} < 0, \ i \in \mathbb{K}, \ t \in [0, h) \tag{4.67}$$

subject to the LMI boundary conditions

$$Q_0 > 0, \ I_x'(Q_0 - Q_h)I_x > 0 \tag{4.68}$$

the optimal solution of the convex programming problem

$$\varrho_{2rob}^2 = \inf_{Q(\cdot)} \left\{ \mathbf{tr}(J_c' Q_h^{-1} J_c) + \mathbf{tr}(J_d' Q_0^{-1} J_d) : (4.67)\text{–}(4.68) \right\} \tag{4.69}$$

provides the matrix gain $L = -\hat{X}_0^{-1} V_0'$ constructed with the partitions indicated in (4.9) of matrix $P_0 = Q_0^{-1}$. Then, for all $\lambda \in \Lambda$, the uncertain closed-loop sampled-data system (4.50)–(4.52) is globally asymptotically stable and operates with the guaranteed performance index ϱ_{2rob}^2.

Proof Since each DLMI in the set (4.67) depends linearly on matrices (F_i, G_i), $i \in \mathbb{K}$, then the DLMI

$$\begin{bmatrix} -\dot{Q}(t) + Q(t)F_\lambda' + F_\lambda Q(t) & Q(t)G_\lambda' \\ \bullet & -I \end{bmatrix} < 0, \; t \in [0, h) \tag{4.70}$$

holds for every $\lambda \in \Lambda$. From Theorem 4.4, we conclude that ϱ_{2rob}^2 given in (4.69) is a valid upper bound to the closed-loop transfer function squared norm $||H_{sd}(\zeta, \lambda)||_2^2$ for all $\lambda \in \Lambda$. The proof is concluded. □

A similar result can also be obtained as a corollary of Theorem 4.3. The proof follows the same pattern of that of Corollary 4.2, being thus omitted. It is important to stress, by one side, that these results stem from the fact that the DLMI naturally avoids to expressing the control design conditions through matrix exponential mappings and, by the other side, that only an upper bound of the $\mathcal{H}_2$ performance index is minimized. The next example illustrates the robust control design for sampled-data systems subject to convex bounded parameter uncertainty.

Example 4.2 Consider the sampled-data control system (4.50)–(4.52) with $h = 1.5$ and matrices

$$A = \begin{bmatrix} 0 & 1 & 0 \\ 0 & 0 & 1 \\ a & -5 & -9 \end{bmatrix}, \; B = \begin{bmatrix} 0 \\ 0 \\ b \end{bmatrix}$$

with a pair of uncertain parameters (a, b) belonging to the convex bounded set defined by the vertices, $(a_1, b_1) = (0, 1)$ and $(a_2, b_2) = (9, 2)$. The remaining matrices are

$$E = \begin{bmatrix} 0 \\ 0 \\ 1 \end{bmatrix}, \; E_u = \begin{bmatrix} 0 & 1 \end{bmatrix}, \; C_z' = \begin{bmatrix} 1 \\ 1 \\ 1 \end{bmatrix}, \; D_z = \begin{bmatrix} 1 \end{bmatrix}$$

It is simple to see that matrix A is not Hurwitz stable in the entire parameter uncertainty region because the parameter a is always non-negative needing, by consequence, an effective control action for robust stabilization. With $n_\phi = 64$, we have solved problem (4.69) twice:

(a) For verification purposes, the parameters $(a, b) = (0, 1)$ have been fixed. In this case, Corollary 4.2 provides the optimal state feedback gain

$$L = \begin{bmatrix} -0.8515 & -0.8288 & -0.1370 \end{bmatrix}$$

and the corresponding minimum $\mathcal{H}_2$ performance level $\varrho_2^2 = 2.3187$. We have solved the same problem by calculating the equivalent system and solving the associated Riccati equation that yielded $\varrho_2^2 = 2.0734$. Albeit, this performance level may be attained with a greater n_ϕ (it becomes $\varrho_2^2 = 2.2070$ for $n_\phi = 128$), the price to be paid in terms of computational burden is prohibitive.

(b) The uncertain system with the convex bounded parameter uncertainty defined before has been considered. The robust state feedback matrix gain is

$$L = \begin{bmatrix} -5.5766 & -4.8059 & -0.5034 \end{bmatrix}$$

with guaranteed $\mathcal{H}_2$ performance level $\varrho_{2rob}^2 = 24.1282$.

Notice the enormous difference between the minimum guaranteed cost and the optimal cost associated with the nominal system. This is due to the range of the parameter uncertainty considered. It is true that solving the conditions of Corollary 4.2 is harder but a robust control able to cope with parameter uncertainty can be designed. □

Now, let us move our attention to the second case, in which $E_y \neq 0$. As we know, this case is much more complicated as far as control design is concerned because both matrices H and J_d, quoted in (4.53), depend on the state feedback gain matrix L. Unfortunately, even for a sampled-data system free of parameter uncertainty, we are unable to calculate the optimal state feedback matrix gain using convex programming facilities. However, in principle, a suboptimal solution (whenever exists) can be determined by adopting some *ad hoc* algebraic manipulation that linearizes the problem at the expense of introducing some degree of conservatism. This possibility will not be pursued in this book. Here, more general models, including those with $C_y \neq I$, will be tackled by the adoption of full order dynamic output feedback controllers as we have successfully done previously in Chap. 3.

4.4 Filter Design

Filtering in the context of sampled-data systems is an important issue, mainly if we can cope with parameter uncertainty always present in the transmission channel. The sampled-data system has the standard form

$$\dot{x}(t) = Ax(t) + Ew_c(t), \; x(0) = 0 \tag{4.71}$$

$$y[k] = C_y x[k] + E_y w_d[k] \tag{4.72}$$

$$z_e(t) = C_z x(t) \tag{4.73}$$

where $x(\cdot) : \mathbb{R}_+ \to \mathbb{R}^{n_x}$ is the state, $z_e(\cdot) : \mathbb{R}_+ \to \mathbb{R}^{n_z}$ is the variable to be estimated, and $y[\cdot] : \mathbb{N} \to \mathbb{R}^{n_y}$ denotes the measurement available only at the sampling times. The filter to be designed has the state space realization

$$\dot{\hat{x}}(t) = \hat{A}_c(t)\hat{x}(t), \ \hat{x}(0) = 0 \tag{4.74}$$

$$\hat{x}(t_k) = \hat{A}_d\hat{x}(t_k^-) + \hat{B}_d y[k] \tag{4.75}$$

$$\hat{z}_e(t) = \hat{C}_c(t)\hat{x}(t) \tag{4.76}$$

valid in the time interval $t \in [t_k, t_{k+1}),\ k \in \mathbb{N}$ and $\hat{x}(\cdot) : \mathbb{R}_+ \to \mathbb{R}^{n_x}$ is the state and $\hat{z}_e(\cdot) : \mathbb{R}_+ \to \mathbb{R}^{n_z}$ is the estimation produced by the filter. Observe that this filter depends on two matrices of appropriate dimensions $(\hat{A}_d, \hat{B}_d)$ that define the discrete-time domain of the jumps and two matrix-valued functions $\hat{A}_c(\cdot) : \mathbb{R}_+ \to \mathbb{R}^{n_x \times n_x}$, $\hat{C}_c(\cdot) : \mathbb{R}_+ \to \mathbb{R}^{n_z \times n_x}$ that define the continuous-time evolution of the filter between successive samplings. We want to verify the conditions under which these matrix-valued functions can be set as constant matrices without losing optimality. The connection of the filter to the plant has the hybrid linear system model (4.33)–(4.35) with state variable $\psi(t)' = [x(t)' \ \hat{x}(t)']$ and matrices

$$F = \begin{bmatrix} A & 0 \\ 0 & \hat{A}_c \end{bmatrix}, \ J_c = \begin{bmatrix} E \\ 0 \end{bmatrix}, \ G = \begin{bmatrix} C_z & -\hat{C}_c \end{bmatrix}, \tag{4.77}$$

and

$$H = \begin{bmatrix} I & 0 \\ \hat{B}_d C_y & \hat{A}_d \end{bmatrix}, \ J_d = \begin{bmatrix} 0 \\ \hat{B}_d E_y \end{bmatrix} \tag{4.78}$$

where, to ease the notation, the time dependence of $\hat{A}_c(\cdot)$ and $\hat{C}_c(\cdot)$ has been dropped. The variable $z(t) = G\psi(t) = z_e(t) - \hat{z}_e(t)$ indicates that the output of the hybrid linear system is the estimation error produced by the filter. Our first purpose is to find the optimal time-varying filter (4.74)–(4.76) for the $\mathcal{H}_2$ performance index. Later on, a time-invariant version of the same filter, preserving optimality, is determined. Notice that, based on the measurements available only at the sampling times $\{t_k\}_{k\in\mathbb{N}}$, this filter provides the best estimation of the output $z_e(t)$ for all $t \in [t_k, t_{k+1}), k \in \mathbb{N}$, that is, during the whole inter-sampling time interval.

To this end, given $h > 0$, we need to introduce the following symmetric positive definite matrix-valued function $P(\cdot) : [0, h] \to \mathbb{R}^{2n_x \times 2n_x}$ of the form

$$P(t) = \begin{bmatrix} X(t) & V(t) \\ V(t)' & \hat{X}(t) \end{bmatrix} > 0 \tag{4.79}$$

The four partitions are matrix-valued functions defined in the time interval $[0, h]$ with image in $\mathbb{R}^{n_x \times n_x}$. Let us also denote, with the same partitioning dimensions, the following matrix-valued functions:

$$Q(t) = P(t)^{-1} = \begin{bmatrix} Y(t) & U(t) \\ U(t)' & \hat{Y}(t) \end{bmatrix} > 0 \tag{4.80}$$

$$\Gamma(t) = \begin{bmatrix} I & I \\ 0 & U(t)'Y(t)^{-1} \end{bmatrix} \tag{4.81}$$

It can be verified that similar definitions and partitions have been used in Chap. 3, but for time-invariant matrices instead of matrix-valued functions. Let us define the matrix-valued function $Z(t) = Y(t)^{-1} > 0$ and, from now on, to simplify the notation, the dependence of these functions with respect to time is omitted. Algebraic manipulations put in evidence that

$$\Gamma' P \Gamma = \begin{bmatrix} X & Z \\ Z & Z \end{bmatrix} > 0 \tag{4.82}$$

and $QP = I$ is settled provided that any U non-singular is given. In addition, calculating the time derivative of P yields

$$\begin{aligned} \Gamma' \dot{P} \Gamma &= -\Gamma' P \dot{Q} P \Gamma \\ &= \begin{bmatrix} \dot{X} & \dot{X} + \dot{V} U' Z \\ \bullet & \dot{Z} \end{bmatrix} \end{aligned} \tag{4.83}$$

where the two formulas in the first equality have been used to obtain the second equality in (4.83). In the next remark, some useful matrix calculations to be handled afterward are presented.

Remark 4.3 Each term in the DLMI (4.10) of Theorem 4.1 can be expressed as

$$\Gamma' P F \Gamma = \begin{bmatrix} XA & XA + V\hat{A}_c U' Z \\ ZA & ZA \end{bmatrix}$$

$$G\Gamma = \begin{bmatrix} C_z & C_z - \hat{C}_c U' Z \end{bmatrix}$$

while the positive definiteness $P > 0$ is equivalent to $X > Z > 0$ as an immediate consequence of the Schur complement applied to the second diagonal element of (4.82). The boundary conditions (4.11) can be written as

$$\begin{bmatrix} \Gamma_h' P_h \Gamma_h & \Gamma_h' H' P_0 \Gamma_0 \\ \bullet & \Gamma_0' P_0 \Gamma_0 \end{bmatrix} = \begin{bmatrix} X_h & Z_h & X_0 + C_y' \hat{B}_d' V_0' & Z_0 \\ \bullet & Z_h & X_0 + C_y' \hat{B}_d' V_0' + Z_h U_h \hat{A}_d' V_0' & Z_0 \\ \bullet & \bullet & X_0 & Z_0 \\ \bullet & \bullet & \bullet & Z_0 \end{bmatrix}$$

whereas for the objective function we have $J_c' P_h J_c = E' X_h E$, and adopting the same reasoning as before, it follows that

$$\begin{bmatrix} W_d & J_d' P_0 \Gamma_0 \\ \bullet & \Gamma_0' P_0 \Gamma_0 \end{bmatrix} = \begin{bmatrix} W_d & E_y' \hat{B}_d' V_0' & 0 \\ \bullet & X_0 & Z_0 \\ \bullet & \bullet & Z_0 \end{bmatrix}$$

which, by Schur complement calculation, is equivalent to $Z_0 > 0$ and

$$W_d > J_d' P_0 J_d \Leftrightarrow \begin{bmatrix} W_d & E_y' \hat{B}_d' V_0' \\ \bullet & X_0 - Z_0 \end{bmatrix} > 0 \tag{4.84}$$

These are the key algebraic manipulations that we need in order to design the optimal filter and, in the sequel, the robust filter, both belonging to the same class that we judge to be large enough. It remains to show that they can be expressed through a DLMI and LMIs by a proper definition of a one-to-one change of variables. □

The next theorem is the more important result of this section. Based on it, for precisely known sampled-data systems, the optimal time-varying filter is determined. Actually, elaborating more on that direction, it is shown that the optimal filter is, in fact, time invariant. Another remarkable consequence is that it also allows the determination of a robust filter whenever the plant is subject to convex bounded parameter uncertainty. The theoretical basis needed to handle those situations has been discussed in Remark 4.2.

Theorem 4.5 *Let $h > 0$ be given. If the matrix-valued functions $Z(t)$, $X(t)$, $M_c(t)$, $K_c(t)$, the matrices M_d, K_d, and W_d satisfy the DLMI*

$$\begin{bmatrix} \dot{X} + A'X + XA & XA + A'Z + M_c & C_z' \\ \bullet & \dot{Z} + A'Z + ZA & C_z' - K_c' \\ \bullet & \bullet & -I \end{bmatrix} < 0, \ t \in [0, h) \tag{4.85}$$

subject to the LMI boundary condition

$$\begin{bmatrix} X_h & Z_h & X_0 + C_y' K_d' & Z_0 \\ \bullet & Z_h & X_0 + C_y' K_d' + M_d' & Z_0 \\ \bullet & \bullet & X_0 & Z_0 \\ \bullet & \bullet & \bullet & Z_0 \end{bmatrix} > 0 \tag{4.86}$$

then the solution of the convex programming problem

$$\varrho_2^2 = \inf_{X>Z>0, M_c, K_c, M_d, K_d, W_d} \{\mathbf{tr}(E'X_h E) + \mathbf{tr}(W_d) : (4.84)\text{–}(4.86)\} \tag{4.87}$$

provides the optimal filter whose state space realization is given by

$$\hat{A}_c = (Z - X)^{-1}\big(M_c - \dot{Z}\big), \; \hat{C}_c = K_c \tag{4.88}$$

$$\hat{A}_d = (Z_0 - X_0)^{-1} M_d, \; \hat{B}_d = (Z_0 - X_0)^{-1} K_d \tag{4.89}$$

Proof It follows directly from the generalization of Theorem 4.1 discussed in Remark 4.2, concerning time-varying hybrid systems. First, adopting the partitioning (4.79)–(4.81), due to (4.82), it is clear that $P > 0$ if and only if $X > Z > 0$. Moreover, the DLMI (4.10) rewritten in the equivalent form

$$\begin{bmatrix} \Gamma'(\dot{P} + F'P + PF)\Gamma & \Gamma'G' \\ \bullet & -I \end{bmatrix} < 0, \; t \in [0, h) \tag{4.90}$$

together with the calculations provided in Remark 4.3 and (4.83) is expressed as the DMLI (4.85) with $M_c = V\hat{A}_c U'Z + \dot{V}U'Z + \dot{X}$ and $K_c = \hat{C}_c U'Z$. On the other hand, the boundary conditions (4.11) rewritten as

$$\begin{bmatrix} \Gamma_h' P_h \Gamma_h & \Gamma_h' H' P_0 \Gamma_0 \\ \bullet & \Gamma_0' P_0 \Gamma_0 \end{bmatrix} > 0 \tag{4.91}$$

reproduce, as indicated in Remark 4.3, the LMI (4.86) whenever the relationships $M_d = V_0 \hat{A}_d U_h' Z_h$ and $K_d = V_0 \hat{B}_d$ are adopted. Finally, using again the calculations provided in Remark 4.3, the objective function in (4.12) is expressed in terms of the new matrix-valued functions and matrix variables. Hence, the filter with optimal $\mathcal{H}_2$ performance is determined from the solution of the convex problem (4.87).

It remains to extract the optimal filter from the solution of the problem (4.87). To this end, set the matrix-valued function $U = Y = Z^{-1}$ in which case we have $V = Z - X < 0$ that yields the matrix-valued functions $\hat{C}_c = K_c(U'Z)^{-1} = K_c$ and

$$\begin{aligned} \hat{A}_c &= V^{-1}\big(M_c - \dot{V}U'Z - \dot{X}\big)(U'Z)^{-1} \\ &= (Z - X)^{-1}\big(M_c - \dot{Z}\big) \end{aligned} \tag{4.92}$$

as well as the matrices $\hat{A}_d = V_0^{-1} M_d (U_h' Z_h)^{-1} = (Z_0 - X_0)^{-1} M_d$ and $\hat{B}_d = V_0^{-1} K_d = (Z_0 - X_0)^{-1} K_d$ that all together reproduce the filter state space matrices (4.88) and (4.89), respectively. The proof is concluded. □

The consequence of this theorem is twofold. The first one is the characterization of the optimal filter of the class under consideration, for sampled-data systems by means of a convex programming problem with DLMI and LMIs constraints avoiding exponential and similar maps that destroy convexity. Second, for a sampled-data system free of parameter uncertainty, the result of Theorem 4.5 can be deeply simplified as the next theorem states.

Theorem 4.6 *Let $h > 0$ be given. If the matrix-valued function $X(t)$ and the matrices X_0, X_h, and K_d satisfy the DLMI*

$$\begin{bmatrix} \dot{X} + A'X + XA & C_z' \\ \bullet & -I \end{bmatrix} < 0, \; t \in [0, h) \tag{4.93}$$

subject to the LMI boundary condition

$$\begin{bmatrix} X_h & X_0 + C_y' K_d' \\ \bullet & X_0 \end{bmatrix} > 0 \tag{4.94}$$

then the solution of the convex programming problem

$$\varrho_2^2 = \inf_{X, K_d} \left\{ \mathbf{tr}(E' X_h E) + \mathbf{tr}\left(E_y' K_d' X_0^{-1} K_d E_y\right) : (4.93)\text{–}(4.94) \right\} \tag{4.95}$$

provides the optimal time-invariant filter whose state space realization is given by

$$\hat{A}_c = A, \; \hat{C}_c = C_z \tag{4.96}$$

$$\hat{A}_d = I + X_0^{-1} K_d C_y, \; \hat{B}_d = -X_0^{-1} K_d \tag{4.97}$$

Proof We have to prove that the result of Theorem 4.5 still remains true in this particular situation. Interchanging the first and second rows and columns, it is seen that the DLMI (4.85) can be simplified by eliminating the variables $K_c = C_z$ and $M_c = -A'Z - XA$. Doing this, it is equivalent to the DLMI $\dot{Z} + A'Z + ZA < 0$ and the DLMI (4.93). Looking at the LMI (4.94), it is clear that it is nothing but the LMI (4.86) where the second and fourth rows and columns have been eliminated. This elimination is possible to be done by keeping $X_h > Z_h > Z_0 > 0$, setting $M_d = -X_0 - K_d C_y$, and making matrices Z_h and Z_0 arbitrarily close to the null matrix. Notice that these algebraic manipulations and choice of variables have been done preserving feasibility and without introducing any kind of conservativeness.

On the other hand, the fact that $0 < X_0 - Z_0 < X_0$ makes it clear that the objective function of problem (4.87) must be replaced by that of problem (4.95) whenever $Z_0 > 0$ arbitrarily close to zero is feasible. The consequence is that the matrix-valued function $Z(t)$ subject to the boundary conditions $Z(0) = Z_0$ and $Z(h) = Z_h$ arbitrarily close to zero for all $t \in [0, h]$ is feasible. Taking this fact into account, from (4.88), we obtain $\hat{C}_c(t) = C_z$ and

$$\hat{A}_c(t) = (Z - X)^{-1}\big(- XA - A'Z - \dot{Z}\big)$$
$$= A \tag{4.98}$$

such that the formulas in (4.89) provide $\hat{A}_d = I + X_0^{-1}K_dC_y$ and $\hat{B}_d = -X_0^{-1}K_d$, respectively. Finally, due to (4.94), $X_h > 0$ implies that all feasible solutions to the DLMI (4.93) are such that $X(t) > 0$, $\forall t \in [0, h]$. Hence, the constraint $X > Z > 0$ appearing in problem (4.87) is superfluous since $Z > 0$ is arbitrarily close to zero. The proof is concluded. □

As expected, the optimal sampled-data filter is time invariant and can be designed from the solution of a convex programming problem expressed by a DLMI and an LMI. It is interesting to observe that the optimal filter exhibits the *internal model* structure for the continuous-time part and the *innovation* structure for the discrete-time part that receives and processes the available measurement on each sampling time t_k, $k \in \mathbb{N}$. Indeed, plugging the matrices given in Theorem 4.6 into Eqs. (4.74)–(4.75), it follows that

$$\dot{\hat{x}}(t) = A\hat{x}(t),\ \hat{x}(0) = 0 \tag{4.99}$$
$$\hat{x}(t_k) = \hat{x}(t_k^-) - X_0^{-1}K_d\big(y(t_k) - C_y\hat{x}(t_k^-)\big) \tag{4.100}$$

A second and equally important consequence of Theorem 4.5 is that the design conditions are linear with respect to the matrix parameters, allowing thus the determination of a robust filter for sampled-data systems subject to convex bounded uncertainty. Of course, in this case, there is no sense in talking about a filter with the internal model structure simply because, due to uncertainty, it does not exist. This important result is settled in the next corollary.

Corollary 4.3 *Let $h > 0$ be given. Consider the set of DLMIs*

$$\begin{bmatrix} \dot{X} + A_i'X + XA_i & XA_i + A_i'Z + M_c & C_{zi}' \\ \bullet & \dot{Z} + A_i'Z + ZA_i & C_{zi}' - K_c' \\ \bullet & \bullet & -I \end{bmatrix} < 0,\ i \in \mathbb{K},\ t \in [0, h) \tag{4.101}$$

subject to the LMI-based boundary condition (4.86). The solution of the convex programming problem

$$\varrho_{2rob}^2 = \inf_{X>Z>0, M_c, K_c, M_d, K_d, W_d} \{\mathbf{tr}(E'X_hE) + \mathbf{tr}(W_d) : (4.84), (4.86), (4.101)\} \tag{4.102}$$

provides a robust filter with state space realization (4.88)–(4.89) that operates with the guaranteed performance index ϱ_{2rob}^2, for all parameter convex bounded uncertainty defined by $\lambda \in \Lambda$.

Proof It is quite immediate. Indeed, if the set of DLMIs (4.101) holds for the extreme matrices $\{A_i, C_{zi}\}_{i\in\mathbb{K}}$, then the result of Theorem 4.5 also holds for any convex combination $(A_\lambda, C_{z\lambda})$ with $\lambda \in \Lambda$. The proof is concluded. □

It is simple to verify that the boundary conditions (4.86) depend linearly on the measurement matrix C_y. The objective function (4.102), whenever expressed in terms of a linear objective function and LMIs, exhibits the same property with respect to the matrices E and E_y, respectively. Hence, similarly to Corollary 4.3, there is no difficulty to cope with parameter uncertainty acting on all matrices of the sampled-data system (4.71)–(4.73). Observe that, in general, the robust filter is time-varying. The next example illustrates the results obtained so far.

Remark 4.4 The conditions of Corollary 4.3 are unfeasible unless the matrix A_λ is Hurwitz stable for all $\lambda \in \Lambda$. Indeed, a necessary condition for the inequality (4.101) to be feasible is that $\dot{Z} + A_\lambda' Z + Z A_\lambda < 0$, which is equivalent to say that

$$Z_0 > e^{A_\lambda' h} Z_h e^{A_\lambda h}$$

On the other hand, a necessary condition for inequality (4.86) to be feasible is that the LMI obtained by the elimination of the first and third rows and columns, namely

$$\begin{bmatrix} Z_h & Z_0 \\ Z_0 & Z_0 \end{bmatrix} > 0 \iff Z_h > Z_0 > 0$$

is feasible. However, both inequalities together are feasible if and only if $e^{A_\lambda h}$ is Schur stable that is the same to say that A_λ is Hurwitz stable. This is a well known condition in the context of robust filter design, as the reader can verify in the Bibliography notes at the end of this chapter. It is important to keep in mind that this is true because if matrix A is uncertain, then a filter based on the internal model of the plant does not exist. This interpretation agrees with the result of Theorem 4.6, which does not have this limitation because, for certain systems, the internal model is naturally available. □

Example 4.3 This example illustrates the design of a filter for a certain unstable system. Consider the sampled-data system (4.71)–(4.73) with $h = 1.5$ and matrices

$$A = \begin{bmatrix} 0 & 1 & 0 \\ 0 & 0 & 1 \\ 1 & -5 & -9 \end{bmatrix}, \quad E = \begin{bmatrix} 0 \\ 0 \\ 1 \end{bmatrix}$$

where, clearly, matrix A is not Hurwitz stable. The goal is to estimate $x_2(t)$ using the information provided by the measurement of $x_1(t)$, for all $t \geq 0$. The remaining matrices with a high gain in the discrete-time perturbation channel are

$$C_y = \begin{bmatrix} 1 & 0 & 0 \end{bmatrix}, \quad E_y = \begin{bmatrix} 5 \end{bmatrix}, \quad C_z = \begin{bmatrix} 0 & 1 & 0 \end{bmatrix}$$

With $n_\phi = 64$, we have solved the convex programming problem (4.95) to calculate the optimal filter (4.74)–(4.76) with state space matrices

$$\hat{A}_c(t) = A, \ \hat{C}_c(t) = C_z, \ \hat{A}_d = I - \hat{B}_d C_y, \ \hat{B}_d = \begin{bmatrix} 0.3753 \\ 0.0586 \\ 0.0090 \end{bmatrix}$$

and the corresponding minimum $\mathcal{H}_2$ performance level $\varrho_2^2 = 0.4488$. To evaluate the precision of this solution, we have determined by simulation that the optimal filter produces the squared $\mathcal{H}_2$ estimation error $\varrho_2^2 = 0.4440$ that is quite close to the minimum cost. Figure 4.1 shows in dashed line the output $z_e(t)$ of the system starting from the initial $x(0) = E$ imposed by a unitary continuous-time impulse applied in the channel w_c. The solid lines are two estimated trajectories $\hat{z}_e(t)$ provided by the optimal filter, evolving from the initial condition $\hat{x}(0) = \hat{B}_d C_y E = 0$, due to a unitary continuous-time impulse applied in the channel w_c and $\hat{x}(0) = \hat{B}_d E_y$, which is due to the occurrence of a unitary discrete-time impulse in the channel w_d. It is apparent that the optimal filter is very effective to cope with the lack of information between successive samplings. □

Example 4.4 This example illustrates the design of a robust filter for a system subject to a convex bounded uncertainty. Consider the sampled-data system (4.71)–(4.73) with $h = 0.5$ and matrices

$$A = \begin{bmatrix} 0 & a \\ 1 & -0.5 \end{bmatrix}, \ E = \begin{bmatrix} -2 \\ 1 \end{bmatrix}$$

where $-1.3 \leq a \leq -0.7$ is the uncertain interval. The goal is to estimate the first component of the state variable, namely $x_1(t)$ using the information provided by the measurement of the difference $x_2(t) - x_1(t)$, with a high gain, for all $t \geq 0$, leading to the matrices

$$C_y = \begin{bmatrix} -100 & 100 \end{bmatrix}, \ E_y = \begin{bmatrix} 1 \end{bmatrix}, \ C_z = \begin{bmatrix} 1 & 0 \end{bmatrix}$$

Fig. 4.1 Time evolution of the true and estimated outputs

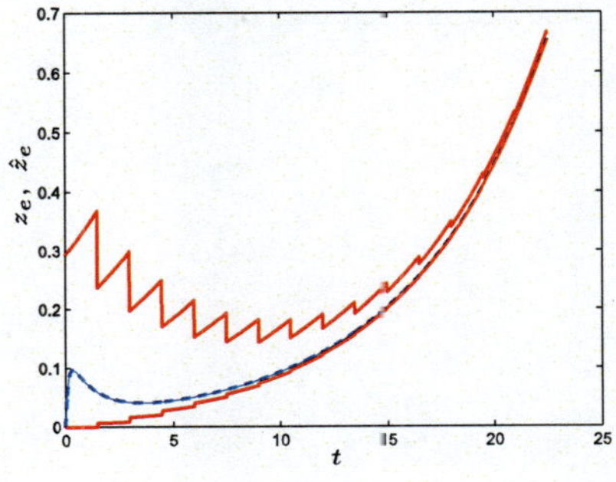

It is important to keep in mind that the robust filter provided by Corollary 4.3 is time-varying with state space realization (4.88)–(4.89). It is natural to implement the piecewise linear procedure with $M_c(t)$ and $K_c(t)$ piecewise constant in each subinterval. However, since $Z(t)$ and $X(t)$ are piecewise linear, then $\hat{A}_c(t)$ has a nonlinear dependence of time in each time subinterval. Hence, for implementation purposes, it is simpler (and more precise, as far as optimality is concerned) to adopt n_ϕ large, typically $n_\phi \geq 32$, in which situation the $\hat{A}_c(t)$ can approximately be considered piecewise constant in each subinterval of length h/n_ϕ. We want to make explicit two complementary situations:

(a) Problem (4.102) has been solved with the time interval $[0, h]$ divided into $n_\phi \in \{1, 2, 4, 8, 16\}$ subintervals. We have determined the associated minimum guaranteed squared $\mathcal{H}_2$ performance index $\varrho_{2rob}^2 = \{6.1345, 3.2149, 2.1240, 1.7241, 1.5758\}$. In each case, the time-varying filter has been calculated but not implemented.
(b) Problem (4.102) has been solved with the time interval $[0, h]$ divided into $n_\phi = 32$ subintervals. It provided the minimum guaranteed squared $\mathcal{H}_2$ performance index $\varrho_{2rob}^2 = 1.5184$ and the associated time-varying filter with piecewise constant approximation. It has been numerically implemented to obtain the exact value of the guaranteed $\mathcal{H}_2$ squared performance $\varrho_{2rob}^2 = 0.6472$. Figure 4.2 shows the output of the plant in blue solid lines together with the output of the filter in red solid lines for all $\lambda \in \Lambda$ and unitary impulses applied simultaneously in each perturbation channel.

This example shows that the robust time-varying filter is very effective to cope with uncertainty. In some sense, the degree of liberty imposed by the uncertain parameters appears to be compensated by the flexibility of the time-varying matrix-valued functions involved. However, the price to be paid is the computation burden needed for calculations and online implementation, which is much higher when compared to the time-invariant optimal filter whose robustness with respect to parameter uncertainty is not guaranteed. Of course, we are interpreting on the basis of only one example solved, in which case, additional numerical experiences are recommended for final verification. □

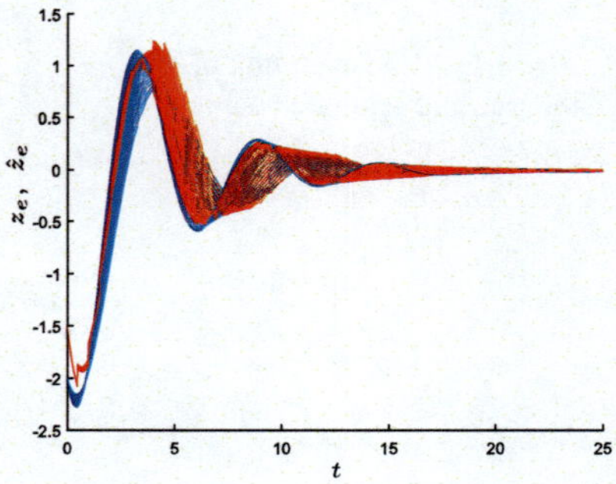

Fig. 4.2 Time evolution of the true and estimated outputs

The class of causal and time-varying filters considered in this section is inspired by several results already available in the literature that deal with Riccati-based equations. For more details, the reader is requested to see the discussion about this point provided in the Bibliography notes at the end of this chapter. Fortunately, making use of DLMIs, the possibility of designing robust filters in the context of sampled-data systems, through convex programming problems, is a reality. It is interesting to put in evidence that, for certain sampled-data systems, among all causal and time-varying filters, the optimal one is time invariant and exhibits the internal model structure with innovation. The same idea is now adopted to cope with dynamic output feedback control design.

4.5 Dynamic Output Feedback Design

The filtering design that we have just presented is surely helpful to prepare the algebraic manipulations needed for the design of full order dynamic output feedback controllers. From the very beginning, we adopt the same one-to-one change of variables that has been adopted in Chap. 3, but now in the context of DLMIs. This change of strategy is necessary because, as the reader can verify, the one-to-one transformation that works well for filtering is not suitable for control design if one wants to express the conditions through DLMIs and LMIs. It is shown that, as for the optimal filter, the optimal output feedback controller of that class is time invariant. For convenience, recall the sampled-data control system of interest

$$\dot{x}(t) = Ax(t) + Bu(t) + Ew_c(t), \; x(0) = 0 \tag{4.103}$$

$$y[k] = C_y x[k] + E_y w_d[k] \tag{4.104}$$

$$z(t) = C_z x(t) + D_z u(t) \tag{4.105}$$

where $x(\cdot) : \mathbb{R}_+ \to \mathbb{R}^{n_x}$ is the state and the other variables remain the same. The full order dynamic output feedback controller has the state space realization

$$\dot{\hat{x}}(t) = \hat{A}_c(t)\hat{x}(t), \; \hat{x}(0) = 0 \tag{4.106}$$

$$\hat{x}(t_k) = \hat{A}_d \hat{x}(t_k^-) + \hat{B}_d y[k] \tag{4.107}$$

$$v(t) = \hat{C}_c(t)\hat{x}(t) \tag{4.108}$$

valid in the time interval $t \in [t_k, t_{k+1})$, $k \in \mathbb{N}$, and $\hat{x}(\cdot) : \mathbb{R}_+ \to \mathbb{R}^{n_x}$ indicates that the plant and the controller share the same dimensions. Notice that the control signal belongs to the continuous-time domain, and it is generated from a measured signal belonging to the discrete-time domain. According to the control structure introduced in the beginning of Chap. 3, we set $u(t) = v(t) + E_u w_c(t)$. Hence, the hybrid linear system that comes out of the feedback connection of the plant and the controller is given by (4.33)–(4.35) with the state variable $\psi(t)' = [x(t)' \; \hat{x}(t)']$ and matrices

$$F = \begin{bmatrix} A & B\hat{C}_c \\ 0 & \hat{A}_c \end{bmatrix}, \; J_c = \begin{bmatrix} E + BE_u \\ 0 \end{bmatrix}, \; G = \begin{bmatrix} C_z & D_z\hat{C}_c \end{bmatrix}, \tag{4.109}$$

and

$$H = \begin{bmatrix} I & 0 \\ \hat{B}_d C_y & \hat{A}_d \end{bmatrix}, \; J_d = \begin{bmatrix} 0 \\ \hat{B}_d E_y \end{bmatrix} \tag{4.110}$$

where, to ease the notation, once again, the time dependence of $\hat{A}_c(\cdot)$ and $\hat{C}_c(\cdot)$ has been dropped. Due to the fact that J_c remains independent of the controller matrices, $E_u = 0$ is set with no loss of generality because its effect can be incorporated in matrix E, by the replacement $E \leftarrow E + BE_u$. Our purpose is to find the optimal controller of this class, whenever the $\mathcal{H}_2$ performance index is concerned. The matrix-valued functions $P(t) = Q(t)^{-1}$, $\Gamma(t)$ with time domain $[0, h]$ and their partitioned forms are given in (4.79)–(4.80), respectively. However, instead of (4.81), a slightly different matrix needed to perform the calculations, namely

$$\Gamma(t) = \begin{bmatrix} I & Y(t) \\ 0 & U(t)' \end{bmatrix} \tag{4.111}$$

is introduced. As usual, to simplify notation, from now on, the dependence of all matrix-valued functions with respect to time is omitted. Simple algebraic manipulations put in evidence that

$$\Gamma' P \Gamma = \begin{bmatrix} X & I \\ I & Y \end{bmatrix} > 0 \tag{4.112}$$

and $QP = I$ is imposed, provided that any matrix-valued function U non-singular is given. In addition, the calculation of the time derivative of P yields

$$\begin{aligned} \Gamma' \dot{P} \Gamma &= -\Gamma' P \dot{Q} P \Gamma \\ &= \begin{bmatrix} \dot{X} & \dot{X}Y + \dot{V}U' \\ \bullet & -\dot{Y} \end{bmatrix} \end{aligned} \tag{4.113}$$

In the next remark, all calculations needed afterward are performed by taking into account the hybrid linear system model under consideration, with state space realization matrices given by (4.109) and (4.110), respectively.

Remark 4.5 Each term of the DLMI (4.10) in Theorem 4.1 is expressed as

$$\Gamma' P F \Gamma = \begin{bmatrix} XA & XAY + XB\hat{C}_c U' + V\hat{A}_c U' \\ A & AY + B\hat{C}_c U' \end{bmatrix}$$

$$G\Gamma = \big[\, C_z \;\; C_z Y + D_z \hat{C}_c U' \,\big]$$

The positive definiteness $P > 0$ is equivalent to $X > Y^{-1} > 0$ as an immediate consequence of the Schur Complement applied to the second diagonal element of (4.112). The same algebraic manipulations, but with different matrices, are required to determine the equivalence

$$P_h > H' P_0 H \Leftrightarrow \begin{bmatrix} X_h & I & X_0 + C_y' \hat{B}_d' V_0' & I \\ \bullet & Y_h & Y_h X_0 + Y_h C_y' \hat{B}_d' V_0' + U_h \hat{A}_d' V_0' & Y_h \\ \bullet & \bullet & X_0 & I \\ \bullet & \bullet & \bullet & Y_0 \end{bmatrix} > 0$$

as well as those needed to linearize the objective function, that is

$$W_c > J_c' P_h J_c \Leftrightarrow W_c > E' X_h E$$

$$W_d > J_d' P_0 J_d \Leftrightarrow \begin{bmatrix} W_d & E_y' \hat{B}_d' V_0' & 0 \\ \bullet & X_0 & I \\ \bullet & \bullet & Y_0 \end{bmatrix} > 0$$

Although tedious, there is no difficulty to obtain these equivalences, provided that the calculations are performed carefully and by following the partition of all involved matrices and matrix-valued functions. □

Remark 4.6 The proof of the next theorem fully depends on the following strategies to determine some matrix variables and matrix-valued functions, without introducing any kind of conservatism. The main mathematical tool to be applied is the variable elimination procedure highlighted in Remark 4.1. The DLMI (4.10) is equivalent to

$$\Gamma' \big(\dot{P} + F' P + P F + G' G \big) \Gamma < 0, \; t \in [0, h)$$

which, by definition, is naturally partitioned into four blocks. By the elimination variable procedure just mentioned, this DLMI holds if and only if the two main diagonal blocks are feasible, that is

$$\dot{X} + A' X + X A + C_z' C_z < 0 \tag{4.114}$$

$$\dot{Z} + (A + B K_c)' Z + Z (A + B K_c) + (C_z + D_z K_c)' (C_z + D_z K_c) < 0 \tag{4.115}$$

where $Z = Y^{-1} > 0$, $K_c = \hat{C}_c U' Z$, and the off-diagonal block determines the matrix-valued function $M_c = V \hat{A}_c U' Z + \dot{V} U' Z + \dot{X}$ by imposing

$$A'Z + X(A + BK_c) + M_c + C_z'(C_z + D_z K_c) = 0$$

in the same time interval. Let us now turn our attention to the boundary conditions (4.11). From Remark 4.5, multiplying that LMI from both sides by diag(I, Z_h, I, Z_0), we obtain

$$\begin{bmatrix} X_h & Z_h & X_0 + C_y' K_d' & Z_0 \\ \bullet & Z_h & X_0 + C_y' K_d' + M_d' & Z_0 \\ \bullet & \bullet & X_0 & Z_0 \\ \bullet & \bullet & \bullet & Z_0 \end{bmatrix} > 0$$

where $K_d = V_0 \hat{B}_d$ and $M_d = V_0 \hat{A}_d U_h' Z_h$. Calculating successively the Schur complement with respect to the fourth block-diagonal element and then, with respect to the first block-diagonal element of the resulting LMI, we obtain a four-block LMI and the variable elimination procedure indicates that it is feasible if and only if the two main diagonal blocks are feasible, that is,

$$\begin{bmatrix} X_h & Z_h & Z_0 \\ \bullet & Z_h & Z_0 \\ \bullet & \bullet & Z_0 \end{bmatrix} > 0, \quad \begin{bmatrix} X_h & X_0 + C_y' K_d' & Z_0 \\ \bullet & X_0 & Z_0 \\ \bullet & \bullet & Z_0 \end{bmatrix} > 0 \tag{4.116}$$

and the off-diagonal block is null, which imposes

$$(X_0 - Z_0 + K_d C_y) + M_d - (X_0 - Z_0 + K_d C_y)(X_h - Z_0)^{-1}(Z_h - Z_0) = 0$$

where the existence of the indicated matrix inversion is assured. The fact that $Z_0 > 0$ is already imposed in the precedent LMI, it is seen that the LMI used to express the objective function becomes

$$\begin{bmatrix} W_d & E_y' K_d' \\ \bullet & X_0 - Z_0 \end{bmatrix} > 0 \tag{4.117}$$

Finally, setting, as usual, $U = Y$ that yields $V = Z - X$, then $K_c = \hat{C}_c$, $M_c = (Z - X)\hat{A}_c + \dot{Z}$, $K_d = (Z_0 - X_0)\hat{B}_d$, and $M_d = (Z_0 - X_0)\hat{A}_d$. Notice that, in general, the couple (K_c, M_c) are matrix-valued functions, whereas the couple (K_d, M_d) are matrices, all with compatible dimensions. □

The relationship among variables and data raised in the last remark makes it clear that a full order dynamic output feedback controller, robust with respect to convex bounded parameter uncertainty, cannot be designed unless some conservatism is introduced. Actually, this claim appears to be valid since it is true in the framework of LTI systems control. Hence, for certain sampled-data control systems, the next theorem gives the optimality conditions that, whenever solved, yield the corresponding time-invariant full order dynamic output feedback controller. It is shown how to determine it by a proper choice of variables preserving optimality, exactly as we have done in the precedent case of optimal filter design. With no surprise, the stabilizing solution of an algebraic Riccati equation is at the core of that result.

Theorem 4.7 *Let $h > 0$ be given and set $Z_0 = \bar{Z} > 0$ where $\bar{Z}$ is the stabilizing solution to the algebraic Riccati equation*

$$A'\bar{Z} + \bar{Z}A - \big(D_z'C_z + B'\bar{Z}\big)'(D_z'D_z)^{-1}\big(D_z'C_z + B'\bar{Z}\big) + C_z'C_z = 0 \tag{4.118}$$

If the matrix-valued function $X(t)$, the symmetric matrices X_0, X_h, W_d, and the matrix K_d satisfy the DLMI

$$\begin{bmatrix} \dot{X} + A'X + XA & C_z' \\ \bullet & -I \end{bmatrix} < 0, \ t \in [0, h) \tag{4.119}$$

subject to the LMI boundary condition

$$\begin{bmatrix} X_h & X_0 + C_y'K_d' & \bar{Z} \\ \bullet & X_0 & \bar{Z} \\ \bullet & \bullet & \bar{Z} \end{bmatrix} > 0 \tag{4.120}$$

then the solution of the convex programming problem

$$\varrho_2^2 = \inf_{X, K_d, W_d} \big\{\mathbf{tr}(E'X_hE) + \mathbf{tr}(W_d) : (4.117), (4.119)\text{–}(4.120)\big\} \tag{4.121}$$

provides the optimal full order dynamic output feedback controller whose state space realization is given by

$$\hat{A}_c = A + B\hat{C}_c \tag{4.122}$$

$$\hat{C}_c = -(D_z'D_z)^{-1}\big(D_z'C_z + B'\bar{Z}\big) \tag{4.123}$$

$$\hat{A}_d = I - \hat{B}_dC_y \tag{4.124}$$

$$\hat{B}_d = (\bar{Z} - X_0)^{-1}K_d \tag{4.125}$$

Proof It follows from the result of Theorem 4.1. The algebraic manipulations provided in Remark 4.6 indicate that the optimal controller follows from the solution of the non-convex problem

$$\inf_{X > Z > 0, K_c, K_d, W_d} \big\{\mathbf{tr}(E'X_hE) + \mathbf{tr}(W_d) : (4.114)\text{–}(4.117)\big\} \tag{4.126}$$

which puts in clear evidence that from the perspective of the objective function to be minimized, the initial condition $Z_0 > 0$ must be minimal (in the matrix sense) and the same is true for the second LMI in (4.116). Calculating the Schur complement twice, the first LMI in (4.116) is equivalent to $X_h > Z_h > Z_0 > 0$. Hence, the less constrained inequality follows from the choice of $Z_h > 0$ and $Z_0 > 0$ arbitrarily closed to minimal matrix $\bar{Z} > 0$. From Remark 2.1, it solves the algebraic Lyapunov equation

$$(A + BK_c)'Z + Z(A + BK_c) + (C_z + D_zK_c)'(C_z + D_zK_c) = 0 \tag{4.127}$$

which, by its turn, admits a minimal feasible solution, that is, the pair $(\bar{K}, \bar{Z})$ with $\bar{Z} > 0$ being the stabilizing solution of the Riccati equation (4.118) and $\bar{K} = -(D_z'D_z)^{-1}(D_z'C_z + B'\bar{Z})$ the correspondent stabilizing gain, such that $Z \geq \bar{Z}$ for all feasible pairs (K_c, Z). In addition, defining $\Xi = X - \bar{Z}$, subtracting the Schur complement of (4.119) to the Riccati equation (4.118), we obtain

$$\dot{\Xi} + A'\Xi + \Xi A + \bar{K}'(D_z'D_z)\bar{K} < 0 \tag{4.128}$$

valid for all $t \in [0, h)$ and subject to the final boundary condition $\Xi(h) = X_h - \bar{Z} > 0$ imposed by (4.120). Consequently, the constraint $X(t) > \bar{Z}$ holds in the whole time interval $[0, h]$. Plugging $Z_0 > 0$ arbitrarily close to $\bar{Z}$ in the objective function and in the second LMI in (4.116), then imposing the DLMI (4.114), the optimal controller design problem is formulated as (4.121).

It remains to calculate the controller state space matrices from the one-to-one change of variables given in Remark 4.6. First, $K_c = \bar{K}$ gives $\hat{C}_c = \bar{K}$ as in (4.123). In addition, setting $Z(t) = Z_0 = Z_h = \bar{Z}$ and $K_c = \bar{K}$ for all $t \in [0, h]$, we have

$$\begin{aligned} M_c &= -A'Z - X(A + BK_c) - C_z'(C_z + D_zK_c) \\ &= -(A + BK_c)'Z - X(A + BK_c) - C_z'(C_z + D_zK_c) + K_c'B'Z \\ &= (Z - X)(A + BK_c) + K_c'\big(D_z'(C_z + D_zK_c) + B'Z\big) \\ &= (Z - X)(A + BK_c) \end{aligned} \tag{4.129}$$

and consequently, taking into account that $\dot{Z} = 0$, we obtain $\hat{A}_c = (Z - X)^{-1}M_c$ reproducing thus (4.122). Finally, $\hat{B}_d = (Z_0 - X_0)^{-1}K_d$ is exactly (4.125) and

$$\begin{aligned} M_d &= Z_0 - X_0 - K_dC_y \\ &= (Z_0 - X_0)\big(I - (Z_0 - X_0)^{-1}K_dC_y\big) \end{aligned} \tag{4.130}$$

together with $\hat{A}_d = (Z_0 - X_0)^{-1}M_d$ reproduces (4.124). The proof is complete. □

This theorem is important for full order dynamic feedback controller design because it was possible to simplify considerably the optimality conditions. The resulting design conditions are easier to solve by the numerical machinery available in the literature to date. Perhaps the most interesting characteristic of the presented solution is the fact that, for certain sampled-data systems, the optimal controller is time invariant, being simple to be implemented in practice. It has the structure of a state feedback controller connected to a filter based on the internal model of the plant with the well known innovation structure. The next example illustrates the closed-loop operation of the optimal controller. The numerical burden needed to implement the solution of the DLMI (4.119) is assessed and discussed.

Example 4.5 This example illustrates the design of a full order dynamic output feedback optimal controller with state space realization (4.106)–(4.108) for the sampled-data system (4.103)–(4.105) with $h = 1.5$ and matrices

$$A = \begin{bmatrix} 0 & 1 & 0 \\ 0 & 0 & 1 \\ 1 & -5 & -9 \end{bmatrix}, \; B = E = \begin{bmatrix} 0 \\ 0 \\ 1 \end{bmatrix}$$

The state space variable $x_1(t)$ is measured by a device that includes an additional external perturbation with high gain. The controlled output has two components, the first one is the first state space variable $x_1(t)$ and the second is $x_2(t) + u(t)$, for all $t \geq 0$, leading to the matrices

$$C_y = \begin{bmatrix} 1\,0\,0 \end{bmatrix}, \; E_y = \begin{bmatrix} 5 \end{bmatrix}, \; C_z = \begin{bmatrix} 1\,0\,0 \\ 0\,1\,0 \end{bmatrix}, \; D_z = \begin{bmatrix} 0 \\ 1 \end{bmatrix}$$

where it is to be noticed that the open-loop system is unstable and matrices C_z and D_z are not orthogonal. Using Matlab routines, we have solved the convex programming problem (4.121) with $n_\phi = 32$ to obtain the optimal controller with state space realization defined by matrices

$$\hat{A}_c = \begin{bmatrix} 0 & 1 & 0 \\ 0 & 0 & 1 \\ -1.4142 & -8.9650 & -9.3236 \end{bmatrix}, \; \hat{C}_c = \begin{bmatrix} -2.4142 \\ -3.9650 \\ -0.3236 \end{bmatrix}'$$

$$\hat{A}_d = \begin{bmatrix} 0.6236 & 0 & 0 \\ -0.0588 & 1 & 0 \\ -0.0090 & 0 & 1 \end{bmatrix}, \; \hat{B}_d = \begin{bmatrix} 0.3764 \\ 0.0588 \\ 0.0090 \end{bmatrix}$$

and the minimum $\mathcal{H}_2$ performance index $\varrho_2^2 = 168.5667$ whose precision is about 1.2% determined by numerical simulation. On the top of Fig. 4.3, we provide the time evolution of the controlled output, $z_1(t)$ in dashed line and $z_2(t)$ in solid line from the application of simultaneous impulses in the w_c and w_d perturbation channels. The effect of the sampling is clear in the second component of $z(t)$. On the bottom of the same figure, for each value of $n_\phi = \{1, 2, 4, 8, 16, 32, 64\}$, we show in dashed line the performance level ϱ_2^2 determined from the optimal solution of problem (4.121) and in solid line its value calculated by simulation of the associated closed-loop system. The convergence of both sequences is apparent. □

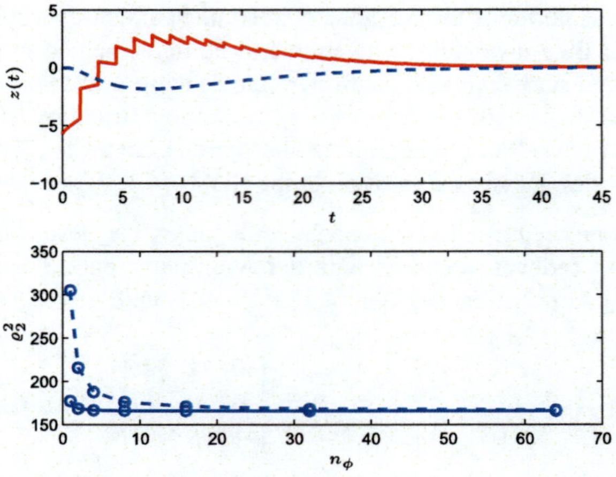

Fig. 4.3 Time evolution and $\mathcal{H}_2$ performance

The difficulty to deal with parameter uncertainty remains in the context of dynamic output feedback control of sampled-data systems. This important aspect is treated in the next chapter, where $\mathcal{H}_\infty$ control design of sampled-data systems is tackled. For the moment being, however, let us make it clear that the result of Theorem 4.7 puts in clear evidence that, for systems with known parameters, the design of that class of controllers is clearly possible due to its simplicity. That makes it possible to handle sampled-data control systems of practical interest with positive impact in real-world problems.

We now move our attention to the design of a pure discrete-time full order output feedback controller with state space realization

$$\hat{x}[k] = \hat{A}_d\hat{x}[k-1] + \hat{B}_d y[k] \tag{4.131}$$

$$v[k] = \hat{C}_d\hat{x}[k-1] + \hat{D}_d y[k] \tag{4.132}$$

valid for all $k \in \mathbb{N}$, with the initial condition $\hat{x}[-1] = 0$. The feedback connection of the plant and this sampled-data controller with $u[k] = v[k] + E_u w_d[k]$ is modeled as a hybrid linear system with state variable $\psi(t)' = [x(t)'\ u(t)'\ \hat{x}(t)']$, where it is clear that, in this case, the full order controller must be such that $\hat{x}(\cdot) : \mathbb{R}_+ \to \mathbb{R}^{n_x+n_u}$, in order to preserve the convexity of the design problem. With a little abuse of notation, the state space model matrices are

$$F \leftarrow \begin{bmatrix} F & 0 \\ 0 & 0 \end{bmatrix},\ J_c \leftarrow \begin{bmatrix} J_c \\ 0 \end{bmatrix},\ G \leftarrow \begin{bmatrix} G & 0 \end{bmatrix} \tag{4.133}$$

where, as usual,

$$F = \begin{bmatrix} A & B \\ 0 & 0 \end{bmatrix},\ J_c = \begin{bmatrix} E \\ 0 \end{bmatrix},\ G = \begin{bmatrix} C_z & D_z \end{bmatrix} \tag{4.134}$$

and

$$H = \begin{bmatrix} I_{xx} + I_u \hat{D}_d C_y I_x' & I_u \hat{C}_d \\ \hat{B}_d C_y I_x' & \hat{A}_d \end{bmatrix}, \quad J_d = \begin{bmatrix} I_u \hat{D}_d E_y + I_u E_u \\ \hat{B}_d E_y \end{bmatrix} \tag{4.135}$$

where

$$I_x = \begin{bmatrix} I \\ 0 \end{bmatrix}, \quad I_u = \begin{bmatrix} 0 \\ I \end{bmatrix}, \quad I_{xx} = I_x I_x' \tag{4.136}$$

Fortunately, the same one-to-one change of variables already used to cope with full order output feedback controller design is once again adopted. In the next remark, a set of useful algebraic manipulations and matrix calculations are presented in order to be applied afterward.

Remark 4.7 The matrix calculations included here follow the same reasoning adopted in the former design case. However, they are used specifically to deal with the present control design problem. For the DLMI, we need to determine

$$\Gamma' P F \Gamma = \begin{bmatrix} XF & XFY \\ F & FY \end{bmatrix}$$

$$G\Gamma = \begin{bmatrix} G & GY \end{bmatrix}$$

while $P > 0$ is equivalent to $X > Y^{-1} > 0$. For the other constraints we need to introduce the one-to-one change of variables $(\hat{A}_d, \hat{B}_d, \hat{C}_d, \hat{D}_d) \Leftrightarrow (M_d, K_d, L_d, D_d)$ given by

$$\begin{bmatrix} M_d - X_0 I_{xx} Y_h & K_d \\ L_d & D_d \end{bmatrix} = \begin{bmatrix} V_0 & X_0 I_u \\ 0 & I \end{bmatrix} \begin{bmatrix} \hat{A}_d & \hat{B}_d \\ \hat{C}_d & \hat{D}_d \end{bmatrix} \begin{bmatrix} U_h' & 0 \\ C_y I_x' Y_h & I \end{bmatrix}$$

which is well defined since V_0 and U_h are, by construction, full rank square matrices. From it, the boundary conditions are transformed into the LMI

$$P_h > H' P_0 H \Leftrightarrow \begin{bmatrix} X_h & I & I_{xx} X_0 + I_x C_y' K_d' & I_{xx} + I_x C_y' D_d' I_u' \\ \bullet & Y_h & M_d' & Y_h I_{xx} + L_d' I_u' \\ \bullet & \bullet & X_0 & I \\ \bullet & \bullet & \bullet & Y_0 \end{bmatrix} > 0 \tag{4.137}$$

and the objective function is expressed by the same set of matrix variables including W_d, satisfying the LMI

$$W_d > J_d' P_0 J_d \Leftrightarrow \begin{bmatrix} W_d & E_u' I_u' X_0 + E_y' K_d' & E_u' I_u' + E_y' D_d' I_u' \\ \bullet & X_0 & I \\ \bullet & \bullet & Y_0 \end{bmatrix} > 0 \tag{4.138}$$

It is important to stress that all constraints have been adequately converted into linear constraints after the adoption of the proposed one-to-one change of variables. The fact that the dimensions of the square matrices V_0 and U_h are $(n_x + n_u) \times (n_x + n_u)$ is essential to achieve this goal. □

The full order output feedback controller (4.131)–(4.132) does not have the structure usually adopted to deal with pure discrete-time control systems, including those stemmed from the application of the equivalence procedure to some sampled-data control system. The main difference is that, at any time $k \in \mathbb{N}$, the measurement is available and modifies the current state and the control signal. See and compare with (3.60)–(3.61). Another interesting aspect to be raised is the order $n_x + n_u > n_x$ of the controller that is greater than the order of the plant. Hence, in general, it is expected the cancelation of n_u poles and zeros.

Theorem 4.8 *Let $h > 0$ be given. If the matrix-valued functions $X(t)$, $Y(t)$, the symmetric matrices X_0, X_h, Y_0, Y_h, W_d, the matrices M_d, K_d, L_d, and D_d satisfy the DLMIs*

$$\begin{bmatrix} \dot{X} + F'X + XF & G' \\ \bullet & -I \end{bmatrix} < 0, \ t \in [0, h) \tag{4.139}$$

$$\begin{bmatrix} -\dot{Y} + FY + YF' & YG' \\ \bullet & -I \end{bmatrix} < 0, \ t \in [0, h) \tag{4.140}$$

and the LMI boundary condition (4.137), then the solution to the convex programming problem

$$\varrho_2^2 = \inf_{X > Y^{-1} > 0, M_d, K_d, L_d, D_d, W_d} \{\mathbf{tr}(J_c' X_h J_c) + \mathbf{tr}(W_d) : (4.137)\text{–}(4.140)\} \tag{4.141}$$

provides the optimal full order dynamic output feedback controller whose state space realization matrices are yielded by the one-to-one transformation provided in Remark 4.7.

Proof Following the calculations included in Remark 4.7, it is seen that, with respect to the indicated set of matrix variables, the LMIs (4.137)–(4.138) are equivalent to the boundary conditions and are used to build the objective function of the problem stated in Theorem 4.1. In addition, the time derivative of $XY + VU' = I$ provides the identity $\dot{X}Y + \dot{V}U' = -X\dot{Y} - V\dot{U}'$ and using once again the matrix manipulations given in Remark 4.7, the DLMI (4.10) multiplied to the right by Γ and to the left by its transpose yields

$$\begin{bmatrix} \dot{X} + F'X + XF + G'G & -X\dot{Y} - V\dot{U}' + F' + XFY + G'GY \\ \bullet & -\dot{Y} + FY + YF' + YG'GY \end{bmatrix} < 0 \tag{4.142}$$

which is satisfied if and only if the two main blocks reproduced in (4.139) and (4.140) hold and the off-diagonal block is set to zero, that is,

$$\begin{aligned} V\dot{U}' &= X\big(-\dot{Y} + FY\big) + (F + YG'G)' \\ &= VU'(F + YG'G)' + X\big(-\dot{Y} + FY + YF' + YG'GY\big) \\ &= VU'(F + YG'G)' - XR \end{aligned} \tag{4.143}$$

where, to obtain the second equality, we have used the identity $XY + VU' = I$, and denoted $-R = -\dot{Y} + FY + YF' + YG'GY$. Since, by assumption, the inverse of V exists, then, from the same equality, we obtain

$$\dot{U} = \Big(F + YG'G + R\big(Y - X^{-1}\big)^{-1}\Big)U \tag{4.144}$$

which, whenever the matrix-valued functions X and Y are available, is a mere time-varying linear differential equation. Hence, setting $U_0 = Y_0$ that imposes $V_0 = Y_0^{-1} - X_0$, the forward integration of (4.144) provides $U_h = U(h)$. The conclusion is that the one-to-one transformation is well defined, and whenever the convex programming problem (4.141) is solved, it yields the optimal controller with state space matrices $(\hat{A}_d, \hat{B}_d, \hat{C}_d, \hat{D}_d)$. The positivity of the solution is imposed through the constraints $X > Y^{-1} > 0$, completing thus the proof. □

It is interesting to compare the result of the last theorem with the one that we have already presented in Chap. 3, concerning full order output feedback controller design based on the equivalent system. Toward that goal, the starting point is the observation that any feasible solution to the DMLI (4.139) satisfies

$$X_0 > F_d' X_h F_d + G_d' G_d \tag{4.145}$$

and any feasible solution to the DLMI (4.140) expressed in terms of its inverse satisfies a similar inequality that, by Schur complement, is rewritten as the LMI

$$\begin{bmatrix} Y_0 & Y_0 F_d' & Y_0 G_d' \\ \bullet & Y_h & 0 \\ \bullet & \bullet & I \end{bmatrix} > 0 \tag{4.146}$$

where

$$F_d = e^{Fh}, \; G_d' G_d = \int_0^h e^{F't} G' G e^{Ft} dt \tag{4.147}$$

Hence, based on these relationships, there is no surprise in verifying that the control design problem (4.141) can be alternatively stated as in the next corollary.

Corollary 4.4 *Let $h > 0$ be given. The solution to the convex programming problem with the symmetric matrix X_0, X_h, Y_0, Y_h, W_d, the matrix M_d, K_d, L_d, and D_d variables*

$$\varrho_2^2 = \inf\{\mathbf{tr}(J_c' X_h J_c) + \mathbf{tr}(W_d) : (4.137)–(4.138), (4.145)–(4.146)\} \tag{4.148}$$

provides the optimal full order dynamic output feedback controller, whose state space realization matrices are yielded by the one-to-one transformation provided in Remark 4.7.

Proof We have already established that any feasible solution to the DLMI (4.139) satisfies the LMI (4.145). Adopting the same reasoning to the inverse of the feasible solutions of (4.140), it is seen that

$$Y_0^{-1} > F_d' Y_h^{-1} F_d + G_d' G_d \tag{4.149}$$

which can be readily converted into the equivalent LMI (4.146). Now, observe that the matrix variable $Y_0 > 0$ appears only on the main diagonal of the LMIs (4.137) and (4.138), which means that, as far as these constraints are concerned, it must be maximal, which implies that the inequality (4.149) becomes arbitrarily tight. In other words, the optimal solution of problem (4.148) can be assumed to satisfy

$$-\dot{Y} + FY + YF' + YG'GY = 0, \ Y(0) = Y_0, \ Y(h) = Y_h \tag{4.150}$$

Two conclusions can be drawn. First, expressing (4.150) in terms of $Z = Y^{-1} > 0$, subtracting the result from the Schur complement of (4.139), it follows that $X > Z > 0$, because the LMI (4.137) imposes $X_h > Y_h^{-1} > 0$. Second, the determination of the controller matrices depends on the change of variables that is well defined by setting $V_0 = Y_0^{-1} - X_0$ and calculating $U(h) = U_h$ by forward integration of $\dot{U} = (F + YG'G)U$ from the initial condition $U(0) = Y_0$, since (4.150) implies that $R = 0$, see (4.144). The proof is concluded. □

This corollary puts in clear perspective that the optimal full order output feedback controller can be determined through a convex programming problem with linear objective function and LMIs constraints, exclusively. The inconvenience of this result is the fact that the controller matrices still depend on the determination of matrices (V_0, U_h) that are calculated by forward integration of a time-varying linear differential equation. The next example illustrates the results obtained so far.

Example 4.6 In this example, we want to compare the methods presented to solve the same control design problem. The open-loop sampled-data system is defined by $h = 1.0$ and matrices

$$A = \begin{bmatrix} 0 & 1 \\ -1 & 0 \end{bmatrix}, \ B = E = \begin{bmatrix} 0 \\ 1 \end{bmatrix}, \ E_u = \begin{bmatrix} 0 \end{bmatrix}$$

Table 4.2 Optimal $\mathcal{H}_2$ performance—ϱ_2^2

n_ϕ	2	4	8	16	32	64
$d = 0$	4.89	2.95	2.24	1.75	1.41	1.20
$d = 1$	6.56	3.91	3.10	2.80	2.69	2.65

$$C_y = \begin{bmatrix} 1 & 0 \end{bmatrix}, \ E_y = \begin{bmatrix} d \end{bmatrix}, \ C_z = \begin{bmatrix} 1 & 0 \\ 0 & 0 \end{bmatrix}, \ D_z = \begin{bmatrix} 0 \\ d \end{bmatrix}$$

where $d \in \{0, 1\}$ is a given parameter. The following calculations have been performed:

(a) Problem (4.141) has been solved to calculate the entries of Table 4.2. Under the adopted precision, it appears that for $n_\phi = 64$, the solution for the case $d = 1$ is close to the optimum.

$$\mathcal{C}_h(\zeta) = \frac{-1.222\zeta + 1.607}{\zeta + 0.7549}, \ \varrho_2^2 = 1.20 \Longleftarrow d = 0$$

$$\mathcal{C}_h(\zeta) = \frac{-0.04417\zeta^2 + 0.339\zeta}{\zeta^2 - 0.267\zeta + 0.2176}, \ \varrho_2^2 = 2.65 \Longleftarrow d = 1$$

It is interesting to notice that the near-optimal controller corresponding to $d = 0$ is of order one even though we are dealing with a second order plant. At least, the cancelation of a pole–zero pair always occurred.

(b) For comparison purposes, we have solved problem (3.73) stated in Theorem 3.1 with the same precision. We have obtained the following optimal controllers with the respective $\mathcal{H}_2$ performance.

$$\mathcal{C}_h(\zeta) = \frac{-1.636\zeta^2 + 1.829\zeta - 0.02193}{\zeta^2 + 0.8194\zeta - 0.01008}, \ \varrho_2^2 = 0.91 \Longleftarrow d = 0$$

$$\mathcal{C}_h(\zeta) = \frac{-0.05626\zeta^2 + 0.369\zeta + 0.0002554}{\zeta^2 - 0.1369\zeta + 0.1662}, \ \varrho_2^2 = 2.61 \Longleftarrow d = 1$$

It confirmed our first impression that the near-optimal controller corresponding to $d = 0$ designed in item (a) is not close enough to the optimum. On the contrary, for $d = 1$, both controllers, although with different transfer functions, perform similarly as far as the $\mathcal{H}_2$ performance index is concerned.

From the numerical viewpoint, comparing the results of Theorem 4.8, Corollary 4.4, and Theorem 3.1, there is no doubt that the last one is preferable. Indeed, it does not depend on DLMIs and works with a full order controller of order $n_x = 2$, whereas the others deal with controllers of order $n_x + n_u = 3$ and this can make an expressive difference, even if the order of the plant is not large. □

This chapter ends with a warning concerning the optimization of the $\mathcal{H}_2$ performance in the context of filtering and control. For nominal systems, when possible, working with equivalent systems is always preferable than the direct manipulation of DLMIs because, in general, higher computational effort is involved. However, DLMIs are important to cope with parameter uncertainty, in the context of filter and state feedback control design, where equivalent systems cannot be determined and convexity is preserved since the handling of exponential or similar non-convex mappings is avoided.

4.6 Bibliography Notes

The main goal of this chapter was to put in clear evidence the importance of DLMIs in the study of stability and $\mathcal{H}_2$ performance of sampled-data systems subject to parameter uncertainty. The presented results are based on two theoretical properties of LMIs that could be applied to DLMI, namely *variable elimination* and a *one-to-one change of variables*. Both have been presented early in [46]. The former procedure simplifies the involved LMIs and DLMIs without introducing any kind of conservatism. The last, by its turn, converts complicated constraints into linear ones.

This chapter treats sampled-data systems design problems with $\mathcal{H}_2$ performance, exclusively. For sampled-data system with known parameters, the results of [41] as well as those of [36] are reobtained from equivalent system calculation and DLMIs. It is important to recognize the presence of similar ideas related to equivalent systems in those celebrated references. The structure of the filter proposed in [49] has been adopted, and for it, the optimal and the robust filters have been fully characterized. In particular, the robust stability condition raised in [21] for LTI systems filter design is, once again, present in the context of sampled-data robust filtering, see also [25]. Example 4.4, inspired by that reference, illustrates this relevant theoretical aspect of the design problem.

The classes of filters and full order output feedback controllers considered have been inspired by Geromel et al. [27] and Sun et al. [49]. The most important feature is the simplification of the optimality conditions, obtained by the variable elimination procedure applied together with the search for minimal and maximal (in matrix sense) variables, that could be determined *a priori*, preserving optimality. This kind of theoretical manipulation is new in the literature to date and appears to play a central role in sampled-data systems design, mainly as far as numerical efficiency and reduction of the computational burden are concerned.

The results of [27] for the design of a pure discrete-time full order output feedback controller have been revisited. New aspects, see [37], as for instance, a simpler procedure for controller construction and the solution of several academic examples, have been added for comparison and illustration.

Chapter 5
$\mathcal{H}_\infty$ Filtering and Control

5.1 Introduction

The $\mathcal{H}_\infty$ theory was essential for the development of robust control and filter design. The Small Gain Theorem is a central result that enables one to cope with model and parameter uncertainty of the important norm bounded class. Our intention is to present this theory but restricted to the framework of sampled-data systems. Several aspects of analysis and control and filtering design are presented and discussed in detail, given particular attention to the design conditions expressed through convex programming problems with DLMIs and LMIs constraints. The $\mathcal{H}_\infty$ performance index is formally defined and it is shown how to calculate it efficiently. Afterwards, the design problems involving filter and controllers are tackled. Several illustrative examples are solved and discussed.

5.2 $\mathcal{H}_\infty$ Performance Analysis

Let us consider again the following open-loop sampled-data system with state space realization

$$\dot{x}(t) = Ax(t) + Bu(t) + Ew_c(t), \; x(0) = 0 \tag{5.1}$$

$$z(t) = C_z x(t) + D_z u(t) \tag{5.2}$$

$$u(t) = w_d[k], \; t \in [t_k, t_{k+1}), \; \forall k \in \mathbb{N} \tag{5.3}$$

where, as before, $x(\cdot) : \mathbb{R}_+ \to \mathbb{R}^{n_x}$, $u(\cdot) : \mathbb{R}_+ \to \mathbb{R}^{n_u}$, $w_c(\cdot) : \mathbb{R}_+ \to \mathbb{R}^{r_c}$ and $z(\cdot) : \mathbb{R}_+ \to \mathbb{R}^{n_z}$ are the state, the control, the exogenous input, and the controlled output of the continuous-time process, respectively. In open-loop operation, the sampled-data control belongs to the time-invariant set $\mathbb{U}$ composed by all piecewise constant

J. C. Geromel, *Differential Linear Matrix Inequalities*,
https://doi.org/10.1007/978-3-031-29754-0_5

trajectories fully defined by the exogenous discrete-time sensor perturbation $w_d[\cdot] : \mathbb{N} \to \mathbb{R}^{r_d}$. For completeness, it is appropriate to repeat here the definition of $\mathcal{H}_\infty$ performance index already introduced in Chap. 3, that is:

Definition 5.1 The $\mathcal{H}_\infty$ performance index associated with the sampled-data system (5.1)–(5.3) is given by

$$\varrho_\infty^2 = \sup_{0 \neq (w_c, w_d) \in \mathcal{L}_2 \times \ell_2} \frac{||z||_2^2}{||w_c||_2^2 + ||w_d||_2^2} \tag{5.4}$$

where $z(t) : \mathbb{R}_+ \to \mathbb{R}^q$ is its response to the continuous-time $w_c(t) : \mathbb{R}_+ \to \mathbb{R}^{r_c}$ and to the discrete-time $w_d[k] : \mathbb{N} \to \mathbb{R}^{r_d}$ exogenous perturbations, assuming that they are not null, simultaneously.

The difficulty we have to evaluate the $\mathcal{H}_\infty$ performance is apparent. It stems from the fact that we have to manipulate signals $(w_c, w_d) \in \mathcal{L}_2 \times \ell_2$ in both domains, continuous and discrete-time, simultaneously. For this reason, in this case, the equivalent system cannot be adopted, unless we agree with the determination of a mere lower bound to ϱ_∞^2. Hence, we need to solve the problem (5.4), and for that, the hybrid system model appears to be the most suitable. For the sampled-data system (5.1)–(5.3) the hybrid model state space representation is of the form

$$\dot{\psi}(t) = F\psi(t) + J_c w_c(t) \tag{5.5}$$

$$z(t) = G\psi(t) \tag{5.6}$$

$$\psi(t_k) = H\psi(t_k^-) + J_d w_d[k] \tag{5.7}$$

where $t \in [t_k, t_{k+1})$ and $k \in \mathbb{N}$. Moreover, as we have verified in the last chapter, the indicated matrices are given by

$$F = \begin{bmatrix} A & B \\ 0 & 0 \end{bmatrix}, \; G = \begin{bmatrix} C_z & D_z \end{bmatrix}, \; J_c = \begin{bmatrix} E \\ 0 \end{bmatrix}, \; J_d = \begin{bmatrix} 0 \\ I \end{bmatrix} \tag{5.8}$$

and $H = \text{diag}(I, 0)$. Starting from zero initial condition $\psi(0) = 0$, the exogenous disturbances $(w_c, w_d) \in \mathcal{L}_2 \times \ell_2$ are injected in the respective channel, yielding the associated controlled output $z(t)$, for all $t \in \mathbb{R}_+$. Afterwards, based on this preliminary discussion, the right hand side of (5.4) is evaluated, in order to characterize the indicated supremum. The next theorem summarizes this procedure as one of the most important and central results of this chapter. Notice that it states a general result valid for any hybrid linear system with state space representation of the form (5.5)–(5.7). Naturally, it can be verified that the proof follows closely to the one of Theorem 4.2.

Theorem 5.1 *Let $h > 0$ be given. The hybrid linear system (5.5)–(5.7) is globally asymptotically stable with ϱ_∞^2 performance index if and only if the differential inequality*

$$\dot{P}(t) + F'P(t) + P(t)F + \gamma^{-2}P(t)J_cJ_c'P(t) + G'G < 0, \; t \in [0, h) \tag{5.9}$$

subject to the boundary conditions

$$P_h > 0, \; \gamma^2 I > J_d' P_0 J_d, \; P_h > H'\Big(P_0^{-1} - \gamma^{-2}J_dJ_d'\Big)^{-1}H \tag{5.10}$$

is feasible. In the affirmative case, the equality

$$\varrho_\infty^2 = \inf_{\gamma>0, P(\cdot)} \{\gamma^2 : (5.9)\text{-}(5.10)\} \tag{5.11}$$

holds.

Proof To prove sufficiency, let us consider that the hybrid linear system (5.5)–(5.7) evolves from an arbitrary initial condition $\psi(0) \in \mathbb{R}^{n_x}$. The cost function denoted $V_c(\psi, t)$ is well defined in the time interval $t \in [0, h)$. Since, by assumption, the differential inequality (5.9) admits a solution, the Hamilton–Jacobi–Bellman inequality

$$\sup_{w_c \in \mathcal{L}_2} \left\{ \frac{\partial V_c'}{\partial \psi}(F\psi + J_c w_c) + \psi' G' G \psi - \gamma^2 w_c' w_c \right\} + \frac{\partial V_c}{\partial t} < 0 \tag{5.12}$$

holds for the quadratic cost function $V_c(\psi, t) = \psi' P(t) \psi$ for all $t \in [0, h)$ which, by simple integration yields

$$\begin{aligned} V_c(\psi(t_k), t_k) > \sup_{w_c \in \mathcal{L}_2} \bigg\{ \int_{t_k}^{t_{k+1}} \Big(z(t)'z(t) - \gamma^2 w_c(t)' w_c(t) \Big)\, dt + \\ + V_c(\psi(t_{k+1}^-), t_{k+1}) \bigg\} \end{aligned} \tag{5.13}$$

Now, taking into account that $P_h > 0$ implies $P_0 > 0$, define the positive definite quadratic function $v_d(\psi) = \psi' P_0 \psi$. Due to the time-invariant nature of the hybrid system, the function $V_c(\psi, t) = \psi' P(t - t_k)\psi$ remains feasible in any subsequent time interval, $k \in \mathbb{N}$, provided that the same boundary conditions (5.10) are imposed. Hence, by construction, we have $V_c(\psi(t_k), t_k) = v_d(\psi(t_k))$. Calculating the Schur complement, the boundary conditions (5.10) can be converted into the equivalent inequality

$$\begin{bmatrix} P_h & 0 \\ 0 & \gamma^2 I \end{bmatrix} > \begin{bmatrix} H' \\ J_d' \end{bmatrix} P_0 \begin{bmatrix} H & J_d \end{bmatrix} \tag{5.14}$$

which when multiplied to the left by $[\psi(t_{k+1}^-)' \; w_d[k+1]']$ and to the right by its transpose, together with the jump equation $\psi(t_{k+1}) = H\psi(t_{k+1}^-) + J_d w_d[k+1]$, yield

$$
\begin{aligned}
V_c(\psi(t_{k+1}^-), t_{k+1}) &= \psi(t_{k+1}^-)' P_h \psi(t_{k+1}^-) \\
&> \psi(t_{k+1})' P_0 \psi(t_{k+1}) - \gamma^2 w_d[k+1]' w_d[k+1] \\
&= v_d(\psi(t_{k+1})) - \gamma^2 w_d[k+1]' w_d[k+1]
\end{aligned} \tag{5.15}
$$

which is valid for all $w_d \in \ell_2$. Putting together the inequalities (5.13) and (5.15) we obtain

$$
\begin{aligned}
v_d(\psi(t_k)) > \sup_{(w_c, w_d) \in \mathcal{L}_2 \times \ell_2} & \left\{ \int_{t_k}^{t_{k+1}} \left(z(t)'z(t) - \gamma^2 w_c(t)' w_c(t) \right) dt - \right. \\
& \left. - \gamma^2 w_d[k+1]' w_d[k+1] + v_d(\psi(t_{k+1})) \right\}
\end{aligned} \tag{5.16}
$$

Due to the fact that this inequality holds, in particular, for any initial condition $\psi(0) \neq 0$ and for $(w_c, w_d) = (0, 0)$, then it implies that there exists $\varepsilon > 0$ sufficiently small such that $v_d(\psi(t_{k+1})) \leq (1 - \varepsilon) v_d(\psi(t_k))$ assuring that the sequence $v_d(\psi(t_k))$ converges to zero as $k \in \mathbb{N}$ goes to infinity, that is, the hybrid system is globally asymptotically stable. On the other hand, for all $0 \neq (w_c, w_d) \in \mathcal{L}_2 \times \ell_2$, the telescoping sum of (5.16) yields

$$
\begin{aligned}
v_d(\psi(0)) - \lim_{k \to \infty} v_d(\psi(t_k)) &> \sum_{k=0}^{\infty} \left\{ \int_{t_k}^{t_{k+1}} \left(z(t)'z(t) - \gamma^2 w_c(t)' w_c(t) \right) dt - \right. \\
& \qquad \left. - \gamma^2 w_d[k+1]' w_d[k+1] \right\} \\
&= \int_0^{\infty} \left(z(t)'z(t) - \gamma^2 w_c(t)' w_c(t) \right) dt - \\
& \qquad - \gamma^2 \sum_{k=0}^{\infty} w_d[k+1]' w_d[k+1]
\end{aligned} \tag{5.17}
$$

Hence, setting $\psi(t_0^-) = 0$ and $w_d[0] = 0$ to impose $\psi(0) = \psi(t_0) = H\psi(t_0^-) + J_d w_d[0] = 0$ and keeping in mind that, due to asymptotic stability the indicated limit is zero, it is readily concluded that

$$
||z||_2^2 < \gamma^2 \left(||w_c||_2^2 + ||w_d||_2^2 \right), \ \forall \, 0 \neq (w_c, w_d) \in \mathcal{L}_2 \times \ell_2 \tag{5.18}
$$

and, by consequence, the upper bound $\varrho_\infty^2 < \gamma^2$ holds. By Definition 5.1 the smallest upper bound given by the optimal solution to problem (5.11) equals the $\mathcal{H}_\infty$ performance index ϱ_∞^2 as it is shown in the necessity part of the proof.

Necessity follows from Bellman's Principle of Optimality applied to the hybrid linear system. To this end, let us define the cost-to-go function

$$V_c(\xi, t_k) = \sup_{(w_c,w_d)\in\mathcal{L}_2\times\ell_2} \left\{ \int_{t_k}^{\infty} \Big(\psi(t)'G'G\psi(t) - \gamma^2 w_c(t)'w_c(t)\Big)\, dt - \right.$$
$$\left. -\gamma^2 \sum_{n=k}^{\infty} w_d[n+1]'w_d[n+1] \right\} \tag{5.19}$$

where for $t \geq t_k$, the trajectory $\psi(t)$ is provided by the linear, time-invariant differential equation $\dot{\psi}(t) = F\psi(t) + J_c w_c(t)$, subject to the initial condition $\psi(t_k) = \xi$ and the jump equation $\psi(t_{k+1}) = H\psi(t_{k+1}^-) + J_d w_d[k+1]$, $\forall k \geq n \in \mathbb{N}$. Clearly, it is positive definite because $(w_c, w_d) = (0, 0) \in \mathcal{L}_2 \times \ell_2$.

Assuming that the hybrid system is globally asymptotically stable and the $\mathcal{H}_\infty$ performance index ϱ_∞^2 is well defined and finite, our goal is to prove that it is equal to the smallest value of γ^2 such that the stationary version of the cost-to-go function $v_d(\psi(t_k))$ satisfies the inequality (5.16) arbitrarily close to equality. Moreover, it has the form $v_d(\xi) = \xi'S\xi$ for some $S > 0$, that is, the stationary cost-to-go function is quadratic. Indeed, to prove this claim, by induction, let us assume that this actually occurs at sampling time t_{k+1}, and split the whole time interval $t_k \mapsto t_{k+1}$ in two successive events, namely $t_k \mapsto t_{k+1}^-$ and $t_{k+1}^- \mapsto t_{k+1}$. For the second part of the time interval, we first point out that the quadratic programming problem

$$\sup_{w_d}\{v_d(H\psi + J_d w_d) - \gamma^2 w_d' w_d\} = \psi' H' \Big(S^{-1} - \gamma^{-2} J_d J_d'\Big)^{-1} H\psi \tag{5.20}$$

admits a finite optimal solution with respect to w_d if and only if the objective is strictly concave, which means that $\gamma^2 I > J_d' S J_d$, and the equality (5.20) holds. Based on this, set $P_h > 0$ satisfying

$$P_h > H'\Big(S^{-1} - \gamma^{-2} J_d J_d'\Big)^{-1} H \tag{5.21}$$

arbitrarily close to the matrix on the right hand side of (5.21). Concerning the first part of the time interval, particularizing (5.20) for $\psi = \psi(t_{k+1}^-)$, $w_d = w_d[k+1]$ and plugging the result into the right hand side of (5.16), we have

$$\begin{aligned} v_d(\psi(t_k)) > \sup_{w_c\in\mathcal{L}_2} & \left\{ \int_{t_k}^{t_{k+1}} \Big(z(t)'z(t) - \gamma^2 w_c(t)'w_c(t)\Big)\, dt + \right. \\ & \left. + \psi(t_{k+1}^-)' P_h \psi(t_{k+1}^-) \right\} \\ = \ & \psi(t_k)' P(0) \psi(t_k) \end{aligned} \tag{5.22}$$

where the equality follows from the well known result on LTI systems $\mathcal{H}_\infty$ Theory which states that the supremum indicated in (5.22) is given by the quadratic function $V_c(\psi, t) = \psi' P(t)\psi$ evaluated at $t = 0$ with $P(t)$ being the positive definite solution to the differential Riccati equation that follows from (5.9) taken arbitrarily close to equality subject to the final boundary condition $P(h) = P_h$. Hence, from (5.22), the smallest cost is quadratic and is reached by setting $S = P(0) = P_0$ which together with (5.21) imply that the boundary conditions (5.10) hold. Finally, since $\psi(0) = 0$ requires $w_d[0] = 0$ and the worst signals w_c and w_d depend linearly on $\psi(\cdot)$, the asymptotic stability assures that $(w_c, w_d) \in \mathcal{L}_2 \times \ell_2$, concluding thus the proof. □

There are several aspects that should be assessed. First, if we set $\gamma > 0$ arbitrarily large, the inequalities (5.9) and (5.10) reduce to those of Theorem 4.1, corresponding to the $\mathcal{H}_2$ performance index calculation. The second one concerns the differential inequality (5.9) which, by calculation of Schur Complements is converted to an equivalent DLMI, that is

$$\begin{bmatrix} \dot{P}(t) + F'P(t) + P(t)F & P(t)J_c & G' \\ \bullet & -\mu I & 0 \\ \bullet & \bullet & -I \end{bmatrix} < 0, \ t \in [0, h) \tag{5.23}$$

with respect to the variables $P(t) \in \mathbb{R}^{n_x \times n_x}$ for all $t \in [0, h)$ and $\mu = \gamma^2 > 0$. On the other hand, similar algebraic manipulations allow us to rewrite the boundary conditions (5.10) as the LMI

$$\begin{bmatrix} P_h & H'P_0 & 0 \\ \bullet & P_0 & P_0 J_d \\ \bullet & \bullet & \mu I \end{bmatrix} > 0 \tag{5.24}$$

and from them, two important consequences arise. Since $P_h > 0$ then any feasible solution to the DLMI (5.23) and the LMI (5.24) is positive definite in the whole time interval $t \in [0, h]$. And, in addition, the determination of the $\mathcal{H}_\infty$ performance is better accomplished by solving, instead of (5.11), the convex programming problem

$$\varrho_\infty^2 = \inf_{\mu, P(\cdot)} \{\mu : (5.23)\text{-}(5.24)\} \tag{5.25}$$

for instance, by the numerical method based on piecewise linear approximation successfully applied many times up to now.

Remark 5.1 Filter and output feedback control design consider time-varying models with matrices (F, G) being time dependent but fully defined in the time interval $[0, h]$. This means that $F(t) = F(t - t_k)$ and $G(t) = G(t - t_k)$ for all $k \in \mathbb{N}$. Under this assumption, it can be seen that the result of Theorem 5.1 remains entirely valid. This is true because the quadratic function $V_c(\psi, t) = \psi' P(t - t_k)\psi$ still can be adopted for all $k \in \mathbb{N}$. The same holds for the stationary cost-to-go quadratic and positive definite function $v_d(\xi)$. □

It is important to stress that Theorem 5.1 has been stated and it is valid for any hybrid linear system with state space realization (5.5)–(5.7). Of course, it can be particularized to cope with the open-loop sampled-data system (5.1)–(5.3) by setting the matrices as in (5.8) and $H = \text{diag}(I, 0)$, which is responsible only for imposing continuity to the state variable of the sampled-data system. In this case, the jump is governed by the discrete-time exogenous perturbation $w_d \in \ell_2$. The next example illustrates the result of Theorem 5.1 by comparing it with the upper and lower bounds to the $\mathcal{H}_\infty$ performance index provided in Chap. 3.

Example 5.1 At this point it is natural and desirable to compare the exact value of the $\mathcal{H}_\infty$ performance index with the lower and upper bounds provided in Lemma 3.4 and Lemma 3.5, respectively. For each value of $h > 0$, we have solved problem (5.25) to calculate $\varrho_\infty = \sqrt{\mu}$ as a function of $h > 0$, adopting the piecewise linear approximation procedure with the time interval $[0, h]$ split into $n_\phi = 32$ subintervals. For comparison purposes we have considered the same fifth order sampled-data system with two inputs and two outputs already treated in Example 3.8. For completeness, the matrices of the state space realization (5.1)–(5.3), are given

$$A = \begin{bmatrix} 0 & 1 & 0 & 0 & 0 \\ 0 & 0 & 1 & 0 & 0 \\ 0 & 0 & 0 & 1 & 0 \\ 0 & 0 & 0 & 0 & 1 \\ -3 & -5 & -8 & -5 & -3 \end{bmatrix}, \quad B = \begin{bmatrix} 1 & 0 \\ 1 & 0 \\ 1 & 0 \\ 1 & 0 \\ 1 & 0 \end{bmatrix}, \quad E = \begin{bmatrix} 0 & 1 \\ 0 & 1 \\ 0 & 1 \\ 0 & 1 \\ 0 & 1 \end{bmatrix}$$

and

$$C_z = \begin{bmatrix} 1 & 0 & 0 & 0 & 0 \\ 0 & 1 & 0 & 0 & 0 \end{bmatrix}, \quad D_z = \begin{bmatrix} 1 & 0 \\ 0 & 1 \end{bmatrix}$$

which together with $H = \text{diag}(I, 0)$ determine the state space realization of the hybrid linear system (5.5)–(5.7). Figure 5.1 shows in dotted line the exact value of the performance index ϱ_∞ and in solid lines its lower and upper bounds, as a function of the sampling period in the time interval $h \in (0, 10]$. As expected, for small values of the sampling period, the bounds virtually coincide with the true

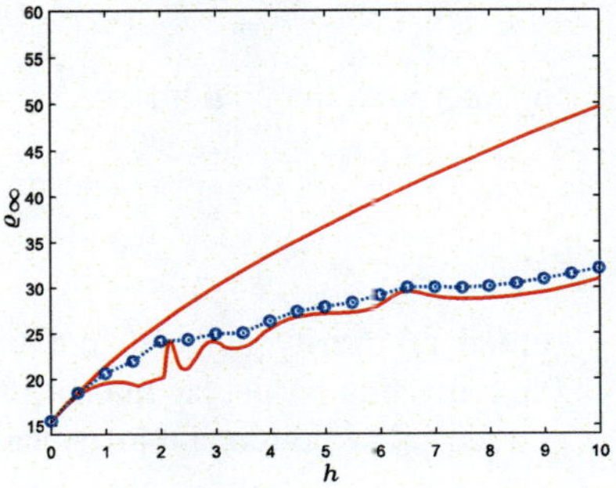

Fig. 5.1 The index ϱ_∞ and its bounds as a function of h

value of ϱ_∞. In this region, the continuous-time as well as the discrete-time models represent the sampled-data system within some reasonable precision. Outside of this region, in this example, the true $\mathcal{H}_\infty$ performance index is much closer to the lower bound than to the upper bound. This means that the proposed upper bound is useless since it cannot replace ϱ_∞ with an adequate precision. □

This example puts in clear evidence the importance of Theorem 5.1. The upper bound is a crude approximation of ϱ_∞ and so, it cannot be used for analysis and synthesis unless one accepts significant differences between them. The approximation provided by the lower bound is much better (in this single example) but it cannot be adopted as well, since the optimization of a lower bound does not provide any guarantee for performance improvement. Fortunately, the result of Theorem 5.1 is amenable as far as filter or control synthesis are concerned.

In the context of $\mathcal{H}_\infty$ performance index calculation, let us tackle two particular and natural situations of interest. The first one raises by imposing $w_d = 0 \in \ell_2$ such as the sampled-data system becomes a pure continuous-time system with state space realization

$$\dot{x}(t) = Ax(t) + Ew_c(t), \ x(0) = 0 \tag{5.26}$$

$$z(t) = C_z x(t) \tag{5.27}$$

which means that the equality $\varrho_\infty = ||C_z(sI - A)^{-1}E||_\infty$ must be provided by Theorem 5.1 as well by setting $J_d = 0$. Actually, by doing this, the differential inequality (5.9) remains unchanged but the boundary conditions (5.10) become $P_h > 0$ and $P_h > H'P_0H$ with $P_0 > 0$. It is well known that any $\gamma > 0$ such that $||C_z(sI - A)^{-1}E||_\infty < \gamma$ ensures the existence of matrix $X_\infty > 0$ satisfying the algebraic Riccati inequality

$$A'X_\infty + X_\infty A + \gamma^{-2} X_\infty E E' X_\infty + C_z'C_z < 0 \tag{5.28}$$

which allows us to introduce the following partitioned matrix-valued function with constant first main block diagonal, that is

$$P(t) = \begin{bmatrix} X_\infty & V(t) \\ V(t)' & \hat{X}(t) \end{bmatrix} > 0, \ t \in [0, h) \tag{5.29}$$

together with the final matrix

$$P_h > \begin{bmatrix} X_\infty & 0 \\ 0 & 0 \end{bmatrix} \tag{5.30}$$

settled arbitrarily close to the one appearing on the right hand side of (5.30). Plugging this particular matrix-valued function into (5.9) it is readily seen that it is a feasible solution and provides at $t = 0$ the initial matrix

$$P_0 = \begin{bmatrix} X_\infty & V_0 \\ V_0' & \hat{X}_0 \end{bmatrix} > 0 \tag{5.31}$$

where the block matrices V_0 and $\hat{X}_0$ are of appropriated dimensions. Simple algebraic substitution shows that $P_h - H'P_0H > 0$ making it clear that the boundary condition is strictly fulfilled. The minimum value of $\gamma > 0$ preserving feasibility of this solution equals the $\mathcal{H}_\infty$ norm of the continuous-time system (5.26)–(5.27). It is important to notice that matrices B and D_z are superfluous in these calculations because, in this particular situation, we have imposed $u(t) = w_d[k] = 0,\ t \in [t_k, t_{k+1}),\ \forall k \in \mathbb{N}$.

Let us now turn our attention to the second situation of interest by imposing $w_c = 0 \in \mathcal{L}_2$ so that the sampled-data system becomes

$$\dot{x}(t) = Ax(t) + Bu(t),\ x(0) = 0 \tag{5.32}$$

$$z(t) = C_z x(t) + D_z u(t) \tag{5.33}$$

$$u(t) = w_d[k],\ t \in [t_k, t_{k+1}),\ \forall k \in \mathbb{N} \tag{5.34}$$

and we observe that since $u \in \mathbb{U}$, then the equivalent discrete-time system can be determined yielding $\varrho_\infty = ||C_{zd}(\zeta I - A_d)^{-1}B_d + D_{zd}||_\infty$. Recall that for any $\gamma > 0$ such that the inequality $||C_{zd}(\zeta I - A_d)^{-1}B_d + D_{zd}||_\infty < \gamma$ holds, there exists a matrix $X_\infty > 0$ satisfying the LMI

$$\begin{bmatrix} A_d'X_\infty A_d - X_\infty & A_d'X_\infty B_d & C_{zd}' \\ \bullet & B_d'X_\infty B_d - \gamma^2 I & D_{zd}' \\ \bullet & \bullet & -I \end{bmatrix} < 0 \tag{5.35}$$

Hence, our goal is to use this matrix to construct a feasible solution to the conditions of Theorem 5.1 with $J_c = 0$, which reduces the Riccati differential inequality (5.9) to a Lyapunov differential inequality that always admits a feasible solution. The boundary conditions (5.10) remain unchanged. Now, set the initial and final matrices satisfying

$$P_0 < \begin{bmatrix} X_\infty & 0 \\ 0 & \gamma^2 I \end{bmatrix},\ P_h > \begin{bmatrix} X_\infty & 0 \\ 0 & 0 \end{bmatrix} \tag{5.36}$$

arbitrarily close to the ones appearing on the right hand side of each inequality. Keeping in mind that $J_d = [0\ I]'$, the conditions (5.10) are satisfied since with the positive definite matrices in (5.36), we have $J_d'P_0J_d < \gamma^2 I$ which enables us to verify immediately that

$$\begin{bmatrix} I \\ 0 \end{bmatrix}' \left(P_0^{-1} - \gamma^{-2}J_dJ_d'\right)^{-1} \begin{bmatrix} I \\ 0 \end{bmatrix} < X_\infty \tag{5.37}$$

which when multiplied to the left by $[I\ 0]'$ of compatible dimensions and to the right by its transpose produces

$$H'\left(P_0^{-1} - \gamma^{-2} J_d J_d'\right)^{-1} H \le \begin{bmatrix} I \\ 0 \end{bmatrix} X_\infty \begin{bmatrix} I \\ 0 \end{bmatrix}' \tag{5.38}$$

making it clear that the choices in (5.36) satisfy the boundary conditions (5.10). Finally, remembering that any feasible solution of the Lyapunov differential inequality (5.9) satisfies

$$P_0 > e^{F'h} P_h e^{Fh} + \int_0^h e^{F't} G' G e^{Ft} dt \tag{5.39}$$

then, using again the matrices given in (5.36), this inequality becomes

$$\begin{aligned} \begin{bmatrix} X_\infty & 0 \\ 0 & \gamma^2 I \end{bmatrix} &> e^{F'h} \begin{bmatrix} X_\infty & 0 \\ 0 & 0 \end{bmatrix} e^{Fh} + \int_0^h e^{F't} G' G e^{Ft} dt \\ &= \begin{bmatrix} A_d' \\ B_d' \end{bmatrix} X_\infty \begin{bmatrix} A_d' \\ B_d' \end{bmatrix}' + \begin{bmatrix} C_{zd}' \\ D_{zd}' \end{bmatrix} \begin{bmatrix} C_{zd}' \\ D_{zd}' \end{bmatrix}' \end{aligned} \tag{5.40}$$

which is nothing else than the inequality one obtains by performing the Schur Complement with respect to the last row and column in the LMI (5.35). As before, the same conclusion can be drawn, that is, a feasible solution to the conditions of Theorem 5.1 is built based on the $\mathcal{H}_\infty$ norm of a pure discrete-time linear system.

These results, albeit expected are important to make it clear that Theorem 5.1 provides the exact $\mathcal{H}_\infty$ performance index for the sampled-data system under consideration, involving perturbations defined in continuous-time, discrete-time and both domains, simultaneously. It is also important to stress that the conditions of Theorem 5.1 are equivalently expressed by the DLMI (5.23) and the LMI (5.24), making (5.25) a convex programming problem. In the next section the corresponding state feedback control design is tackled. Particular attention is given in order to keep the problems to be solved jointly convex.

5.3 State Feedback Design

As usual, let us consider again a sampled-data control system with the following state space realization

$$\dot{x}(t) = Ax(t) + Bu(t) + Ew_c(t), \; x(0) = 0 \tag{5.41}$$

$$z(t) = C_z x(t) + D_z u(t) \tag{5.42}$$

$$u(t) = Lx[k] + (E_u + LE_y)w_d[k], \ t \in [t_k, t_{k+1}), \ \forall k \in \mathbb{N} \tag{5.43}$$

obtained by setting $C_y = I$ in the general model (3.1)–(3.3) in order to make explicit that the whole state is available for feedback. The associated hybrid linear system is defined by matrices F, G and J_c given in (5.8) and

$$H = \begin{bmatrix} I & 0 \\ L & 0 \end{bmatrix}, \ J_d = \begin{bmatrix} 0 \\ E_u + LE_y \end{bmatrix} \tag{5.44}$$

As we already know, the matrix E_y whenever is not null causes non-convexity that makes the corresponding state feedback control design problem hard to solve, unless some degree of conservatism in terms of suboptimality is accepted. Some strategies to cope with this difficulty exist, but none of them is detailed in this book, and so, from now on in this section, it is assumed that $E_y = 0$.

Theorem 5.2 *Let $h > 0$ be given. Consider that together with the DLMI*

$$\begin{bmatrix} -\dot{Q}(t) + Q(t)F' + FQ(t) & Q(t)G' & J_c \\ \bullet & -I & 0 \\ \bullet & \bullet & -\mu I \end{bmatrix} < 0, \ t \in [0, h) \tag{5.45}$$

subject to the LMI boundary conditions

$$\begin{bmatrix} Q_h & Q_h \begin{bmatrix} I \\ 0 \end{bmatrix} \\ \bullet & W \end{bmatrix} > 0, \ \begin{bmatrix} W & \begin{bmatrix} W & K' \end{bmatrix} & 0 \\ \bullet & Q_0 & J_d \\ \bullet & \bullet & \mu I \end{bmatrix} > 0 \tag{5.46}$$

the optimal solution of the convex programming problem

$$\varrho_\infty^2 = \inf_{\mu, W, K, Q(\cdot)} \{\mu : (5.45)\text{-}(5.46)\} \tag{5.47}$$

provides the matrix gain $L = KW^{-1}$. Then, the closed-loop sampled-data control system (5.41)–(5.43) is globally asymptotically stable and operates with minimum ϱ_∞^2 performance index.

Proof The proof is similar to that of Theorem 4.3. The boundary conditions (5.10) can be split into the LMIs given in (5.46). In addition, taking into account that $Q_0 > 0$ implies $Q(t) > 0$ for all $t \in [0, h]$ then defining $P(t) = Q(t)^{-1} > 0$, the DLMI (5.45) is readily shown to be equivalent to the Riccati differential inequality (5.9). The proof is concluded. □

The convex programming problem (5.47) is very similar to its $\mathcal{H}_2$ performance index calculation counterpart stated in Theorem 4.3. Actually, it is simple to see that imposing $\mu > 0$ arbitrarily large all constraints of both problems coincide. In this sense, Theorem 5.2 is a genuine generalization of Theorem 4.3 with the only difference that they have different objective functions. As before, it is possible to simplify the LMIs (5.46) by eliminating the matrix variables W and K as the next Theorem indicates. The proof does not follow the one of Theorem 4.4 but it is accomplished by applying the matrix variable elimination procedure introduced in the beginning of Chap. 4 as it has been done up to now, in several occasions.

Theorem 5.3 *Let $h > 0$ be given. Consider that together with the DLMI*

$$\begin{bmatrix} -\dot{Q}(t) + Q(t)F' + FQ(t) & Q(t)G' & J_c \\ \bullet & -I & 0 \\ \bullet & \bullet & -\mu I \end{bmatrix} < 0, \; t \in [0, h) \tag{5.48}$$

subject to the LMI boundary conditions

$$\begin{bmatrix} Q_0 & J_d \\ \bullet & \mu I \end{bmatrix} > 0, \; I_x'\big(Q_0 - Q_h\big)I_x > 0 \tag{5.49}$$

the optimal solution of the convex programming problem

$$\varrho_\infty^2 = \inf_{\mu, Q(\cdot)} \{\mu : (5.48)\text{-}(5.49)\} \tag{5.50}$$

provides the matrix gain $L = U_0'Y_0^{-1}$ constructed with the partitions of matrix $Q_0 = P_0^{-1}$. Then, the closed-loop sampled-data control system (5.41)–(5.43) is globally asymptotically stable and operates with minimum ϱ_∞^2 performance index.

Proof Denoting the four-block partition, with appropriate dimensions, of the following positive definite matrices

$$P_h = \begin{bmatrix} X_h & V_h \\ V_h' & \hat{X}_h \end{bmatrix}, \; Q_0 = P_0^{-1} = \begin{bmatrix} Y_0 & U_0 \\ U_0' & \hat{Y}_0 \end{bmatrix} \tag{5.51}$$

the boundary conditions (5.10), together with (5.44), are rewritten in the equivalent form

$$\begin{bmatrix} X_h & V_h & I & L' & 0 \\ \bullet & \hat{X}_h & 0 & 0 & 0 \\ \bullet & \bullet & Y_0 & U_0 & 0 \\ \bullet & \bullet & \bullet & \hat{Y}_0 & E_u \\ \bullet & \bullet & \bullet & \bullet & \mu I \end{bmatrix} > 0 \tag{5.52}$$

which, after calculation of the Schur Complement with respect to the second, third and fifth rows and columns, provide $L = U_0'Y_0^{-1}$ and the LMIs obtained by eliminating the first and the forth rows and columns from (5.52), respectively. Doing this we obtain $\mu > 0$, $Q_0 > \mu^{-1}J_dJ_d'$ and $Y_0 > [I\ 0]P_h^{-1}[I\ 0]'$. Remembering that

$I_x = [I\ 0]'$ the last inequality can be written as $I_x' Q_0 I_x > I_x' Q_h I_x$ showing that they can be written as indicated in (5.49). The remaining part of the proof is obvious being thus omitted. □

It is interesting to compare the result of Theorem 4.4 and that of Theorem 5.3. Clearly, for $\mu > 0$ large enough both share the same set of convex constraints. However, the main difference lies in the formulae of each state feedback gain, that must be identical, namely $L = -\hat{X}_0^{-1} V_0'$ and $L = U_0' Y_0^{-1}$. Indeed, they are equal because the (2, 1) block of the equation $P_0 P_0^{-1} - I = 0$ being $V_0' Y_0 + \hat{X}_0 U_0' = 0$ readily shows that $-\hat{X}_0^{-1} V_0' = U_0' Y_0^{-1}$. The formula indicated in Theorem 5.3 appears to be more convenient to be used since the state feedback gain is calculated directly from the partition of the matrix variable $Q_0 > 0$ and not from its inverse $P_0 > 0$. The next example illustrates the theoretical results presented so far.

Remark 5.2 As we have already mentioned many times, the $\mathcal{H}_\infty$ performance index is strongly related to robustness in the context of norm bounded uncertainty. Indeed, consider the sampled-data system with state space realization

$$\dot{x}(t) = A_\Delta x(t) + B_\Delta u(t),\ x(0) = x_0$$
$$u(t) = Lx[k],\ t \in [t_k, t_{k+1}),\ \forall k \in \mathbb{N}$$

where $A_\Delta = A + E\Delta C_z$ and $B_\Delta = B + E\Delta D_z$. Our goal is to determine a state feedback gain $L \in \mathbb{R}^{n_u \times n_x}$ and the minimum value of γ^2 such that the closed-loop system remains stable for all uncertainty such that $||\Delta||_2^2 \leq \gamma^{-2}$. The hybrid linear system associated with it has the form

$$\dot{\psi}(t) = F_\Delta \psi(t)$$
$$z(t) = G\psi(t)$$
$$\psi(t_k) = H\psi(t_k^-)$$

where $F_\Delta = F + J_c \Delta G$ and all other matrices are taken as before. As we know, for each norm bounded Δ given, this uncertain hybrid linear system is globally asymptotically stable if and only if there exists a feasible solution to the DLMI

$$\dot{P}(t) + F_\Delta' P(t) + P(t) F_\Delta + G'G < 0,\ t \in [0, h)$$

subject to the boundary conditions $P_h > 0$ and $P_h > H' P_0 H$. Simple algebraic manipulation that we have applied many times before, ensure that the existence of a solution to this inequality, for all uncertainty, can be enforced by imposing the existence of a solution to the differential Riccati inequality

$$\dot{P}(t) + F' P(t) + P(t) F + \gamma^{-2} P(t) J_c J_c' P(t) + G'G < 0,\ t \in [0, h)$$

subject to $P_h > 0$ and $P_h > H'P_0H$. This is exactly the result of Theorem 5.1 with $J_d = 0$. As a consequence, the minimum value of $\mu = \gamma^2$ is determined from the solution of the convex programming problem (5.50) with $J_d = 0$. □

Example 5.2 The objectives of this example are twofold. First, we want to illustrate the result of Theorem 5.3 and, second, to put in evidence its relationship with the state feedback robust control discussed in Remark 5.2. To this end, we adopt the data from Example 2.3, repeated here for convenience

$$A = \begin{bmatrix} 0 & 1 & 0 \\ 0 & 0 & 1 \\ 0 & -5 & -9 \end{bmatrix}, \ B = \begin{bmatrix} 0 \\ 0 \\ 1 \end{bmatrix}, \ C_z' = \begin{bmatrix} 1 \\ 1 \\ 1 \end{bmatrix}$$

$E = B$, $D_z = 1$, $E_u = 0$, and $h = 1.5$ the sampling period. In this particular context we have $J_d = 0$. Observe that, on the contrary of what we have done in Example 2.3, the state feedback gain L is to be determined in order to produce the maximal parameter uncertainty set which is provided by the optimal solution of problem (5.50). Indeed, with the time interval $[0, h]$ divided in $n_\phi = 64$ subintervals, its optimal solution provides $\mu = 0.0354$ and the associated state feedback gain

$$L = \begin{bmatrix} -0.8516 & -0.8587 & -0.1138 \end{bmatrix}$$

which means that the closed-loop system remains asymptotically stable for all uncertainty satisfying $||\Delta|| \leq \gamma^{-1} = 1/\sqrt{\mu} = 5.3136$. It is interesting to verify that for the state feedback gain considered in Example 2.3, the closed-loop system asymptotic stability was ensured for a much smaller upper bound of the parameter uncertainty norm, that is $||\Delta|| \leq 2.2657$. To verify that the robustness margin provided by Theorem 5.3 in terms of the minimum value of μ is indeed correct, we have determined the matrix $A_L(\Delta) = A_d(\Delta) + B_d(\Delta)L$ with

$$A_d(\Delta) = e^{A_\Delta h}, \ B_d(\Delta) = \int_0^h e^{A_\Delta t} B_\Delta dt$$

where $A_\Delta = A + E\Delta C_z$ and $B_\Delta = B + E\Delta D_z$ for all $||\Delta|| \leq 5.3136$. Figure 5.2 shows in dashed line the unitary circle and inside it the eigenvalues of matrix $A_L(\Delta)$ for all feasible norm bounded uncertainty which shows that $A_L(\cdot)$ remains Schur stable. The eigenvalue locus having points near the boundary indicates, as far as asymptotically stability is concerned, that the result of Theorem 5.3 is not conservative. □

From the previous results, as far as the $\mathcal{H}_\infty$ performance index is concerned, the optimal state feedback control design is not more difficult to be accomplished than the $\mathcal{H}_2$ performance index. This claim is supported by the fact that the control design problem (5.50) is jointly convex with respect to all variables involved. In addition,

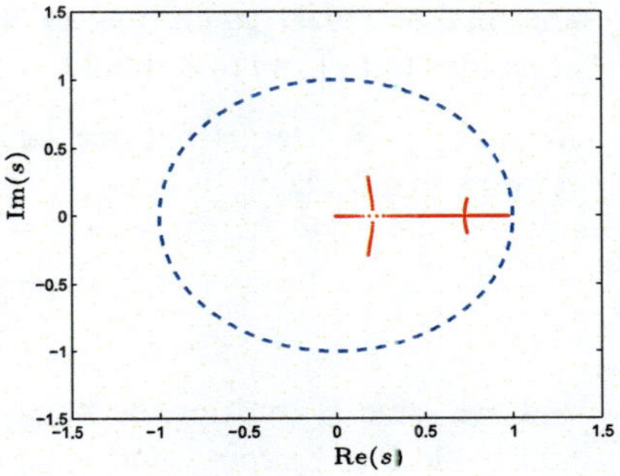

Fig. 5.2 Eigenvalues of the closed-loop uncertain system

the robustness interpretation of the $\mathcal{H}_\infty$ index is very important in practice since parameter uncertainty is always present. We want to stress that when $E_y \neq 0$ the design problem becomes non-convex and quite hard to solve. This situation will be treated in a forthcoming section devoted to dynamic output feedback control design.

5.4 Filter Design

Filter design in the context of $\mathcal{H}_\infty$ performance is a very important issue mainly because robust filters capable of dealing with uncertainty can be determined. The convex nature of the problem to be solved allows the calculation of the optimal filter associated with the highest robustness margin. The convexity of the filter design problem in the context of sampled-data systems is a consequence of the DLMIs that depend implicitly on exponential mappings.

The sampled-data system has the standard form (4.71)–(4.73) and the filter to be designed has the state space realization (4.74)–(4.76). Furthermore, the hybrid linear system that emerges from the connection of the plant and the filter is given by (4.33)–(4.35) with state variable $\psi(t)' = [x(t)' \; \hat{x}(t)']$ and matrices (4.77)–(4.78), which we repeat here for convenience, that is

$$F = \begin{bmatrix} A & 0 \\ 0 & \hat{A}_c \end{bmatrix}, \; J_c = \begin{bmatrix} E \\ 0 \end{bmatrix}, \; G = \begin{bmatrix} C_z & -\hat{C}_c \end{bmatrix} \tag{5.53}$$

and

$$H = \begin{bmatrix} I & 0 \\ \hat{B}_d C_y & \hat{A}_d \end{bmatrix}, \; J_d = \begin{bmatrix} 0 \\ \hat{B}_d E_y \end{bmatrix} \tag{5.54}$$

In addition, the same partitioned matrix-valued variables, namely $P(t) > 0$, $Q(t) = P(t)^{-1} > 0$ and $\Gamma(t)$ with continuous-time domain $[0, h]$ and image $\mathbb{R}^{2n_x \times 2n_x}$ defined in (4.79)–(4.81), together with the matrices related to the initial and final boundary conditions are naturally used. As an obvious consequence, the equalities (4.82) and (4.83) hold.

Remark 5.3 All terms but one of the DLMI (5.23) are given in Remark 4.3. It remains to calculate

$$\Gamma' P J_c = \begin{bmatrix} XE \\ ZE \end{bmatrix}$$

and the same is true for the boundary condition (5.24) that needs the following matrix block to be completed

$$\Gamma_0' P_0 J_d = \begin{bmatrix} V_0 \hat{B}_d E_y \\ 0 \end{bmatrix}$$

With them and the other transformed matrices provided in Remark 4.3, following similar steps and algebraic manipulations adopted previously to tackle the $\mathcal{H}_2$ performance design optimization, the present case of interest is surely handled with no additional difficulty. For this reason, the reader is asked to see the mentioned remark again. □

The content of Remark 5.3 complements that of Remark 4.4 as far as the $\mathcal{H}_\infty$ performance is concerned. If we compare with the $\mathcal{H}_2$ design, the same matrix variables are handled but a slightly different problem must be solved. Fortunately, both are convex programming problems.

Theorem 5.4 *Let $h > 0$ be given. If the matrix-valued functions $Z(t)$, $X(t)$, $M_c(t)$, $K_c(t)$, the matrices M_d, K_d and the scalar $\mu > 0$ satisfy the DLMI*

$$\begin{bmatrix} \dot{X} + A'X + XA & XA + A'Z + M_c & XE & C_z' \\ \bullet & \dot{Z} + A'Z + ZA & ZE & C_z' - K_c' \\ \bullet & \bullet & -\mu I & 0 \\ \bullet & \bullet & \bullet & -I \end{bmatrix} < 0, \ t \in [0, h) \tag{5.55}$$

subject to the LMI boundary condition

$$\begin{bmatrix} X_h & Z_h & X_0 + C_y' K_d' & Z_0 & 0 \\ \bullet & Z_h & X_0 + C_y' K_d' + M_d' & Z_0 & 0 \\ \bullet & \bullet & X_0 & Z_0 & K_d E_y \\ \bullet & \bullet & \bullet & Z_0 & 0 \\ \bullet & \bullet & \bullet & \bullet & \mu I \end{bmatrix} > 0 \tag{5.56}$$

then the solution of the convex programming problem

$$\varrho_\infty^2 = \inf_{\mu, X > Z > 0, M_c, K_c, M_d, K_d} \{\mu : (5.55)\text{-}(5.56)\} \tag{5.57}$$

provides the optimal filter whose state space realization is given by

$$\hat{A}_c = (Z - X)^{-1}(M_c - \dot{Z}), \; \hat{C}_c = K_c \tag{5.58}$$

$$\hat{A}_d = (Z_0 - X_0)^{-1} M_d, \; \hat{B}_d = (Z_0 - X_0)^{-1} K_d \tag{5.59}$$

Proof It follows from Theorem 5.1 by observing that the problem (5.11) can be equivalently formulated as (5.25) subject to the DLMI (5.23) and the LMI (5.24). From the previous partitions it is clear that $P > 0$ if and only if $X > Z > 0$. Moreover, the DLMI (5.23) rewritten in the equivalent form

$$\begin{bmatrix} \Gamma'(\dot{P} + F'P + PF)\Gamma & \Gamma' P J_c & \Gamma' G' \\ \bullet & -\mu I & 0 \\ \bullet & \bullet & -I \end{bmatrix} < 0, \; t \in [0, h) \tag{5.60}$$

becomes the DLMI (5.55) provided that the matrix relationships given in Remarks 4.3 and 5.3 together with the one-to-one change of variables $M_c = V\hat{A}_c U'Z + \dot{V}U'Z + \dot{X}$ and $K_c = \hat{C}_c U'Z$ are adopted. Following the same reasoning, the boundary conditions (5.24) rewritten as

$$\begin{bmatrix} \Gamma_h' P_h \Gamma_h & \Gamma_h' H' P_0 \Gamma_0 & 0 \\ \bullet & \Gamma_0' P_0 \Gamma_0 & \Gamma_0' P_0 J_d \\ \bullet & \bullet & \mu I \end{bmatrix} > 0 \tag{5.61}$$

reproduce the LMI (5.56) whenever the relationships $M_d = V_0 \hat{A}_d U_h' Z_h$ and $K_d = V_0 \hat{B}_d$ are adopted. As a consequence, the optimal $\mathcal{H}_\infty$ filter is determined from the solution of the problem (5.57). Finally, the choice $U = Y = Z^{-1}$ yields $V = Z - X < 0$ and the previous change of variables reproduce the filter state space matrices (5.58) and (5.59), respectively. The proof is concluded. □

In general lines, the same points raised before are also valid in the present case. Indeed, the design conditions given in Theorem 5.4 are expressed by one DLMI and one LMI that are linear with respect to all matrices involved, which is a necessary condition to successfully generalize them to cope with convex bounded uncertainty. In addition, the filter that emerges from Theorem 5.4 is time-varying but fortunately, as stated in the next theorem, the optimal filter can be reduced to a time-invariant one, much easier to be implemented in practice.

Theorem 5.5 *Let $h > 0$ be given. If the matrix-valued function $X(t)$ and the matrices X_0, X_h, and K_d satisfy the DLMI*

$$\begin{bmatrix} \dot{X} + A'X + XA & XE & C_z' \\ \bullet & -\mu I & 0 \\ \bullet & \bullet & -I \end{bmatrix} < 0, \; t \in [0, h) \tag{5.62}$$

subject to the LMI boundary condition

$$\begin{bmatrix} X_h & X_0 + C_y' K_d' & 0 \\ \bullet & X_0 & K_d E_y \\ \bullet & \bullet & \mu I \end{bmatrix} > 0 \tag{5.63}$$

then the solution of the convex programming problem

$$\varrho_\infty^2 = \inf_{\mu, X, K_d} \{\mu : (5.62)\text{-}(5.63)\} \tag{5.64}$$

provides the optimal time-invariant filter whose state space realization is given by

$$\hat{A}_c = A, \; \hat{C}_c = C_z \tag{5.65}$$

$$\hat{A}_d = I + X_0^{-1} K_d C_y, \; \hat{B}_d = -X_0^{-1} K_d \tag{5.66}$$

Proof The first step is to calculate the Schur Complement with respect to the third row and column of DLMI (5.55). Doing this, following the same reasoning of Theorem 4.5, the DLMI (5.55) can be simplified by eliminating the variables $K_c = C_z$ and $M_c = -A'Z - \mu^{-1}XEE'Z - XA$ so as it reduces to $\dot{Z} + A'Z + ZA + \mu^{-1}ZEE'Z < 0$ and the DLMI (5.62). Now, looking at the LMI (5.63) it is clear that it is identical to the LMI (5.56), where the second and fourth rows and columns have been eliminated. This elimination is possible to be done by keeping $X_h > Z_h > Z_0 > 0$, setting $M_d = -X_0 - K_d C_y$, and making matrices Z_h and Z_0 arbitrarily close to the null matrix. Hence, the matrix-valued function $Z(t)$ subject to the boundary conditions $Z(0) = Z_0$ and $Z(h) = Z_h$ arbitrarily close to zero for all $t \in [0, h]$ is feasible. Notice that these algebraic manipulations and choice of variables have been done preserving feasibility and without introducing any kind of conservativeness.

Taking this fact into account, from (5.58), we obtain $\hat{C}_c(t) = C_z$ and

$$\begin{aligned} \hat{A}_c(t) &= (Z - X)^{-1}\big(-XA - A'Z - \mu^{-1}XEE'Z - \dot{Z}\big) \\ &= A \end{aligned} \tag{5.67}$$

so as the formulas in (5.59) provide $\hat{A}_d = I + X_0^{-1} K_d C_y$ and $\hat{B}_d = -X_0^{-1} K_d$, respectively. Finally, due to (5.63), $X_h > 0$ implies that all feasible solutions to the DLMI (5.62) are such that $X(t) > 0, \; \forall t \in [0, h]$. Hence, the constraint $X > Z > 0$ appearing in problem (5.57) is superfluous, since $Z > 0$ is arbitrarily close to zero. The proof is concluded. □

As before, the observer structure based on the internal model of the plant is found. From the matrices (5.65)–(5.66) and the filter state space realization (4.74)–(4.75), we have

$$\dot{\hat{x}}(t) = A\hat{x}(t), \ \hat{x}(0) = 0 \tag{5.68}$$

$$\hat{x}(t_k) = \hat{x}(t_k^-) - X_0^{-1} K_d \big(y(t_k) - C_y \hat{x}(t_k^-) \big) \tag{5.69}$$

which is identical to the structure of the $\mathcal{H}_2$ filter but with the respective gain calculated according to the performance index of interest. The next corollary puts in evidence that convex bounded uncertainties can also be handled due to the linearity of the design conditions with respect to all matrix data involved.

Corollary 5.1 *Let $h > 0$ be given. Consider the set of DLMIs*

$$\begin{bmatrix} \dot{X} + A_i'X + XA_i & XA_i + A_i'Z + M_c & XE_i & C_{zi}' \\ \bullet & \dot{Z} + A_i'Z + ZA_i & ZE_i & C_{zi}' - K_c' \\ \bullet & \bullet & -\mu I & 0 \\ \bullet & \bullet & \bullet & -I \end{bmatrix} < 0, \ t \in [0, h) \tag{5.70}$$

with $i \in \mathbb{K}$, subject to the LMI-based boundary condition (5.56). The solution of the convex programming problem

$$\varrho_{\infty rob}^2 = \inf_{\mu, X > Z > 0, M_c, K_c, M_d, K_d} \{\mu : (5.56), (5.70)\} \tag{5.71}$$

provides a robust filter with state space realization (5.58)–(5.59) that operates with the guaranteed performance index $\varrho_{\infty rob}^2$, for all parameter convex bounded uncertainty defined by $\lambda \in \Lambda$,

Proof The proof follows immediately from the fact that if (5.70) holds for the extreme matrices $\{A_i, E_i, C_{zi}\}_{i \in \mathbb{K}}$, then the result of Theorem 5.4 holds for any convex combination $(A_\lambda, E_\lambda, C_{z\lambda})$ with $\lambda \in \Lambda$, concluding thus the proof. □

It can be seen that this result is very close to its $\mathcal{H}_2$ optimization counterpart with the difference that the input matrix E is also supposed to depend on the convex bounded uncertainty. In the same vein, it can also be seen that the result of Corollary 5.1 is applicable only to plants such that A_λ is Hurwitz stable for all $\lambda \in \Lambda$. This limitation is always present in robust filtering design because, in this case, it is not possible to define the internal model of the plant.

However, there is another important situation that a time-varying robust filter can be determined. Actually, let us assume that all matrices of the plant but (C_y, E_y) are exactly known. These matrices define the measurements provided in discrete-time by sensors located in the plant, that is $y[k] = C_y x[k] + E_y w_d[k]$, $k \in \mathbb{N}$. The convex bounded uncertainty, expressed as usual, by the matrix vertices (C_{yi}, E_{yi}), $i \in \mathbb{K}$ may be useful to model, for instance, a fault due to some sensor breakdown. This situation is treated in the next corollary.

Corollary 5.2 *Let $h > 0$ be given. Consider the DLMI (5.55) subject to the set of LMI-based boundary conditions*

$$
\begin{bmatrix} X_h & Z_h & X_0 + C_{yi}' K_d' & Z_0 & 0 \\ \bullet & Z_h & X_0 + C_{yi}' K_d' + M_d' & Z_0 & 0 \\ \bullet & \bullet & X_0 & Z_0 & K_d E_{yi} \\ \bullet & \bullet & \bullet & Z_0 & 0 \\ \bullet & \bullet & \bullet & \bullet & \mu I \end{bmatrix} > 0, \; i \in \mathbb{K} \tag{5.72}
$$

The solution of the convex programming problem

$$
\varrho_{\infty rob}^2 = \inf_{\mu, X > Z > 0, M_c, K_c, M_d, K_d} \{\mu : (5.55), (5.72)\} \tag{5.73}
$$

provides a robust filter with state space realization (5.58)–(5.59) that operates with the guaranteed performance index $\varrho_{\infty rob}^2$*, for all parameter convex bounded uncertainty defined by* $\lambda \in \Lambda$*,*

Proof The fact that (5.72) holds for each extreme matrix $\{C_{yi}, E_{yi}\}_{i \in \mathbb{K}}$ implies that the result of Theorem 5.4 holds for any convex combination of these matrices, from what the proof follows. □

Even though the internal mode of the plant is available since the triple of matrices (A, C_z, E) is exactly known, unfortunately, the robust filter is time-varying. This occurs because the matrix variable $Z(t), t \in [0, h]$ arbitrarily close to zero is not feasible anymore. Indeed, this is an immediate consequence of the fact that Z_h arbitrarily close to zero is not feasible to the set of LMIs (5.72). To put this claim in a correct perspective, the reader is asked to verify the consequences whenever the variables M_c and M_d cannot be eliminated. The next example illustrates the result of Corollary 5.2.

Example 5.3 Consider a sampled-data system defined by matrices

$$
A = \begin{bmatrix} 0 & 1 & 0 \\ 0 & 0 & 1 \\ -1 & -5 & -9 \end{bmatrix}, \; E = \begin{bmatrix} 1 \\ 2 \\ 3 \end{bmatrix}, \; C_z' = \begin{bmatrix} 0 \\ 1 \\ 0 \end{bmatrix}
$$

and $h = 0.5$ the sampling period whose output matrices belong to an uncertain set defined by $(N = 2)$ two vertices

$$
C_{y1} = \begin{bmatrix} 1 & 0 & 0 \\ 0 & 0 & 1 \end{bmatrix}, \; E_{y1} = \begin{bmatrix} 5 \\ 5 \end{bmatrix}
$$

$$
C_{y2} = \begin{bmatrix} 1 & 0 & 1 \\ 1 & 0 & -1 \end{bmatrix}, \; E_{y2} = \begin{bmatrix} 2 \\ 2 \end{bmatrix}
$$

Two sensors are coupled to the plant and provide the online measurements $y_j[k] = C_{yj} x[k] + E_{yj} w_d[k]$ for each $j \in \mathbb{K}$, respectively. Sensor "1" measures the first and third state variables separately with low precision due to the additive noise of

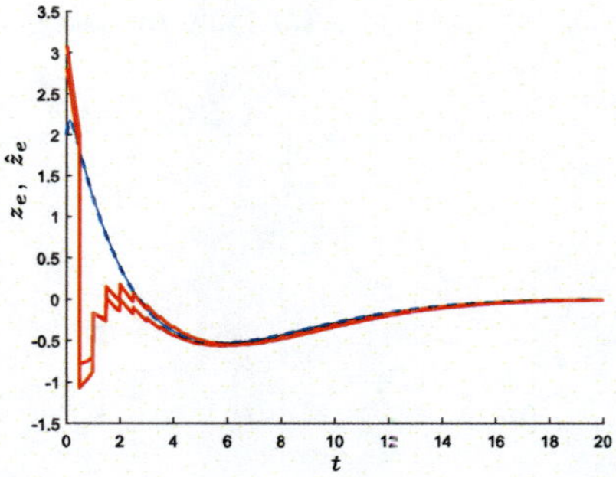

Fig. 5.3 Time evolution of the true and estimated outputs

relatively high intensity. Sensor "2" is more precise as a consequence of an additive noise of relatively low intensity. Our goal is to design a robust filter with minimum $\mathcal{H}_\infty$ guaranteed performance, whenever it is fed with $y[k] = \sum_{j\in\mathbb{K}} \lambda_j y_j[k]$, $k \in \mathbb{N}$ for any $\lambda \in \Lambda$. This means that it must perform well whenever $y[k] = y_1[k]$, $k \in \mathbb{K}$ (sensor "2" fails) as well as $y[k] = y_2[k]$, $k \in \mathbb{K}$ (sensor "1" fails).

Figure 5.3 shows a trajectory of the plant (dashed line) and two estimated trajectories (in solid line) provided by the robust filter, assuming that one sensor fails. The robust time-varying filter has been determined from the convex programming problem (5.73), solved by the piecewise linear approximation method with $n_\phi = 32$. The filter and the plant evolve from initial conditions, yielded by the application of unitary impulses in the perturbation channels w_c and w_d, simultaneously. Even though the sampling period is large and the measurement noise intensities are relatively high, the robust filter performs very well. □

We now move our attention to control design. The idea is to generalize the results presented in Chap. 4 to cope with $\mathcal{H}_\infty$ performance. As expected, in general lines, similar results keeping intact the convexity of the involved control design problems are presented in the next section.

5.5 Dynamic Output Feedback Design

The design of dynamic output feedback controllers in the $\mathcal{H}_\infty$ framework follows closely the same path of its $\mathcal{H}_2$ counterpart. Hence, the sampled-data control system under consideration is given in (4.103)–(4.105) and the full order dynamic output feedback controller to be designed appears in (4.106)–(4.108) where it is important to remember that the control signal $u(t) = v(t) + E_u w_c(t)$ is supposed to be corrupted by the continuous-time actuator perturbation w_c, but as discussed in Chap. 4, to ease the notation, the assumption $E_u = 0$ can be imposed with no loss of generality by replacing $E \leftarrow E + BE_u$.

The feedback connection of the plant and the controller produces a closed-loop system modeled as the hybrid linear system (4.33)–(4.35) with the state variable

$\psi(t)' = [x(t)' \ \hat{x}(t)']$ and matrices

$$F = \begin{bmatrix} A & B\hat{C}_c \\ 0 & \hat{A}_c \end{bmatrix}, \ J_c = \begin{bmatrix} E \\ 0 \end{bmatrix}, \ G = \begin{bmatrix} C_z & D_z\hat{C}_c \end{bmatrix}, \tag{5.74}$$

and

$$H = \begin{bmatrix} I & 0 \\ \hat{B}_d C_y & \hat{A}_d \end{bmatrix}, \ J_d = \begin{bmatrix} 0 \\ \hat{B}_d E_y \end{bmatrix} \tag{5.75}$$

where, as usual, the time dependence of $\hat{A}_c(\cdot)$ and $\hat{C}_c(\cdot)$ has been dropped. The matrix-valued functions $P(t) = Q(t)^{-1}$ with time domain $[0, h]$ and their partitioned forms are given in (4.79)–(4.80), respectively. In addition, together with the matrix-valued function $\Gamma(t)$, defined in the same time interval, they satisfy the equalities (4.112) and (4.113). In the next remark some useful relationships needed to handle the control design problem under consideration are presented. Prior to that, the reader is requested to see with particular care the material included in Remarks 4.5 and 4.6.

Remark 5.4 To prove the next theorem we need the matrix-valued function $\Gamma'\dot{P}\Gamma$ given in (4.113) and $\Gamma'PF\Gamma$, $G\Gamma$ given in Remark 4.5, used to rewrite the differential inequality (5.9) as a DLMI by the adoption of a one-to-one change of variables. To this end, it remains to calculate

$$\Gamma' P J_c = \begin{bmatrix} XE \\ E \end{bmatrix}$$

and the matrices that enable us to rewrite the boundary condition (5.10) as an LMI

$$\Gamma_h' H' P_0 \Gamma_0 = \begin{bmatrix} X_0 + C_y'\hat{B}_d'V_0' & I \\ Y_h X_0 + Y_h C_y \hat{B}_d' V_0' + U_h \hat{A}_d' V_0' & Y_h \end{bmatrix}$$

$$\Gamma_0' P_0 J_d = \begin{bmatrix} V_0 \hat{B}_d E_y \\ 0 \end{bmatrix}$$

from the adoption of the same one-to-one change of variables but evaluated at the initial and final instants of the time interval $[]0, h]$. The main feature of the next result is that the optimal full order dynamic output controller is time-invariant and, by consequence, easy to be implemented in practice. □

Remark 5.5 To simplify the proof of the next theorem, let us make some matrix calculations. We present again a one-to-one change of variables identical to the one introduced in Remark 4.6 that together with the variable elimination procedure provides the desired results. The differential inequality (5.9) is equivalent to

$$\Gamma'\big(\dot{P} + F'P + PF + \gamma^{-2}PJ_cJ_c'P + G'G\big)\Gamma < 0,\ t \in [0, h)$$

which, when partitioned in four blocks puts in evidence, by the elimination variable procedure, that it holds if and only if the two main diagonal blocks are feasible, that is

$$\dot{X} + A'X + XA + \gamma^{-2}XEE'X + C_z'C_z < 0$$
$$\dot{Z} + (A + BK_c)'Z + Z(A + BK_c) + \gamma^{-2}ZEE'Z + (C_z + D_zK_c)'(C_z + D_zK_c) < 0$$

where $Z = Y^{-1} > 0$, $K_c = \hat{C}_cU'Z$ and the off-diagonal block determines the matrix-valued function $M_c = V\hat{A}_cU'Z + \dot{V}U'Z + \dot{X}$ by imposing

$$A'Z + X(A + BK_c) + M_c + \gamma^{-2}XEE'Z + C_z'(C_z + D_zK_c) = 0$$

in the same time interval. Let us now turn our attention to the boundary conditions (5.10), which are equivalent to the LMI (5.24). Partitioning the matrices variables of that LMI in four blocks and multiplying the result from both sides by $\mathrm{diag}(I, Z_h, I, Z_0, I)$, yield

$$\begin{bmatrix} X_h & Z_h & X_0 + C_y'K_d' & Z_0 & 0 \\ \bullet & Z_h & X_0 + C_y'K_d' + M_d' & Z_0 & 0 \\ \bullet & \bullet & X_0 & Z_0 & K_dE_y \\ \bullet & \bullet & \bullet & Z_0 & 0 \\ \bullet & \bullet & \bullet & \bullet & \mu I \end{bmatrix} > 0$$

where $K_d = V_0\hat{B}_d$ and $M_d = V_0\hat{A}_dU_h'Z_h$. Calculating successively the Schur Complement with respect to the fifth, fourth and first block-diagonal element, we obtain a four-block LMI and the variable elimination procedure indicates that it is feasible if and only if the two main diagonal blocks are feasible, that is

$$\begin{bmatrix} X_h & Z_h & Z_0 & 0 \\ \bullet & Z_h & Z_0 & 0 \\ \bullet & \bullet & Z_0 & 0 \\ \bullet & \bullet & \bullet & \mu I \end{bmatrix} > 0,\ \begin{bmatrix} X_h & X_0 + C_y'K_d' & Z_0 & 0 \\ \bullet & X_0 & Z_0 & K_dE_y \\ \bullet & \bullet & Z_0 & 0 \\ \bullet & \bullet & \bullet & \mu I \end{bmatrix} > 0$$

and the off-diagonal block is null, which imposes

$$(X_0 - Z_0 + K_dC_y) + M_d - (X_0 - Z_0 + K_dC_y)(X_h - Z_0)^{-1}(Z_h - Z_0) = 0$$

where the existence of the indicated matrix inversion is assured. Finally, setting as usual, $U = Y$, we have $V = Z - X$ and by consequence $K_c = \hat{C}_c$, $M_c = (Z - X)\hat{A}_c + \dot{Z}$, $K_d = (Z_0 - X_0)\hat{B}_d$ and $M_d = (Z_0 - X_0)\hat{A}_d$. Notice that, in general, the couple (K_c, M_c) are matrix-valued functions whereas the couple (K_d, M_d) are matrices, all with compatible dimensions. □

The algebraic manipulations that we have just done in the last two remarks are similar to their $\mathcal{H}_2$ counterparts discussed in Chap. 4. However, there is an important difference represented by the positive scalar $\mu > 0$. As it is seen in the proof of the next theorem, the determination of the optimal $\mathcal{H}_\infty$ output feedback controller is more involved and needs an additional effort since we want to cope with jointly convex problems only. The main difficulty stems from the fact that the algebraic Riccati, being dependent on the variable $\mu > 0$, cannot be solved *a priori* as it was the case of the $\mathcal{H}_2$ design.

Theorem 5.6 *Let $h > 0$ be given and set $Z_0(\mu) = \bar{Z} > 0$, where $\bar{Z}$ is the stabilizing solution to the algebraic Riccati equation*

$$A'\bar{Z} + \bar{Z}A - \left(D_z'C_z + B'\bar{Z}\right)'(D_z'D_z)^{-1}\left(D_z'C_z + B'\bar{Z}\right) + \\ + \mu^{-1}\bar{Z}EE'\bar{Z} + C_z'C_z = 0 \tag{5.76}$$

for each possible value of $\mu > 0$. If the matrix-valued function $X(t)$, the symmetric matrices X_0, X_h, and the matrix K_d satisfy the DLMI

$$\begin{bmatrix} \dot{X} + A'X + XA & XE & C_z' \\ \bullet & -\mu I & 0 \\ \bullet & \bullet & -I \end{bmatrix} < 0, \ t \in [0, h) \tag{5.77}$$

subject to the LMI boundary condition

$$\begin{bmatrix} X_h & X_0 + C_y'K_d' & Z_0(\mu) & 0 \\ \bullet & X_0 & Z_0(\mu) & K_dE_y \\ \bullet & \bullet & Z_0(\mu) & 0 \\ \bullet & \bullet & \bullet & \mu I \end{bmatrix} > 0 \tag{5.78}$$

then the solution of the non-convex programming problem

$$\varrho_\infty^2 = \inf_{\mu, X, K_d} \{\mu : (5.77)\text{-}(5.78)\} \tag{5.79}$$

provides the optimal full order dynamic output feedback controller whose state space realization is given by

$$\hat{A}_c = A + \mu^{-1}EE'\bar{Z} + B\hat{C}_c \tag{5.80}$$

$$\hat{C}_c = -(D_z'D_z)^{-1}\left(D_z'C_z + B'\bar{Z}\right) \tag{5.81}$$

$$\hat{A}_d = I - \hat{B}_dC_y \tag{5.82}$$

$$\hat{B}_d = (\bar{Z} - X_0)^{-1} K_d \tag{5.83}$$

Proof It follows from the result of Theorem 5.1 which states that the optimal $\mathcal{H}_\infty$ follows from the solution of

$$\inf_{\mu, P(\cdot)} \{\mu : (5.23)\text{-}(5.24)\} \tag{5.84}$$

In Remark 5.5 it has been established that the DLMI (5.23) is equivalent to two decoupled DLMIs with respect to X and Z, respectively, plus an equation that yields the matrix-valued function M_c. On the other hand, the LMI (5.24) is found to be equivalent to the LMI (5.78) and $X_h > Z_h > Z_0 > 0$. These conditions indicate that the less constrained inequality follows from the choice of $Z_h > 0$ and Z_0 arbitrarily close to the minimal matrix $\bar{Z} > 0$. From Remark 2.1, it solves the algebraic Riccati equation

$$\begin{aligned}(A + BK_c)'Z + Z(A + BK_c) + (C_z + D_z K_c)'(C_z + D_z K_c) + \\ + \mu^{-1} Z E E' Z = 0\end{aligned} \tag{5.85}$$

which, by its turn, admits a minimal feasible solution, that is, the pair $(\bar{K}, \bar{Z})$ with $\bar{Z} > 0$ being the stabilizing solution of the Riccati equation (5.76) and the corresponding stabilizing matrix gain $\bar{K} = -(D_z' D_z)^{-1}(D_z' C_z + B'\bar{Z})$, such that $Z \geq \bar{Z}$ for all feasible pairs (K_c, Z). In addition, defining $\Xi = X - \bar{Z}$, subtracting the Schur Complement of (5.77) to the Riccati equation (5.76), we obtain

$$\dot{\Xi} + \bar{A}'\Xi + \Xi\bar{A} + \mu^{-1}\Xi E E' \Xi + \bar{K}'(D_z' D_z)\bar{K} < 0 \tag{5.86}$$

valid for all $t \in [0, h)$ with $\bar{A} = A + \mu^{-1} E E' \bar{Z}$ and subject to the final boundary condition $\Xi(h) = X_h - \bar{Z} > 0$ imposed by (5.78). Consequently, the constraint $X(t) > \bar{Z}$ holds in the whole time interval $[0, h]$. Plugging $Z_0 > 0$ arbitrarily close to $\bar{Z}$ into the LMI (5.78), then imposing the DLMI (5.77), the optimal controller design problem is formulated as (5.79).

Our remaining task is to calculate the controller state space matrices from the one-to-one change of variables given in Remark 5.5. First, $K_c = \bar{K}$ gives $\hat{C}_c = \bar{K}$ as in (5.81) and setting $Z(t) = Z_0 = Z_h = \bar{Z}$ and $K_c = \bar{K}$ for all $t \in [0, h]$, we determine

$$\begin{aligned}
M_c &= -A'Z - X(A + BK_c) - \mu^{-1} X E E' Z - C_z'(C_z + D_z K_c) \\
&= -(A + BK_c)'Z - X(A + BK_c) - \mu^{-1} X E E' Z - C_z'(C_z + D_z K_c) + K_c' B' Z \\
&= (Z - X)(A + \mu^{-1} E E' Z + BK_c) + K_c'\big(D_z'(C_z + D_z K_c) + B'Z\big) \\
&= (Z - X)(A + \mu^{-1} E E' Z + BK_c)
\end{aligned} \tag{5.87}$$

and consequently, taking into account that $\dot{Z} = 0$, we obtain $\hat{A}_c = (Z - X)^{-1} M_c$ which is exactly (5.80). Finally, $\hat{B}_d = (Z_0 - X_0)^{-1} K_d$ is (5.83) and

$$\begin{aligned} M_d &= Z_0 - X_0 - K_d C_y \\ &= (Z_0 - X_0)\big(I - (Z_0 - X_0)^{-1} K_d C_y\big) \end{aligned} \tag{5.88}$$

together with $\hat{A}_d = (Z_0 - X_0)^{-1} M_d$ reproduce (5.82). The proof is complete. □

Following the proof of Theorem 5.6, it is clearly seen that the algebraic Riccati equation is absolutely necessary to determine the time-invariant version of the optimal output feedback controller of the class under consideration. This claim is true because the factorization needed to obtain (5.87), that yields (5.80), fully depends on the existence of a stabilizing solution to (5.76). Actually, to obtain M_c we have used (5.76). The consequence is that the control design problem (5.79), being dependent on $Z_0(\mu) = \bar{Z}$, is non-convex, but, fortunately, this undesirable characteristic can be circumvented as indicated in the next corollary.

Corollary 5.3 *Let $h > 0$ be given. If the matrix-valued function $X(t)$, the symmetric matrices X_0, X_h, Y_0 and the matrices K_d, K_0 satisfy the DLMI (5.77) subject to the LMI boundary conditions*

$$\begin{bmatrix} AY_0 + BK_0 + Y_0A' + K_0'B' & E & Y_0C_z' + K_0'D_z' \\ \bullet & -\mu I & 0 \\ \bullet & \bullet & -I \end{bmatrix} < 0 \tag{5.89}$$

and

$$\begin{bmatrix} X_h & X_0 + C_y'K_d' & I & 0 \\ \bullet & X_0 & I & K_dE_y \\ \bullet & \bullet & Y_0 & 0 \\ \bullet & \bullet & \bullet & \mu I \end{bmatrix} > 0 \tag{5.90}$$

then the solution of the convex programming problem

$$\varrho_\infty^2 = \inf_{\mu, X, K_d, Y_0, K_0} \{\mu : (5.77), (5.89)\text{-}(5.90)\} \tag{5.91}$$

provides the optimal full order dynamic output feedback controller, whose state space realization, with $Z_0 = Y_0^{-1} > 0$, is given by

$$\hat{A}_c = A + \mu^{-1} E E' Z_0 + B\hat{C}_c \tag{5.92}$$

$$\hat{C}_c = -(D_z'D_z)^{-1}\big(D_z'C_z + B'Z_0\big) \tag{5.93}$$

$$\hat{A}_d = I - \hat{B}_d C_y \tag{5.94}$$

$$\hat{B}_d = (Z_0 - X_0)^{-1} K_d \tag{5.95}$$

Proof It is fully based on the result of Theorem 5.6. Notice that the LMI (5.90) imposes $Y_0 > 0$. Setting $Z_0 = Y_0^{-1} > 0$ and $K_c = K_0 Y_0^{-1}$, the Schur Complement with respect to the last two rows and columns of the LMI (5.89) yields

$$(A + BK_c)' Z_0 + Z_0 (A + BK_c) + (C_z + D_z K_c)'(C_z + D_z K_c) + \\ + \mu^{-1} Z_0 E E' Z_0 < 0 \tag{5.96}$$

This inequality is identical to (5.85) and shows that the stabilizing solution $\bar{Z} > 0$ of the algebraic Riccati equation (5.76) belongs to its boundary. Consequently, for each $\mu > 0$ the matrix $Y_0 > 0$ arbitrarily close to $\bar{Z}^{-1}$ is feasible, which indicates that problems (5.91) and (5.79) share the same optimal variables (μ, X, K_d) and Y_0 close to $\bar{Z}^{-1}$. The proof is concluded. □

In the numerical context a warning must be taken into account. Indeed, the extraction of the time-invariant controller matrices requires that $Z_0 = \bar{Z}$ be exactly the stabilizing solution of the algebraic Riccati equation (5.76). However, the optimal solution of the problem (5.91) provides $Z_0 = Y_0^{-1} > \bar{Z}$ and close to $\bar{Z}$, depending on the numerical precision adopted by the designer. Hence, the correct numerical precision chosen in this step of the design procedure is essential. This aspect, among others, is illustrated in the next example.

Example 5.4 We want to design the optimal full order dynamic output feedback optimal controller with state space realization (4.106)–(4.108) for the sampled-data system (4.103)–(4.105) with sampling period $h = 0.5$, matrices

$$A = \begin{bmatrix} 0 & 1 & 0 \\ 0 & 0 & 1 \\ -10 & 8 & 1 \end{bmatrix}, \quad B = E = \begin{bmatrix} 0 \\ 0 \\ 1 \end{bmatrix}$$

and the ones corresponding to the measurement and controlled outputs, that is

$$C_y = \begin{bmatrix} 1 & 0 & 0 \end{bmatrix}, \quad E_y = \begin{bmatrix} 1 \end{bmatrix}, \quad C_z = \begin{bmatrix} 1 & 0 & 0 \\ 0 & 0 & 0 \end{bmatrix}, \quad D_z = \begin{bmatrix} 0 \\ 1 \end{bmatrix}$$

being clear that the open-loop system is unstable and matrices C_z and D_z are orthogonal. Using Matlab routines we have solved the convex programming problem (5.91) with $n_\phi = 32$ to obtain the optimal controller with state space defined by matrices

$$\hat{A}_c = \begin{bmatrix} 0 & 1 & 0 \\ 0 & 0 & 1 \\ -10.0499 & -14.8811 & -6.8383 \end{bmatrix}, \quad \hat{C}_c = \begin{bmatrix} -0.0499 \\ -22.8812 \\ -7.8383 \end{bmatrix}'$$

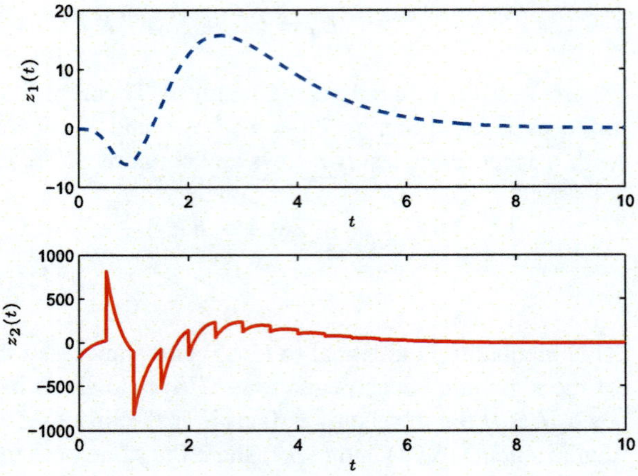

Fig. 5.4 Time evolution of the closed-loop system

$$\hat{A}_d = \begin{bmatrix} 0.0000 & 0 & 0 \\ -3.5674 & 1 & 0 \\ -10.5473 & 0 & 1 \end{bmatrix}, \quad \hat{B}_d = \begin{bmatrix} 1.0000 \\ 3.5674 \\ 10.5473 \end{bmatrix}$$

and the minimum $\mathcal{H}_\infty$ performance index $\varrho_\infty = 569.0225$. The stabilizing solution $\bar{Z} > 0$ of the algebraic Riccati equation (5.76) has been calculated for $\mu = \varrho_\infty^2$ and we have verified that $(4 \times 10^{-4})I > Y_0^{-1} - \bar{Z} > (9 \times 10^{-6})I$ which validates the correctness of the optimal solution found. On the top of Fig. 5.4 we show the time evolution of the controlled output, $z_1(t)$ in dashed line. On the bottom of the same figure we show the time evolution of $z_2(t)$ in solid line. In both cases, simultaneous impulses in the w_c and w_d perturbation channels have been applied. The sampling and the zero order hold effects in the control action (the output $z_2(t)$) are visually clear. The same effect practically disappears in the first controlled output $z_1(t)$. With a relatively large sampling period the designed controller is optimal and very effective in controlling an unstable plant. □

We now tackle the design of a pure discrete-time full order dynamic output feedback controller. The controller state space realization is given in (4.131)–(4.132) and is repeated here for convenience, that is

$$\hat{x}[k] = \hat{A}_d \hat{x}[k-1] + \hat{B}_d y[k] \tag{5.97}$$

$$v[k] = \hat{C}_d \hat{x}[k-1] + \hat{D}_d y[k] \tag{5.98}$$

valid for all $k \in \mathbb{N}$, with the initial condition $\hat{x}[-1] = 0$. It is important to recall that the feedback connection of the plant and this sampled-data controller with $u[k] = v[k] + E_u w_d[k]$ is modeled as a hybrid linear system with state variable $\psi(t)' = [x(t)' \ u(t)' \ \hat{x}(t)']$, where the full order controller must be such that $\hat{x}(\cdot) : \mathbb{R}_+ \to \mathbb{R}^{n_x+n_u}$. The matrices introduced from Eq. (4.133) to (4.136) remain exactly the same. Moreover, many algebraic manipulations, similar to the ones provided in Remark 4.7, are given in the sequel.

Remark 5.6 To proceed, it is important keep in mind that the same four blocks partitioning of several matrices and the same one-to-one change of variables provided in Remark 4.7 and adopted in the context of $\mathcal{H}_2$ control design, still remain valid in the present case of $\mathcal{H}_\infty$ control design. The constraint $P > 0$ is equivalent to $X > Y^{-1} > 0$ and the differential inequality (5.9) multiplied to the left by Γ' and to the right by Γ is equivalent to

$$\begin{bmatrix} \begin{pmatrix} \dot{X} + F'X + XF \\ +\mu^{-1} X J_c J_c' X + G'G \end{pmatrix} & \begin{pmatrix} -X\dot{Y} - V\dot{U}' + F' + XFY \\ +\mu^{-1} X J_c J_c' + G'GY \end{pmatrix} \\ \bullet & \begin{pmatrix} -\dot{Y} + FY + YF' \\ +\mu^{-1} J_c J_c' + YG'GY \end{pmatrix} \end{bmatrix} < 0 \tag{5.99}$$

On the other hand, with the aforementioned one-to-one change of variables that established the relationship $(\hat{A}_d, \hat{B}_d, \hat{C}_d, \hat{D}_d) \Leftrightarrow (M_d, K_d, L_d, D_d)$ the boundary conditions (5.10), rewritten as (5.24), are equivalent to

$$\begin{bmatrix} X_h & I & I_{xx}X_0 + I_x C_y' K_d' & I_{xx} + I_x C_y' D_d' I_u' & 0 \\ \bullet & Y_h & M_d' & Y_h I_{xx} + L_d' I_u' & 0 \\ \bullet & \bullet & X_0 & I & X_0 I_u E_u + K_d E_y \\ \bullet & \bullet & \bullet & Y_0 & I_u E_u + I_u D_d E_y \\ \bullet & \bullet & \bullet & \bullet & \mu I \end{bmatrix} > 0 \tag{5.100}$$

Finally notice that the fact that the dimensions of the square matrices V_0 and U_h are $(n_x + n_u) \times (n_x + n_u)$ is essential to obtain these relationships without including any kind of conservativeness. □

Theorem 5.7 *Let $h > 0$ be given. If the matrix-valued functions $X(t)$, $Y(t)$, the symmetric matrices X_0, X_h, Y_0, Y_h, the matrices M_d, K_d, L_d, and D_d satisfy the DLMIs*

$$\begin{bmatrix} \dot{X} + F'X + XF & XJ_c & G' \\ \bullet & -\mu I & 0 \\ \bullet & \bullet & -I \end{bmatrix} < 0, \; t \in [0, h) \tag{5.101}$$

$$\begin{bmatrix} -\dot{Y} + FY + YF' & J_c & YG' \\ \bullet & -\mu I & 0 \\ \bullet & \bullet & -I \end{bmatrix} < 0, \; t \in [0, h) \tag{5.102}$$

and the LMI boundary condition (5.100), then the solution to the convex programming problem

$$\varrho_\infty^2 = \inf_{\mu, X > Y^{-1} > 0, M_d, K_d, L_d, D_d} \{\mu : (5.100)\text{-}(5.102)\} \tag{5.103}$$

provides the optimal full order dynamic output feedback controller, whose state space realization matrices are yielded by the one-to-one transformation provided in Remark 4.7.

Proof The algebraic manipulations in Remark 5.6 show that the DLMI (5.9) holds if and only if the same is true for the DLMI (5.99), which is satisfied if and only if the two main blocks reproduced in (5.101) and (5.102) hold and the off-diagonal block is set to zero, that is

$$\begin{aligned} V\dot{U}' &= X\big(-\dot{Y} + FY + \mu^{-1}J_cJ_c'\big) + (F + YG'G)' \\ &= VU'(F + YG'G)' + X\big(-\dot{Y} + FY + YF' + \mu^{-1}J_cJ_c' + YG'GY\big) \\ &= VU'(F + YG'G)' - XR \end{aligned} \tag{5.104}$$

where the second equality follows from $XY + VU' = I$, and $-R = -\dot{Y} + FY + YF' + \mu^{-1}J_cJ_c' + YG'GY$. Since, by assumption, the inverse of V exists, we obtain

$$\dot{U} = \Big(F + YG'G + R\big(Y - X^{-1}\big)^{-1}\Big)U \tag{5.105}$$

which, whenever the matrix-valued functions X and Y are available, is a time-varying linear differential equation. Hence, setting $U_0 = Y_0$ which imposes $V_0 = Y_0^{-1} - X_0$, the forward integration of (5.105) provides $U_h = U(h)$. Hence, the one-to-one transformation is well defined and, whenever the convex programming problem (5.103) is solved, it yields the optimal controller with state space matrices $(\hat{A}_d, \hat{B}_d, \hat{C}_d, \hat{D}_d)$. As usual, the positivity of the solution is imposed through the constraints $X > Y^{-1} > 0$, completing thus the proof. □

Certainly, at this point, it is appropriate to investigate, as we have done in the case of $\mathcal{H}_2$ control, if the DLMIs (5.101)–(5.102) can be converted to LMIs by calculating their feasible solutions in the time interval $[0, h]$. Unfortunately, the answer to this question is negative.

First, consider the DLMI (5.102) for some $\mu > 0$ given and assume that there exists a symmetric solution $\bar{Y}$ with appropriate dimensions to the algebraic Riccati equation

$$F\bar{Y} + \bar{Y}F' + \bar{Y}G'G\bar{Y} + \mu^{-1}J_cJ_c' = 0 \tag{5.106}$$

where $\bar{Y}$ is supposed to be symmetric only. Defining $\Xi = Y - \bar{Y}$ and subtracting the Riccati inequality that emerges from the Schur Complement of (5.102) from (5.106), we obtain

$$-\dot{\Xi} + \bar{F}\Xi + \Xi\bar{F}' + \Xi G'G\Xi < 0 \tag{5.107}$$

where $\bar{F} = F + \bar{Y}G'G$. Searching for a non-singular solution, multiplying this inequality both sides by Ξ^{-1} it is seen that Ξ^{-1} satisfies a linear inequality whose solution in the time interval $t \in [0, h]$ provides

$$\Xi(0)^{-1} > e^{\bar{F}'h}\Xi(h)^{-1}e^{\bar{F}h} + \int_0^h e^{\bar{F}'t}G'Ge^{\bar{F}t}dt \tag{5.108}$$

which can be written as an LMI with respect to the variables of interest, namely Y_0 and Y_h, of the following form

$$\begin{bmatrix} Y_0 - \bar{Y} & Y_0\bar{F}_d' - \bar{Y}\bar{F}_d' & Y_0\bar{G}_d' - \bar{Y}\bar{G}_d' \\ \bullet & Y_h - \bar{Y} & 0 \\ \bullet & \bullet & I \end{bmatrix} > 0 \tag{5.109}$$

where the indicated matrices are given by

$$\bar{F}_d = e^{\bar{F}h}, \ \bar{G}_d'\bar{G}_d = \int_0^h e^{\bar{F}'t}G'Ge^{\bar{F}t}dt \tag{5.110}$$

It is interesting to point out that making $\mu \to +\infty$ then $\bar{Y} \to 0$, $\bar{F} \to F$ and the LMI (5.109) becomes the LMI (4.146), the one corresponding to the $\mathcal{H}_2$ case. The existence of the LMI (5.109), depends exclusively on the existence of a symmetric solution to the algebraic Riccati equation (5.106). Taking into account the definitions of matrices F, G and J_c given in (4.133) and (4.134), it can be verified that a possible solution has the form

$$\bar{Y} \leftarrow \begin{bmatrix} \bar{Y} & 0 \\ 0 & 0 \end{bmatrix}, \ \bar{Y} = \begin{bmatrix} \bar{Q} & 0 \\ 0 & 0 \end{bmatrix} \tag{5.111}$$

where the matrix $\bar{Q}$, of appropriate dimensions, is a symmetric solution to the algebraic Riccati equation

$$A\bar{Q} + \bar{Q}A' + \bar{Q}C_z'C_z\bar{Q} + \mu^{-1}EE' = 0 \tag{5.112}$$

whose unique symmetric stabilizing solution exists provided that the associated Hamiltonian matrix does not have eigenvalues lying on the imaginary axis. Under this mild assumption the LMI (5.109) can be constructed.

Let us develop a similar reasoning to cope with the DLMI (5.101). It stems from the determination of a symmetric solution to the algebraic Riccati equation

$$F'\bar{X} + \bar{X}F + \mu^{-1}\bar{X}J_cJ_c'\bar{X} + G'G = 0 \tag{5.113}$$

Due to (4.133), with no loss of generality, we can state that $\bar{X} \leftarrow \text{diag}(\bar{X}, 0)$ but due to (4.134), unfortunately, we notice that $\bar{X} \neq \text{diag}(\bar{R}, 0)$ for some matrix $\bar{R}$, that is,

$\bar{X}$ block diagonal is not a solution. The remedy would be to search a full solution but the matrix

$$F + \mu^{-1} J_c J_c' \bar{X} = \begin{bmatrix} A & B \\ 0 & 0 \end{bmatrix} + \mu^{-1} \begin{bmatrix} EE' & 0 \\ 0 & 0 \end{bmatrix} \bar{X} \tag{5.114}$$

is not asymptotically stable whatever the matrix $\bar{X}$. It has, at least, one eigenvalue in the imaginary axis. Hence, no full stabilizing solution exists. Another possibility is to use the fact that $X^{-1} > 0$ satisfies a DLMI identical to (5.102) which allows us to conclude that (5.101) is imposed by the LMI (5.109) with the variable Y_0 replaced by X_0^{-1} and Y_h replaced by X_h^{-1}. Unfortunately, doing this, the resulting inequality is not an LMI with respect to the variables of interest, namely X_0 and X_h. The conclusion is that the interesting result valid for $\mathcal{H}_2$ control design stated in Corollary 4.4 cannot be generalized to the $\mathcal{H}_\infty$ case. Certainly, this is due to the nonlinear nature of the algebraic Riccati equation that needs to be handled. The next example illustrates the theoretical results obtained so far in this section.

Example 5.5 The same sampled-data system tackled in Example 4.6 is now used to illustrate the optimal $\mathcal{H}_\infty$ controller design provided by Theorem 5.7. We have considered $h = 1.0$ and, for readability reasons, the open-loop system matrices are repeated here

$$A = \begin{bmatrix} 0 & 1 \\ -1 & 0 \end{bmatrix}, \; B = E = \begin{bmatrix} 0 \\ 1 \end{bmatrix}, \; E_u = \begin{bmatrix} 0 \end{bmatrix}$$

$$C_y = \begin{bmatrix} 1 & 0 \end{bmatrix}, \; E_y = \begin{bmatrix} d \end{bmatrix}, \; C_z = \begin{bmatrix} 1 & 0 \\ 0 & 0 \end{bmatrix}, \; D_z = \begin{bmatrix} 0 \\ d \end{bmatrix}$$

where $d \in \{0, 1\}$ is a given parameter used to include, when different of zero, an exogenous measurement perturbation.

Problem (5.103) has been solved to calculate the entries of Table 5.1 that shows the behavior of the minimum $\mathcal{H}_\infty$ performance level with respect to n_ϕ. For $n_\phi = 128$ we have obtained the optimal controllers

$$C_h(\zeta) = \frac{-2.05\zeta + 1.931}{\zeta + 0.898}, \; \varrho_\infty^2 = 0.91 \Longleftarrow d = 0$$

Table 5.1 Optimal $\mathcal{H}_\infty$ performance—ϱ_∞^2

n_ϕ	2	4	8	16	32	64	128
$d = 0$	23.28	7.44	3.60	2.06	1.39	1.07	0.91
$d = 1$	27.81	10.20	6.69	5.43	4.89	4.64	4.54

$$\mathcal{C}_h(\zeta) = \frac{-0.3482\zeta^2 + 0.8969\zeta}{\zeta^2 + 0.3023\zeta + 0.008861}, \quad \varrho_\infty^2 = 4.54 \Longleftarrow d = 1$$

Clearly, it would be necessary to increase n_ϕ in order to obtain a controller closer to the optimum one. It is interesting to notice that the controller corresponding to $d = 0$ is of first order, while the plant is of second order. Since the optimal controller is, by construction, of order $n_x + n_u = 3$, at least one pair of pole/zero is canceled as clearly occurred in the case corresponding to $d = 1$. □

This example puts in clear evidence the fact that the design of full order output feedback controllers in the discrete-time domain, with state space representation (5.97)–(5.98) has a high computational cost. This is due to several reasons. First, the controller has order $n_x + n_u$, but at least n_u poles and zeros are canceled such that the order of the final controller is reduced with no loss of generality, that is, preserving optimality. Second, the numerical method adopted, from the very beginning, to deal with DLMIs is simple to implement, but it lacks efficiency, which makes the treatment of problems involving sampled-data systems of moderate and high dimensions very difficult or even impossible.

5.6 Bibliography Notes

This chapter is a natural generalization of the previous one in the sense that $\mathcal{H}_\infty$ control design problems constitute a counterpart of $\mathcal{H}_2$ control design. It arises in the important context of robust control, facing norm bounded and convex bounded parameter uncertainty. There are several books that tackle this theme in the context of LTI systems, as, for instance, [14] and the references therein. In the context of sampled-data system, the books [13, 33] and the references therein, are strongly recommended. They present the same set of results but from a more classical point of view, expressed through Lyapunov and Riccati equations. This book presents the same class of results in the new framework of DLMIs.

The definition of the $\mathcal{H}_\infty$ performance has appeared in many references, see the book [33] and the paper [49]. Its importance stems from the fact that it takes into account, simultaneously, perturbations defined in discrete and continuous-time domains. All results presented in this chapter are based on Theorem 5.1, where the $\mathcal{H}_\infty$ performance index is evaluated by means of DLMIs together with LMIs. This is the major result of reference [27].

The state feedback design follows the $\mathcal{H}_2$ counterpart with no major additional difficulty, see also [26] where a design strategy based on the existence of a solution to an algebraic Riccati equation is proposed. From the theoretical and numerical implementation viewpoints the results are very similar. The full order linear filter design has two important consequences, namely the full characterization of the time-invariant filter and the determination of the robust filter to face parameters uncertainty. These results are new in the present context.

Compared with the state feedback matrix gain and the linear filter, the full order sampled-data controller design is much more involved because we have to deal with algebraic Riccati equations. In the case of time-invariant full order controllers, the algebraic Riccati equation has been converted to an equivalent LMI preserving feasibility and optimality. Unfortunately, the same was not possible to be done for the pure discrete-time full order controller. Actually, it has been shown that, in that case, no simplification is possible and the optimality conditions, expressed through DLMIs must be directly handled and solved. In general, this causes important impact on the computational burden involved. The numerical examples solved puts in clear evidence the necessity to develop more powerful numerical methods to deal with DLMIs, see on that respect [3], and [31].

Chapter 6
Markov Jump Linear Systems

6.1 Introduction

Markov jump linear systems (MJLSs) is an important class of dynamic systems that has been deeply studied in the past years. The main reason is because MJLS models adequately, with precision, many practical applications of particular interest and importance. Nevertheless, concerning sampled-data control of MJLS, very few results are available in the literature.

This chapter aims to design linear filters and controllers in both contexts of $\mathcal{H}_2$ and $\mathcal{H}_\infty$ performance optimization. As a by-product, several aspects of robust filtering and control are assessed and discussed. The same classes of filters and controllers are dealt with by generalizing the procedures already established in the previous chapters for deterministic dynamic systems. It is important to make clear that the stochastic nature of MJLS naturally reflects the need for more involved calculations to obtain the aforementioned performance indices as well as filter and control design conditions. The reader is requested to see the Bibliography notes of this chapter where useful information and references about Markov linear systems are provided and briefly discussed.

6.2 MJLS Model

Consider the set $\mathbb{K}$, with N (finite) elements. The triple $(\Omega, \mathcal{F}, \mathcal{P})$ denotes a complete probability space equipped with the filtration $\{\mathcal{F}_t, t \in \mathbb{R}_+\}$. The Markov state $\theta(t) : \mathbb{R}_+ \to \mathbb{K}$ is a continuous-time Markov chain with values in the finite set $\mathbb{K}$. The matrix $\Lambda \in \mathbb{R}^{N \times N}$ with elements λ_{ij} such that $\lambda_{ij} \geq 0,\ i \neq j$, and $\sum_{j \in \mathbb{K}} \lambda_{ij} = 0, \forall i \in \mathbb{K}$, is denominated *transition rate matrix*. For any $\varepsilon \geq 0$, it defines the transition probabilities

J. C. Geromel, *Differential Linear Matrix Inequalities*,
https://doi.org/10.1007/978-3-031-29754-0_6

$$\mathcal{P}(\theta(t+\varepsilon) = j|\theta(t) = i) = \begin{cases} 1 + \lambda_{ii}\varepsilon + o(\varepsilon) \ , \ i = j \\ \lambda_{ij}\varepsilon + o(\varepsilon) \quad , \ i \neq j \end{cases} \tag{6.1}$$

with $\lim_{\varepsilon\to 0^+} o(\varepsilon)/\varepsilon = 0$. It defines the time evolution of the continuous-time Markov chain starting from the initial condition $\theta(0) = \theta_0 \in \mathbb{K}$, such that $\mathcal{P}(\theta_0 = i) = \pi_{0i} \geq 0, \ i \in \mathbb{K}$. In the continuous-time context, we have to handle different stochastic processes of the general form $\{\zeta(t) \in \mathbb{R}^n : t \in \mathbb{R}_+\}$ that belong to the already mentioned probability space. Its norm is defined as

$$||\zeta||_2^2 = \int_0^\infty \mathcal{E}\{\zeta(t)'\zeta(t)\}dt \tag{6.2}$$

where $\mathcal{E}\{\cdot\}$ is the mathematical expectation with respect to $\mathcal{P}$. With a slight abuse of notation, the set of all stochastic processes with finite norm, that is, $||\zeta||_2 < \infty$, is denoted by $\mathcal{L}_2$. The discrete-time counterpart follows the same pattern. The norm of a stochastic process of the general form $\{\zeta[k] \in \mathbb{R}^n : k \in \mathbb{N}\}$ is defined as

$$||\zeta||_2^2 = \sum_{k=0}^{\infty} \mathcal{E}\{\zeta[k]'\zeta[k]\} \tag{6.3}$$

and those of finite norm, that is, $||\zeta||_2 < \infty$, compose the set ℓ_2. Let us keep in mind that $\zeta[k] = \zeta(t_k)$ for all $k \in \mathbb{N}$ and that, in general, the sampling times $\{t_k = kh\}_{k\in\mathbb{N}}$ are evenly spaced by $h > 0$ time unities. In the next sections we use these concepts to build models that are described by continuous-time linear differential equations depending on the parameter $\theta(t) \in \mathbb{K}, \forall t \in \mathbb{R}_+$, defined by the previously introduced Markov chain. Of course, this is done in the framework of sampled-data systems control and filtering with $\mathcal{H}_2$ and $\mathcal{H}_\infty$ performance indices.

6.3 $\mathcal{H}_2$ Performance Analysis and Design

Let us introduce the MJLS model to be dealt with later in this section. It operates in open-loop, has the following state space realization

$$\dot{x}(t) = A_{\theta(t)}x(t) + B_{\theta(t)}u(t) + E_{\theta(t)}w_c(t) \tag{6.4}$$

$$z(t) = C_{z\theta(t)}x(t) + D_{z\theta(t)}u(t) \tag{6.5}$$

$$u(t) = w_d[k], \ t \in [t_k, t_{k+1}), \ \forall k \in \mathbb{N} \tag{6.6}$$

and evolves from a given initial condition (x_0, θ_0) at $t = t_0 = 0$ where $x(0) = 0 \in \mathbb{R}^{n_x}$ is given and $\theta(0) = \theta_0 \in \mathbb{K}$ with probability $\mathcal{P}(\theta_0 = i) = \pi_{0i} \geq 0, \ i \in \mathbb{K}$. As before, the variables $x(\cdot) : \mathbb{R}_+ \to \mathbb{R}^{n_x}$, $u(\cdot) : \mathbb{R}_+ \to \mathbb{R}^{n_u}$, $w_c(\cdot) : \mathbb{R}_+ \to \mathbb{R}^{r_c}$, and $z(\cdot) : \mathbb{R}_+ \to \mathbb{R}^{n_z}$ are the state, the control, the exogenous perturbation, and the controlled output of the continuous-time process, respectively. It is important to observe that in open-loop operation, the control signal depends on the exogenous discrete-time sensor perturbation $w_d[\cdot] : \mathbb{N} \to \mathbb{R}^{r_d}$, exclusively. At each $t \in \mathbb{R}_+$, the Markov state $\theta(t) \in \mathbb{K}$ defines a specific operation mode characterized by matrices $(A_i, B_i, E_i, C_{zi}, D_{zi}), i \in \mathbb{K}$. In the sequel, the $\mathcal{H}_2$ performance index is introduced.

Definition 6.1 The $\mathcal{H}_2$ performance index associated with the sampled-data Markov jump linear system (6.4)–(6.6) is given by

$$\varrho_2^2 = \sum_{i=1}^{r_c+r_d} \mathcal{E}\{||z_i||_2^2\} \tag{6.7}$$

where $z(t) = z_i(t), 1 \leq i \leq r_c$, is the output due to a continuous-time impulse $\delta(t)$ in the i-th channel of the input $w_c(t)$ and $z(t) = z_i(t), r_c + 1 \leq i \leq r_c + r_d$, is the output due to a discrete-time impulse $\delta[k]$ in the i-th channel of the input $w_d[k]$. Moreover, the indicated mathematical expectation is performed with respect to the initial Markov state $\theta_0 \in \mathbb{K}$ with probability $\mathcal{P}(\theta_0 = i) = \pi_{0i} \geq 0,\ i \in \mathbb{K}$.

The initial state of the Markov chain $\theta(0) = \theta_0 \in \mathbb{K}$ is a random variable but the initial condition $x(0) = 0 \in \mathbb{R}^{n_x}$ and the impulsive exogenous signals $w_c(t)$ and $w_d[k]$ are deterministic. Interesting enough is the fact that whenever $N = 1$ the MJLS reduces to a deterministic sampled-data system and Definition 6.1 reduces to Definition 4.1 that characterizes the $\mathcal{H}_2$ performance index of the sampled-data system (4.1)–(4.3).

At this point, to be used later in this chapter to deal with sampled-data systems analysis and design, we need to introduce the so-called hybrid Markov jump linear system (HMJLS) which has the general form

$$\dot{\psi}(t) = F_{\theta(t)}\psi(t) + J_{c\theta(t)}w_c(t) \tag{6.8}$$

$$z(t) = G_{\theta(t)}\psi(t) \tag{6.9}$$

$$\psi(t_k) = H_{\theta(t_k)}\psi(t_k^-) + J_{d\theta(t_k)}w_d[k] \tag{6.10}$$

where $t \in [t_k, t_{k+1}),\ k \in \mathbb{N}$, and the indicated Markov state dependent matrices are of compatible dimensions. Several aspects involving this hybrid system must be analyzed carefully. First, the HMJLS we consider is general enough such that the sampled-data systems of interest obey (6.8)–(6.10) provided that the Markov state dependent matrices are properly calculated. Second, it is important to verify what happens to the Markov state $\theta(t)$ at the jump instants $\{t_k\}_{k\in\mathbb{K}}$. This last and very important aspect is addressed in the next remark.

Remark 6.1 The sampling process, which defines the sampling instants $\{t_k\}_{k\in\mathbb{K}}$, is essentially deterministic and independent of the process that defines the Markov state $\theta(t)$ for each $t \in \mathbb{R}_+$. They can be determined as follows:

- For instance, if we consider, as usual, sampling times evenly spaced $t_{k+1}-t_k = h$ with $h > 0$, the sampling instants are given by $t_k = kh$, $k \in \mathbb{N}$.
- A single time evolution of the Markov chain can be determined as follows: at time $t = t_0 = 0$, the initial Markov state $\theta(0) = \theta_0 \in \mathbb{K}$ is determined randomly from the fact that $\mathcal{P}(\theta_0 = i) = \pi_{0i} \geq 0$, $i \in \mathbb{K}$, is known. The Markov state $\theta(t)$ spends a time interval (waiting time) d_i in the mode $i \in \mathbb{K}$, defined by an exponential distribution with mean $1/|\lambda_{ii}|$, and jumps to another mode $j \neq i \in \mathbb{K}$ according to the probability $\mathcal{P}(\theta(d_i + \varepsilon) = j|\theta(d_i) = i) = \lambda_{ij}/|\lambda_{ii}|$ with $\varepsilon > 0$ arbitrarily small. This calculation is repeated in the whole time interval of interest and allows the implementation of a Monte Carlo procedure to evaluate numerically the mathematical expectations involved.

Calling $\{\tau_k\}_{k\in\mathbb{N}}$ the time instants when the jumps on the Markov state occur, it is clear that they are random variables independent of the sampling instants $\{t_k\}_{k\in\mathbb{N}}$. As a consequence, the sentence $\tau_p \neq t_q$ for all pairs $(p, q) \in \mathbb{N} \times \mathbb{N}$ is true almost surely. Or saying it differently, with probability one, jump instants do not coincide with sampling instants. Hence, for any $k \in \mathbb{N}$, taking into account (6.1), if $\theta(t_k) = i \in \mathbb{K}$, then $\theta(t_k^-) = \theta(t_k) = \theta(t_k^+) = i \in \mathbb{K}$ almost surely. □

Finally, as we have done before for sampled-data deterministic systems, the conversion of continuous-time and discrete-time impulse effects into initial conditions is simple and useful as far as the performance index (6.7) is concerned. For $k = 0$, the jump equation (6.10) reads $\psi(t_0) = H_{\theta_0}\psi(t_0^-) + J_{d\theta_0}w_d[0]$, which implies that the effect of a continuous-time impulse $\delta(t)$, occurring $t = 0^-$, injected in the i-th channel of the exogenous disturbance w_c is identical to the effect of an initial condition $\psi(0)$ equal to the i-th column of the matrix $H_{\theta_0}J_{c\theta_0}$. From the same reasoning, the effect of a discrete-time impulse $\delta[k]$, occurring at $k = 0$, injected in the i-th channel of the exogenous disturbance w_d is identical to the effect of an initial condition $\psi(0)$ equal to the i-th column of the matrix $J_{d\theta_0}$.

In general, in the framework of sampled-data MJLS, we have to handle some differential inequalities, a bit different from the ones we have treated up to now. The next lemma puts in evidence an important property of one of them.

Lemma 6.1 *Let $h > 0$ be given. Any feasible solution to the coupled DLMIs*

$$\dot{P}_i(t) + F_i' P_i(t) + P_i(t)F_i + G_i'G_i < -\sum_{j\in\mathbb{K}} \lambda_{ij}P_j(t), \ t \in [0, h) \tag{6.11}$$

subject to the final boundary conditions $P_i(h) = P_{ih} > 0, \forall i \in \mathbb{K}$, is positive definite in the whole time interval $[0, h]$, that is, $P_i(t) > 0, \forall i \in \mathbb{K}$ and $t \in [0, h]$.

Proof Denote $\bar{F}_i = F_i+(\lambda_{ii}/2)I$, $i \in \mathbb{K}$. Assuming that $P_i(t)$ is a feasible solution, thanks to linearity, there exist norm bounded matrices $W_i(t) > 0$ for all $i \in \mathbb{K}$ and $t \in [0, h]$ such that the result of Lemma 2.1 yields

$$P_i(t) = e^{\bar{F}_i'(h-t)}P_{ih}e^{\bar{F}_i(h-t)}$$

$$+\int_t^h e^{\bar{F}_i'(\xi-t)}\Big(G_i'G_i + W_i(t) + \sum_{i\neq j\in\mathbb{K}} \lambda_{ij}P_j(t)\Big)e^{\bar{F}_i(\xi-t)}d\xi \tag{6.12}$$

Based on this equality, it is simple to verify that the sequence of symmetric matrices $R_i^n(t), i \in \mathbb{K}$, satisfying

$$R_i^{n+1}(t) = e^{\bar{F}_i'(h-t)}P_{ih}e^{\bar{F}_i(h-t)} + \int_t^h e^{\bar{F}_i'(\xi-t)}\Big(G_i'G_i + W_i(t) + \sum_{i\neq j\in\mathbb{K}} \lambda_{ij}R_j^n(t)\Big)e^{\bar{F}_i(\xi-t)}d\xi \tag{6.13}$$

for all $t \in [0, h]$ and $n \in \mathbb{N}$ is monotonically non-decreasing, in the sense that $R_i^{n+1}(t) \geq R_i^n(t) > 0$ for all $n \in \mathbb{N}$, provided that $R_i^0(t) > 0$ is taken arbitrarily close to the null matrix, for $i \in \mathbb{K}$ and $t \in [0, h]$. Let us now determine (if any) an upper bound $||R_i^n(t)|| \leq \ell_i(t)$ independent of $n \in \mathbb{N}$. Denoting $||\bar{F}_i|| = \alpha_i$, $||G_i'G_i + W_i(t)|| \leq \beta_i$ for all $i \in \mathbb{K}$ and $t \in [0, h]$, (6.13) provides

$$\begin{aligned}\|R_i^{n+1}(t)\| &\leq e^{2\alpha_i(h-t)}\ell_i(h) + \int_t^h e^{2\alpha_i(\xi-t)}\Big(\beta_i + \sum_{i\neq j\in\mathbb{K}} \lambda_{ij}\ell_j(t)\Big)d\xi \\ &= \ell_i(t)\end{aligned} \tag{6.14}$$

with the final boundary condition $\ell_i(h) = ||P_i(h)||$ for every $i \in \mathbb{K}$. The equality in the last equation imposes that the upper bound satisfies the set of coupled linear differential equations

$$\dot{\ell}_i(t) + 2\alpha_i\ell_i(t) + \beta_i = -\sum_{i\neq j\in\mathbb{K}} \lambda_{ij}\ell_j(t) \tag{6.15}$$

subject to the given final boundary conditions. Being linear, it always admits a unique and norm bounded solution. Denoting $\max_{t\in[0,h]} ||\ell_i(t)\,| = b_i$, the conclusion is that $||R_i^n(t)|| \leq b_i < \infty$ for all $n \in \mathbb{N}$, $i \in \mathbb{K}$ and $t \in [0, h]$. These calculations imply that the non-decreasing sequence $\{R_i^n(t)\}_{n\in\mathbb{N}}$ converges, that is, $\lim_{n\to\infty} R_i^n(t) = P_i(t)$ and by consequence $P_i(t) > 0$ for all $i \in \mathbb{K}$ in the whole time interval $t \in [0, h]$. The proof is concluded. □

Lemma 6.1 states that all matrix-valued functions $P_i(t), i \in \mathbb{K}$, are positive definite in the whole time interval $[0, h]$ provided that $P_i(h) = P_{ih} > 0, i \in \mathbb{K}$. This is an essential property for the determination of the inverses $P_i(t)^{-1} > 0$ in the same time interval and at the core that allows the development of the next results. Before proceeding, to ease the presentation, we need to introduce a more compact notation concerning conditional expectation, that is,

$$\mathcal{E}^{\nu(t)}\{\cdot\} = \mathcal{E}\{\cdot|\nu(t)\} \tag{6.16}$$

where, throughout this chapter, if nothing is contrarily mentioned, we always adopt $\nu(t) = (\psi(t), \theta(t)) = (\xi, i) \in \mathbb{R}^{n_\psi} \times \mathbb{K}$, for some $t \in \mathbb{R}_+$ of interest. The next theorem deals with a general HMJLS with state space realization of the form (6.8)–(6.10).

Theorem 6.1 *Let $h > 0$ be given. The hybrid Markov jump linear system defined by (6.8)–(6.10) is mean square stable with ϱ_2^2 performance index if and only if the coupled DLMIs*

$$\dot{P}_i(t) + F_i' P_i(t) + P_i(t) F_i + G_i' G_i < -\sum_{j \in \mathbb{K}} \lambda_{ij} P_j(t), \ t \in [0, h) \tag{6.17}$$

subject to the LMI boundary conditions

$$P_{ih} > 0, \ P_{ih} > H_i' P_{i0} H_i \tag{6.18}$$

for all $i \in \mathbb{K}$ are feasible. In the affirmative case, the equality

$$\varrho_2^2 = \inf_{P_i(\cdot)} \left\{ \sum_{i \in \mathbb{K}} \pi_{0i} \left(\mathbf{tr}(J_{ci}' P_{ih} J_{ci}) + \mathbf{tr}(J_{di}' P_{i0} J_{di}) \right) : \ (6.17)\text{–}(6.18) \right\} \tag{6.19}$$

holds.

Proof Let us first prove the sufficiency. Consider that the HMJLS (6.8)–(6.10) evolves from an arbitrary initial condition $\psi(0) \neq 0$ and it is free of external perturbations, that is, $w_c = 0$ and $w_d = 0$. Assuming that the coupled DLMIs (6.17) admit a solution in the time interval $t \in [0, h)$, the quadratic cost function $V_\theta(\psi, t) = \psi' P_\theta(t) \psi$ satisfies the Hamilton–Jacobi–Bellman inequalities

$$\frac{\partial V_i'}{\partial \psi} F_i \psi + \psi' G_i G_i \psi + \frac{\partial V_i}{\partial t} < -\sum_{j \in \mathbb{K}} \lambda_{ij} V_j, \ t \in [0, h) \tag{6.20}$$

for all $i \in \mathbb{K}$. Hence, the infinitesimal generator or equivalently Dynkin's formula applied to the cost function enables us to conclude that (6.20) can be rewritten as

$$V_i(\xi, t_k) > \mathcal{E}^{\nu(t_k)} \left\{ \int_{t_k}^{t_{k+1}} z(t)' z(t) dt + V_{\theta(t_{k+1})}(\psi(t_{k+1}^-), t_{k+1}) \right\} \tag{6.21}$$

because the solution of (6.17) is continuous in the time interval $[0, h)$ and $\theta(t_{k+1}^-) = \theta(t_{k+1})$ with probability one (almost surely) for all $k \in \mathbb{N}$, see Remark 6.1. Since the hybrid system is time-invariant, the function $V_i(\psi, t) = \psi' P_i(t - t_k)\psi$, $i \in \mathbb{K}$, remains feasible in any subsequent time interval, $k \in \mathbb{N}$, provided that the same boundary conditions (6.18) are imposed. Now, taking into account that $P_{ih} > 0, \forall i \in \mathbb{K}$, implies that $P_{i0} > 0, \forall i \in \mathbb{K}$ as well, let us define the positive definite quadratic function $v_\theta(\psi) = \psi' P_\theta(0)\psi$. By construction, we have $V_i(\psi, t_k) =$

$v_i(\psi)$ for all $i \in \mathbb{K}$, and using the jump equation $\psi(t_{k+1}) = H_{\theta(t_{k+1})}\psi(t_{k+1}^-)$ together with the boundary condition, denoting $j = \theta(t_{k+1}) \in \mathbb{K}$, we also have

$$\begin{aligned} V_j(\psi(t_{k+1}^-), t_{k+1}) &= \psi(t_{k+1}^-)' P_j(h)\psi(t_{k+1}^-) \\ &> \psi(t_{k+1}^-)' H_j' P_j(0) H_j \psi(t_{k+1}^-) \\ &= \psi(t_{k+1})' P_j(0)\psi(t_{k+1}) \\ &= v_{\theta(t_{k+1})}(\psi(t_{k+1})) \end{aligned} \tag{6.22}$$

Hence, from these algebraic manipulations, it is readily seen that the quadratic function $v_\theta(\psi)$ satisfies

$$v_i(\xi) > \mathcal{E}^{\nu(t_k)}\left\{\int_{t_k}^{t_{k+1}} z(t)'z(t)dt + v_{\theta(t_{k+1})}(\psi(t_{k+1}))\right\} \tag{6.23}$$

and, based on it, two conclusions can be drawn. The first one is that there exists a scalar $\varepsilon > 0$ small enough such that $\mathcal{E}^{\nu(t_k)}\{v_{\theta(t_{k+1})}(\psi(t_{k+1}))\} \le (1-\varepsilon)v_{\theta(t_k)}(\psi(t_k))$ meaning that the sequence $\mathcal{E}\{v_{\theta(t_k)}(\psi(t_k))\}$ converges to zero as $k \in \mathbb{N}$ goes to infinity indicating that the HMJLS under consideration is mean square stable. On the other hand, the same inequality yields the telescoping sum

$$\begin{aligned} \mathcal{E}\left\{\int_0^\infty z(t)'z(t)dt\right\} &= \mathcal{E}\left\{\sum_{k\in\mathbb{N}} \mathcal{E}^{\nu(t_k)}\left\{\int_{t_k}^{t_{k+1}} z(t)'z(t)dt\right\}\right\} \\ &< \mathcal{E}\left\{\sum_{k\in\mathbb{N}} \left(v_{\theta(t_k)}(\psi(t_k)) - v_{\theta(t_{k+1})}(\psi(t_{k+1}))\right)\right\} \\ &= \mathcal{E}\left\{v_{\theta(0)}(\psi(0))\right\} \\ &= \sum_{i\in\mathbb{K}} \pi_{0i}\psi(0)' P_{i0}\psi(0) \end{aligned} \tag{6.24}$$

where, in order to get the last equality, we have used the fact already proven that the system is mean square stable. We are now in position to evaluate the performance index ϱ_2^2 given in (6.7). Actually, as discussed before, the effect on the controlled output of a continuous-time impulse occurring at time $t = 0^-$ in the j-th channel of w_c is identical to the effect of the initial condition $\psi(0) = H_{\theta_0} J_{c\theta_0} e_{cj}$, whereas the effect on the controlled output of a discrete-time impulse occurring at time $k = 0$ in the j-th channel of w_d is identical to the effect of the initial condition $\psi(0) = J_{d\theta_0} e_{dj}$. The vectors $e_{cj} \in \mathbb{R}^{r_c}$ and $e_{dj} \in \mathbb{R}^{r_d}$ have the j-th component equal to one and zero in the others. Hence, summing up all these contributions, we obtain

$$
\begin{aligned}
\varrho_2^2 &= \sum_{j=1}^{r_c+r_d} \mathcal{E}\{||z_j||_2^2\} \\
&< \sum_{j=1}^{r_c}\sum_{i\in\mathbb{K}} \pi_{0i} e_{cj}' J_{ci}' H_i' P_{i0} H_i J_{ci} e_{cj} + \sum_{j=1}^{r_d}\sum_{i\in\mathbb{K}} \pi_{0i} e_{dj}' J_{di}' P_{i0} J_{di} e_{dj} \\
&= \sum_{i\in\mathbb{K}} \pi_{0i} \mathbf{tr}(J_{ci}' H_i' P_{i0} H_i J_{ci}) + \sum_{i\in\mathbb{K}} \pi_{0i} \mathbf{tr}(J_{di}' P_{i0} J_{di}) \\
&\leq \sum_{i\in\mathbb{K}} \pi_{0i} \left(\mathbf{tr}(J_{ci}' P_{ih} J_{ci}) + \mathbf{tr}(J_{di}' P_{i0} J_{di})\right)
\end{aligned} \tag{6.25}
$$

from which, we conclude that each feasible solution to the problem stated in the right hand side of (6.19) yields an upper bound to the $\mathcal{H}_2$ performance index ϱ_2^2. The smallest one, provided by the optimal solution to problem (6.19), equals the $\mathcal{H}_2$ performance index introduced in Definition 6.1, as it is shown in the necessity part of the proof.

Necessity follows from Bellman's Principle of Optimality applied to the hybrid linear system. To this end, let us define the positive cost-to-go function

$$
V_i(\xi, t_k) = \mathcal{E}^{\nu(t_k)}\left\{\int_{t_k}^{\infty} \psi(t)' G_{\theta(t)}' G_{\theta(t)} \psi(t) dt\right\} \tag{6.26}
$$

where for $t \geq t_k$, the trajectory $\psi(t)$ is provided by the linear differential equation $\dot{\psi}(t) = F_{\theta(t)}\psi(t)$, subject to the initial condition $\psi(t_k) = \xi$, and the jump equation $\psi(t_{k+1}) = H_{\theta(t_{k+1})}\psi(t_{k+1}^-)$. Assuming that the hybrid Markov jump linear system is mean square stable and the $\mathcal{H}_2$ performance index ϱ_2^2 is well defined and finite, we have to prove that the stationary version of the cost-to-go function, namely $v_\theta(\psi)$ satisfies the inequality (6.23) arbitrarily close to equality and it is quadratic, that is, it has the form $v_\theta(\psi) = \psi' S_\theta \psi$ for some $S_i > 0$, $i \in \mathbb{K}$. Indeed, to prove this claim by induction, let us assume that this actually occurs at sampling time t_{k+1} and split the whole time interval $t_k \mapsto t_{k+1}$ in two successive events, namely $t_k \mapsto t_{k+1}^-$ and $t_{k+1}^- \mapsto t_{k+1}$. For the second part of the time interval, let us set $P_{ih} > 0$ satisfying $P_{ih} > H_i' S_i H_i$, but arbitrarily close to equality, for all $i \in \mathbb{K}$. Concerning the first part of the time interval, the right hand side of (6.23) yields

$$
\begin{aligned}
v_i(\xi) &> \mathcal{E}^{\nu(t_k)}\left\{\int_{t_k}^{t_{k+1}} z(t)'z(t)dt + \psi(t_{k+1})' S_{\theta(t_{k+1})} \psi(t_{k+1})\right\} \\
&= \mathcal{E}^{\nu(t_k)}\left\{\int_{t_k}^{t_{k+1}} z(t)'z(t)dt + \psi(t_{k+1}^-)' P_{\theta(t_{k+1})}(h) \psi(t_{k+1}^-)\right\} \\
&= \xi' P_i(0) \xi
\end{aligned} \tag{6.27}
$$

for all $i \in \mathbb{K}$, where the last equality follows from well known calculations involving a pure MJLS system with a quadratic criterion. Indeed, the right hand side of (6.27) is given by $V_{\theta(t)}(\psi(t), t) = \psi(t)' P_{\theta(t)}(t)\psi(t)$ evaluated at $t = 0$ with $P_i(t)$, being the positive definite solution to the coupled differential Lyapunov equations that follow from (6.17) taken arbitrarily close to equality, subject to the final boundary conditions $P_i(h) = P_{ih}, \forall i \in \mathbb{K}$. Hence, from (6.27), the smallest cost is quadratic and is reached by setting $S_i = P_i(0) = P_{i0}, \forall i \in \mathbb{K}$, meaning that the boundary conditions (6.18) hold. Finally, evolving from $t_0 = 0$, we obtain the exact evaluation $||z||_2^2 = \mathcal{E}\{v_{\theta(0)}(\psi(0))\}$ yielding thus the exact value of the $\mathcal{H}_2$ performance index ϱ_2^2, concluding thus the proof. □

The theorem has just presented a general result in the context of HMJLS. The $\mathcal{H}_2$ performance index determination associated with the sampled-data MJLS (6.4)–(6.6) constitutes a particular case well defined by matrices

$$F_i = \begin{bmatrix} A_i & B_i \\ 0 & 0 \end{bmatrix}, \; G_i = \begin{bmatrix} C_{zi} & D_{zi} \end{bmatrix}, \; J_{ci} = \begin{bmatrix} E_i \\ 0 \end{bmatrix}, \; J_{di} = \begin{bmatrix} 0 \\ I \end{bmatrix} \tag{6.28}$$

and $H_i = \text{diag}(I, 0)$, with compatible dimensions for all $i \in \mathbb{K}$. For each Markov mode $i \in \mathbb{K}$, they are identical to the ones already given in Chap. 4, more specifically, in (4.8). Due to this similarity, robustness with respect to convex bounded parameter uncertainty can be addressed with no additional difficulty. Indeed, if matrices $(F_i, G_i), i \in \mathbb{K}$, are not precisely known but belong to polyhedral convex sets $\Delta_i, i \in \mathbb{K}$, then the coupled DLMIs

$$\begin{bmatrix} \dot{P}_i(t) + F_i' P_i(t) + P_i(t) F_i + \sum_{j\in\mathbb{K}} \lambda_{ij} P_j(t) & G_i' \\ \bullet & -I \end{bmatrix} < 0, \; t \in [0, h) \tag{6.29}$$

imposed in all vertices remain valid in the whole uncertainty sets $\Delta_i, i \in \mathbb{K}$. Doing this, the minimum guaranteed $\mathcal{H}_2$ performance index ϱ_{2rob}^2 is readily determined. It is important to stress that it is possible to take into account convex bounded parameter uncertainty only because the coupled DLMIs (6.29) depend linearly on the pair of matrices (F_i, G_i) for all $i \in \mathbb{K}$.

Comparing the conditions of Theorem 6.1 with the ones of Theorem 4.1, it is apparent that the boundary conditions are identical because (6.18) must hold for each $i \in \mathbb{K}$, independently. The only coupling term appears in the DLMIs (6.17) whose feasible solutions $P_i(t), i \in \mathbb{K}$, are positive definite in the whole time interval $[0, h]$. In many situations, we also need to express those DMLIs in terms of the inverses $Q_i(t) = P_i(t)^{-1} > 0, i \in \mathbb{K}$. To this end, the key algebraic manipulation is synthesized by the inequalities

$$P_i(t)^{-1} \left(\sum_{j\in\mathbb{K}} \lambda_{ij} P_j(t) \right) P_i(t)^{-1} = \sum_{i\neq j\in\mathbb{K}} \lambda_{ij} \Big(Q_i(t) Q_j(t)^{-1} Q_i(t) - Q_i(t) \Big)$$

$$< \sum_{i \neq j \in \mathbb{K}} \lambda_{ij} S_{ij}(t), \ i \in \mathbb{K} \tag{6.30}$$

where we have used the relations $\lambda_{ij} \geq 0, i \neq j \in \mathbb{K}$, $\lambda_{ii} = -\sum_{i \neq j \in \mathbb{K}} \lambda_{ij}$, and

$$\begin{bmatrix} S_{ij}(t) + Q_i(t) & Q_i(t) \\ \bullet & Q_j(t) \end{bmatrix} > 0, \ \forall i \neq j \in \mathbb{K} \times \mathbb{K} \tag{6.31}$$

which holds in the time interval $[0, h]$. Actually, this can be readily verified by calculating the Schur Complement with respect to the second row and column of the LMIs (6.31). Inversely, there exists $S_{ij}(t)$ satisfying (6.31) such that (6.30) holds arbitrarily close to equality. Finally, it is immediate to see that $S_{ij}(t)$ is, in general, sign indefinite. As it is clear afterward, this simple algebraic manipulation plays an important role in reproducing to sampled-data MJLS control design the corresponding results already presented in Chap. 4. Throughout, to ease the presentation, whenever possible, the time dependence of matrix-valued functions is dropped.

6.3.1 State Feedback Design

The sampled-data MJLS to be considered in order to design a state feedback controller has the following state space realization:

$$\dot{x}(t) = A_{\theta(t)} x(t) + B_{\theta(t)} u(t) + E_{\theta(t)} w_c(t) \tag{6.32}$$

$$z(t) = C_{z\theta(t)} x(t) + D_{z\theta(t)} u(t) \tag{6.33}$$

$$u(t) = L_{\theta[k]} x[k] + E_{u\theta[k]} w_d[k], \ t \in [t_k, t_{k+1}), \ \forall k \in \mathbb{N} \tag{6.34}$$

and evolves from a given initial condition (x_0, θ_0) at $t = t_0 = 0$, where $x(0) = 0 \in \mathbb{R}^{n_x}$ is given and $\theta(0) = \theta_0 \in \mathbb{K}$ with probability $\mathcal{P}(\theta_0 = i) = \pi_{0i} \geq 0, \ i \in \mathbb{K}$. Notice that it incorporates the feature that the whole state variable is available for feedback. Looking at the HMJLS, all involved matrices are given in (6.28) with the exception of

$$H_i = \begin{bmatrix} I & 0 \\ L_i & 0 \end{bmatrix}, \ J_{di} = \begin{bmatrix} 0 \\ E_{ui} \end{bmatrix}, \ i \in \mathbb{K} \tag{6.35}$$

with appropriate dimensions. Our goal is to design a set of mode dependent matrix gains $L_i, i \in \mathbb{K}$, such that the $\mathcal{H}_2$ performance index of the closed-loop system is minimized. As before, those matrix gains are determined through the global optimal solution of a jointly convex programming problem expressed by LMIs and DLMIs.

Theorem 6.2 *Let $h > 0$ be given. Consider that together with the coupled DLMIs*

$$\begin{bmatrix} -\dot{Q}_i + Q_i F_i' + F_i Q_i + \sum_{i \neq j \in \mathbb{K}} \lambda_{ij} S_{ij} & Q_i G_i' \\ \bullet & -I \end{bmatrix} < 0, \ t \in [0, h) \tag{6.36}$$

subject to the LMI boundary conditions

$$\begin{bmatrix} Q_{ih} & Q_{ih} \begin{bmatrix} I \\ 0 \end{bmatrix} \\ \bullet & W_i \end{bmatrix} > 0, \ \begin{bmatrix} W_i & [W_i \ K_i'] \\ \bullet & Q_{i0} \end{bmatrix} > 0 \tag{6.37}$$

the optimal solution of the convex programming problem with respect to the matrix variables $Q_i(t)$, W_i, K_i for all $i \in \mathbb{K}$ and $S_{ij}(t)$, for all $\forall i \neq j \in \mathbb{K} \times \mathbb{K}$

$$\varrho_2^2 = \inf \left\{ \sum_{i \in \mathbb{K}} \pi_{0i} \left(\mathbf{tr}(J_{ci}' Q_{ih}^{-1} J_{ci}) + \mathbf{tr}(J_{di}' Q_{i0}^{-1} J_{di}) \right) : \ (6.31), \ (6.36)\text{–}(6.37) \right\} \tag{6.38}$$

provides the matrix gains $L_i = K_i W_i^{-1}$, $i \in \mathbb{K}$. Then, the closed-loop sampled-data control system (6.32)–(6.34) is mean square stable and operates with minimum ϱ_2^2 performance index.

Proof First notice that, for each $i \in \mathbb{K}$, the equivalence between the boundary condition (6.37) and (6.18) has been already proven in Theorem 4.3, being thus omitted. Furthermore, since $Q_{i0} > 0$ and $Q_{ih} > 0$, then, similarly to what has been proven in Lemma 6.1, $Q_i(t) > 0$ for all $t \in [0, h]$ and consequently $P_i(t) = Q_i(t)^{-1} > 0$ is well defined in the mentioned time interval. The coupled DLMIs (6.17) are equivalent to (6.29). Multiplying (6.29) both sides by $\text{diag}(P_i(t)^{-1}, I)$, using the fact that the time derivative $-\dot{Q}_i(t) = P_i(t)^{-1} \dot{P}_i(t) P_i(t)^{-1}$ holds, then from the discussion at the end of the previous section, it follows that (6.29) is equivalent to (6.36) and (6.31). Finally, the objective function of problem (6.19) yields the one of (6.38) by simple substitution. The proof is concluded by direct application of Theorem 6.1. □

Observe that (6.31) contains $N(N-1)$ DLMIs, where N is the number of modes of the Markov chain. This number may be large, so the computational burden involved in the numerical solution of (6.38) is, in general, expressive. However, those DLMIs cannot be removed or simplified because they are necessary to represent the coupling terms of the original DLMIs (6.17).

Another point of importance is the fact that the boundary conditions of Theorem 6.1 are those already treated in Chap. 4 but imposed to each $i \in \mathbb{K}$ separately, since all coupling terms appear only in the DLMIs. The consequence is that all algebraic manipulations, whenever restricted to the boundary conditions, remain valid. This is the case of the simpler result, in terms of matrix variables, reported in the next theorem.

Theorem 6.3 *Let $h > 0$ be given. Consider that together with the coupled DLMIs*

$$\begin{bmatrix} -\dot{Q}_i + Q_i F_i' + F_i Q_i + \sum_{i \neq j \in \mathbb{K}} \lambda_{ij} S_{ij} & Q_i G_i' \\ \bullet & -I \end{bmatrix} < 0, \ t \in [0, h) \tag{6.39}$$

subject to the LMI boundary conditions

$$Q_{i0} > 0, \ I_x'(Q_{i0} - Q_{ih}) I_x > 0 \tag{6.40}$$

the optimal solution of the convex programming problem with respect to the matrix variables $Q_i(t)$, for all $i \in \mathbb{K}$ and $S_{ij}(t)$, for all $\forall i \neq j \in \mathbb{K} \times \mathbb{K}$

$$\varrho_2^2 = \inf \left\{ \sum_{i \in \mathbb{K}} \pi_{0i} \left(\mathbf{tr}(J_{ci}' Q_{ih}^{-1} J_{ci}) + \mathbf{tr}(J_{di}' Q_{i0}^{-1} J_{di}) \right) : (6.31), \ (6.39)\text{–}(6.40) \right\} \tag{6.41}$$

provides the matrix gains $L_i = U_{i0}' Y_{i0}^{-1}$, constructed with the partitions of matrices $Q_{i0} = P_{i0}^{-1}, \forall i \in \mathbb{K}$. Then, the closed-loop sampled-data control system (6.32)–(6.34) is mean square stable and operates with minimum ϱ_2^2 performance index.

Proof The reasoning of the proof is identical to that of Theorem 6.2. The only difference is the treatment of the boundary conditions and the gain determination that are handled as in Theorem 4.4. The present formula for the state feedback gains avoids the calculation of the inverse

$$P_{i0} = Q_{i0}^{-1} = \begin{bmatrix} X_{i0} & V_{i0} \\ V_{i0}' & \hat{X}_{i0} \end{bmatrix} \tag{6.42}$$

numerically. Indeed, as it has been already shown, the four-block matrix inverse of the partitioned matrix Q_{i0} provides the equality $L_i = -\hat{X}_{i0}^{-1} V_{i0}' = U_{i0}' Y_{i0}^{-1}$, for all $i \in \mathbb{K}$, completing thus the proof. □

The last two theorems can be generalized to cope with convex bounded parameter uncertainty acting on matrices $(F_i, G_i), i \in \mathbb{K}$, because they appear linearly on the DLMIs (6.36) and (6.39). However, in the case of sampled-data MJLS, for the same reason, the transition rate matrix $\Lambda \in \mathbb{R}^{N \times N}$ can be uncertain, that is, $\Lambda \in \text{co}\{\Lambda_1, \cdots, \Lambda_M\}$ where the vertices are matrices of transition rate type. In this case, as we know, the mentioned DLMIs must be imposed in each vertex of the uncertain domain.

Example 6.1 This example is inspired by Example 4.2. Consider the sampled-data MJLS (6.32)–(6.34) with $h = 1.5$ and a Markov chain with $N = 2$ modes and initial probabilities $\mathcal{P}(\theta_0 = 1) = \pi_{01} = 0.25$, $\mathcal{P}(\theta_0 = 2) = \pi_{02} = 0.75$. The transition rate matrix and the mode dependent matrices are given by $(A_i, B_i) = (A(a_i), B(b_i)), i \in \mathbb{K}$, where $(a_1, b_1) = (0, 1)$, $(a_2, b_2) = (9, 2)$, and

$$\Lambda = \begin{bmatrix} -2 & 2 \\ 1 & -1 \end{bmatrix}, \ A(a) = \begin{bmatrix} 0 & 1 & 0 \\ 0 & 0 & 1 \\ a & -5 & -9 \end{bmatrix}, \ B(b) = \begin{bmatrix} 0 \\ 0 \\ b \end{bmatrix}$$

The remaining matrices are

$$E = \begin{bmatrix} 0 \\ 0 \\ 1 \end{bmatrix}, \ E_u = \begin{bmatrix} 1 \end{bmatrix}, \ C_z' = \begin{bmatrix} 1 \\ 1 \\ 1 \end{bmatrix}, \ D_z = \begin{bmatrix} 1 \end{bmatrix}$$

The system jumps from one mode to the other, according to the Markov chain defined by the transition rate matrix. Both matrices $A_i, i \in \mathbb{K}$, are not Hurwitz stable meaning that an effective control action is needed for mean square stabilization. For illustration, we have solved problem (6.41) twice:

(a) With $n_\phi = 8$, we have calculated the state feedback gains

$$\begin{bmatrix} L_1 \\ L_2 \end{bmatrix} = \begin{bmatrix} -4.4436 & -4.4089 & -0.4703 \\ -5.0113 & -4.1733 & -0.4372 \end{bmatrix}$$

and the corresponding minimum $\mathcal{H}_2$ performance level $\varrho_2^2 = 34.1446$. It is to be remarked that a solution to the problem (6.41) has been found with a small number of subintervals of the time interval $[0, h]$. Even though, due to feasibility, mean square stability is guaranteed.

(b) For comparison with $n_\phi = 32$, we have obtained the state feedback gains

$$\begin{bmatrix} L_1 \\ L_2 \end{bmatrix} = \begin{bmatrix} -4.4436 & -4.4039 & -0.4736 \\ -5.0509 & -4.2124 & -0.4449 \end{bmatrix}$$

with the associated $\mathcal{H}_2$ performance level $\varrho_2^2 = 13.6719$. The difference on the minimum cost indicates that we are likely far from the optimal solution. However, adopting $n_\phi = 64$, the computational burden becomes prohibitive in terms of calculation time.

Compared with the sampled-data control design of LTI systems tackled in Chap. 4, now, the number of matrix variables and the number of LMIs are much greater. This happens due to the number of Markov modes $i \in \mathbb{K}$ $(N > 1)$ and the number of slack variables S_{ij} $(N(N-1))$ that needs to be included in order to keep the problem expressed by (coupled) DLMIs. □

There is an important issue in the context of sampled-data MJLS called *mode independent design*. As the name indicates, it consists of designing only one state feedback gain matrix L, such that instead of (6.34), the state feedback control becomes $u(t) = Lx[k] + E_{u\theta[k]}w_d[k], \ t \in [t_k, t_{k+1}), \ \forall k \in \mathbb{N}$. The rationale behind this proposal is the fact that the current Markov mode does not need to be measured anymore, as far as the control implementation is concerned. Unfortunately, in order to do that some conservatism must be introduced. Indeed, in Theorem 6.2, if the final boundary conditions (6.37) are rewritten with $W_i = W$ and $K_i = K$, then $L_i = K_i W_i^{-1} = KW^{-1} = L$, for all $i \in \mathbb{K}$. Doing this, the implementation is simplified, but, in general, only an upper bound of the $\mathcal{H}_2$ performance index is minimized.

6.3.2 Filter Design

The sampled-data MJLS of interest is the one already considered in Chap. 4 but now depending on the Markov modes, that is,

$$\dot{x}(t) = A_{\theta(t)}x(t) + E_{\theta(t)}w_c(t), \ x(0) = 0 \tag{6.43}$$

$$y[k] = C_{y\theta[k]}x[k] + E_{y\theta[k]}w_d[k] \tag{6.44}$$

$$z_e(t) = C_{z\theta(t)}x(t) \tag{6.45}$$

where $x(\cdot) : \mathbb{R}_+ \to \mathbb{R}^{n_x}$ is the state, $z_e(\cdot) : \mathbb{R}_+ \to \mathbb{R}^{n_z}$ is the variable to be estimated, and $y[\cdot] : \mathbb{N} \to \mathbb{R}^{n_y}$ denotes the measurement available only at the sampling times. The filter to be designed is mode dependent and has the state space realization

$$\dot{\hat{x}}(t) = \hat{A}_{c\theta(t)}(t)\hat{x}(t), \ \hat{x}(0) = 0 \tag{6.46}$$

$$\hat{x}(t_k) = \hat{A}_{d\theta[k]}\hat{x}(t_k^-) + \hat{B}_{d\theta[k]}y[k] \tag{6.47}$$

$$\hat{z}_e(t) = \hat{C}_{c\theta(t)}(t)\hat{x}(t) \tag{6.48}$$

valid in the time interval $t \in [t_k, t_{k+1})$, $k \in \mathbb{N}$, and $\hat{x}(\cdot) : \mathbb{R}_+ \to \mathbb{R}^{n_x}$ is the state and $\hat{z}_e(\cdot) : \mathbb{R}_+ \to \mathbb{R}^{n_z}$ is the estimation produced by the filter. The filter depends on two sets of matrices of appropriate dimensions $(\hat{A}_{di}, \hat{B}_{di}), i \in \mathbb{K}$, that define the discrete-time domain of the jumps and two sets of matrix-valued functions $\hat{A}_{ci}(\cdot) : \mathbb{R}_+ \to \mathbb{R}^{n_x \times n_x}$, $\hat{C}_{ci}(\cdot) : \mathbb{R}_+ \to \mathbb{R}^{n_z \times n_x}$ for $i \in \mathbb{K}$ that define the continuous-time evolution of the filter between successive samplings. We want to verify the conditions under which these matrix-valued functions can be set as constant matrices without losing optimality. The connection of the filter to the plant has the hybrid linear system model (6.8)–(6.10) with state variable $\psi(t)' = [x(t)' \ \hat{x}(t)']$ and matrices

$$F_i = \begin{bmatrix} A_i & 0 \\ 0 & \hat{A}_{ci} \end{bmatrix}, \ J_{ci} = \begin{bmatrix} E_i \\ 0 \end{bmatrix}, \ G_i = \begin{bmatrix} C_{zi} & -\hat{C}_{ci} \end{bmatrix}, \tag{6.49}$$

and

$$H_i = \begin{bmatrix} I & 0 \\ \hat{B}_{di} C_{yi} & \hat{A}_{di} \end{bmatrix}, \quad J_{di} = \begin{bmatrix} 0 \\ \hat{B}_{di} E_{yi} \end{bmatrix} \tag{6.50}$$

where, to ease the notation, the time dependence of $\hat{A}_{ci}(\cdot)$ and $\hat{C}_{ci}(\cdot)$, for all $i \in \mathbb{K}$, has been dropped. The variable $z(t) = G_{\theta(t)}\psi(t) = z_e(t) - \hat{z}_e(t)$ indicates that the output of the hybrid linear system is the estimation error produced by the filter. Notice that, based on the measurements available only at the sampling times $\{t_k\}_{k\in\mathbb{N}}$, this filter provides the best estimation of the output $z_e(t)$ for all $t \in [t_k, t_{k+1})$, $k \in \mathbb{N}$, that is, during the whole inter-sampling time interval. The reader is requested to review Eqs. (4.79)–(4.83), as well as the algebraic manipulations given in Remark 4.3. They are valid for sampled-data MJLS whenever the subscript $i \in \mathbb{K}$ is included in all of them. With that complement, they are exhaustively used in this section.

Remark 6.2 The key result is related to the coupling terms

$$\Gamma_i' P_j \Gamma_i = \begin{bmatrix} X_j & X_j + V_j U_i' Y_i^{-1} \\ \bullet & \Psi_{ij} \end{bmatrix}, \quad (i, j) \in \mathbb{K} \times \mathbb{K}$$

where

$$\begin{aligned} \Psi_{ij} &= X_j + V_j U_i' Y_i^{-1} + Y_i^{-1} U_i V_j' + Y_i^{-1} U_i \hat{X}_j U_i' Y_i^{-1} \\ &= \left(X_j - V_j \hat{X}_j^{-1} V_j'\right) + \left(V_j + Y_i^{-1} U_i \hat{X}_j\right) \hat{X}_j^{-1} \left(V_j + Y_i^{-1} U_i \hat{X}_j\right)' \end{aligned}$$

It is interesting to observe that $\Psi_{ii} = Z_i$ and the inequalities

$$\begin{aligned} \Psi_{ij} &\geq X_j - V_j \hat{X}_j^{-1} V_j' \\ &= Y_j^{-1} = Z_j, \ i \neq j \end{aligned}$$

hold for any choice of matrices U_i. However, for the particular choice $U_i = Y_i$, the equality $Y_i V_i + U_i \hat{X}_i = 0$ implies that $V_i + \hat{X}_i = 0$ and consequently the minimal (in matrix terms) value $\Psi_{ij} = Z_j$ is attained. Hence, taking into account that $U_i = Y_i$ yields $V_i = Z_i - X_i$, then we can unexpectedly verify that equality

$$\Gamma_i' P_j \Gamma_i = \begin{bmatrix} X_j & Z_j \\ \bullet & Z_j \end{bmatrix}, \quad (i, j) \in \mathbb{K} \times \mathbb{K}$$

holds. The linearity of this matrix is at the core of the algebraic manipulations needed to characterize the optimal filter. It is interesting to notice that this formula, valid for $i \in \mathbb{K}$, remains valid for all $j \neq i \in \mathbb{K}$. □

The main difficulty when dealing with sampled-data MJLS is the coupling term appearing in the set of DLMIs (6.17) because the i-th DLMI depends on $P_i(t)$, but the coupling term depends on all $P_j,\ j \in \mathbb{K}$. So, any one-to-one change of variables applied to the i-th DLMI must work for all others. The previous remark shows, and it brings to light, that this is indeed possible to be done provided that the particular change of variables given is adopted. Important is the fact that this can be done without any kind of conservatism.

Theorem 6.4 *Let $h > 0$ be given. If the matrix-valued functions $Z_i(t)$, $X_i(t)$, $M_{ci}(t)$, $K_{ci}(t)$ and the matrices M_{di}, K_{di}, and W_{di} satisfy the DLMI*

$$\begin{bmatrix} \begin{pmatrix} \dot{X}_i + A_i'X_i + X_iA_i \\ +\sum_{j\in\mathbb{K}} \lambda_{ij}X_j \end{pmatrix} & X_iA_i + A_i'Z_i + M_{ci} & C_{zi}' \\ \bullet & \begin{pmatrix} \dot{Z}_i + A_i'Z_i + Z_iA_i \\ +\sum_{j\in\mathbb{K}} \lambda_{ij}Z_j \end{pmatrix} & C_{zi}' - K_{ci}' \\ \bullet & \bullet & -I \end{bmatrix} < 0,\ t \in [0, h) \tag{6.51}$$

subject to the LMI boundary conditions

$$\begin{bmatrix} X_{ih} & Z_{ih} & X_{i0} + C_{yi}'K_{di}' & Z_{i0} \\ \bullet & Z_{ih} & X_{i0} + C_{yi}'K_{di}' + M_{di}' & Z_{i0} \\ \bullet & \bullet & X_{i0} & Z_{i0} \\ \bullet & \bullet & \bullet & Z_{i0} \end{bmatrix} > 0 \tag{6.52}$$

for all $i \in \mathbb{K}$, then the solution of the convex programming problem with respect to $X_i > Z_i > 0$, $M_{ci}, K_{ci}, M_{di}, K_{di}, W_{di}$ for all $i \in \mathbb{K}$

$$\varrho_2^2 = \inf\left\{\sum_{i\in\mathbb{K}} \pi_{0i}\left(\mathbf{tr}(E_i'X_{ih}E_i) + \mathbf{tr}(W_{di})\right) : (6.51)–(6.52)\right\} \tag{6.53}$$

provides the optimal filter whose state space realization is given by

$$\hat{A}_{ci} = (Z_i - X_i)^{-1}\Big(M_{ci} - \dot{Z}_i - \sum_{j\in\mathbb{K}} \lambda_{ij}Z_j\Big),\ \hat{C}_{ci} = K_{ci} \tag{6.54}$$

$$\hat{A}_{di} = (Z_{i0} - X_{i0})^{-1}M_{di},\ \hat{B}_{di} = (Z_{10} - X_{i0})^{-1}K_{di} \tag{6.55}$$

Proof It is a direct consequence of Theorem 6.1. First, adopting the partitioning (4.79)–(4.81), due to (4.82), it is clear that $P_i > 0$ if and only if $X_i > Z_i > 0$ for all $i \in \mathbb{K}$. Moreover, the DLMI (6.17) rewritten in the equivalent form

$$\begin{bmatrix} \Gamma_i'\left(\dot{P}_i + F_i'P_i + P_iF_i + \sum_{j\in\mathbb{K}} \lambda_{ij}P_j\right)\Gamma_i & \Gamma_i'G_i' \\ \bullet & -I \end{bmatrix} < 0,\ t \in [0, h) \tag{6.56}$$

together with the calculations provided in (4.83) and Remark 4.3 is expressed as the DMLI (6.51) with $M_{ci} = V_i \hat{A}_{ci} U_i' Z_i + \dot{V}_i U_i' Z_i + \dot{X}_i + \sum_{j \in \mathbb{K}} \lambda_{ij}(X_j + V_j U_i' Z_i)$, $K_{ci} = \hat{C}_{ci} U_i' Z_i$ and, according to Remark 6.2, the second main diagonal block being

$$\dot{Z}_i + A_i' Z_i + Z_i A_i + \sum_{j \in \mathbb{K}} \lambda_{ij} \Psi_{ij} < 0 \tag{6.57}$$

At this point, we observe that the quantity $\sum_{j \in \mathbb{K}} \lambda_{ij} \Psi_{ij}$ admits the minimal element $\sum_{j \in \mathbb{K}} \lambda_{ij} Z_j$, which is attained by the particular choice $U_i = Y_i = Z_i^{-1}$, $i \in \mathbb{K}$. Clearly, this can be done preserving feasibility and optimality. On the other hand, the boundary condition (6.18) rewritten as

$$\begin{bmatrix} \Gamma_{ih}' P_{ih} \Gamma_{ih} & \Gamma_{ih}' H_i' P_{i0} \Gamma_{i0} \\ \bullet & \Gamma_{i0}' P_{i0} \Gamma_{i0} \end{bmatrix} > 0, \ i \in \mathbb{K} \tag{6.58}$$

reproduces, as indicated in Remark 4.3, the LMI (6.52) where the relationships $M_{di} = V_{i0} \hat{A}_{di} U_{ih}' Z_{ih}$ and $K_{di} = V_{i0} \hat{B}_{di}$ have been used. Finally, using again the calculations provided in Remark 4.3, the objective function in (4.12) is expressed in terms of the new matrix-valued functions and matrix variables. Consequently, the filter with optimal $\mathcal{H}_2$ performance is determined from the solution of the convex programming problem (6.53). Let us now extract the optimal filter state space realization from the solution of problem (6.53). This task is accomplished by adopting the only possible choice $U_i = Y_i = Z_i^{-1}$, in which case we have $V_i = Z_i - X_i < 0$, $i \in \mathbb{K}$, leading to the matrix-valued functions $\hat{C}_{ci} = K_{ci}(U_i' Z_i)^{-1} = K_{ci}$ and

$$\begin{aligned} \hat{A}_{ci} &= V_i^{-1} \left(M_{ci} - \dot{V}_i U_i' Z_i - \dot{X}_i - \sum_{j \in \mathbb{K}} \lambda_{ij} Z_j \right) (U_i' Z_i)^{-1} \\ &= (Z_i - X_i)^{-1} \left(M_{ci} - \dot{Z}_i - \sum_{j \in \mathbb{K}} \lambda_{ij} Z_j \right) \end{aligned} \tag{6.59}$$

as well as the matrices $\hat{A}_{di} = V_{i0}^{-1} M_{di} (U_{ih}' Z_{ih})^{-1} = (Z_{i0} - X_{i0})^{-1} M_{di}$ and $\hat{B}_{di} = V_{i0}^{-1} K_{di} = (Z_{i0} - X_{i0})^{-1} K_{di}$, which all together reproduce the filter state space matrices (6.54) and (6.55), respectively. The proof is concluded. □

From the theoretical viewpoint, the last result is somewhat surprising. The differential matrix inequality (6.56) is highly nonlinear, but there exists a precise choice of the free variables, namely $U_i = Y_i$, $i \in \mathbb{K}$, that is the most favorable possible in the sense that it preserves feasibility and optimality. In other words, that choice does not include any kind of conservatism in the calculations, preserving thus the necessary and sufficient character of Theorem 6.1. This result is important, since the filter design problem (6.53) to be solved is convex and its global optimal solution can be calculated by the numerical procedure discussed in Chap. 2. We are now in position to simplify the result of this theorem considerably.

Theorem 6.5 *Let $h > 0$ be given. If the matrix-valued functions $X_i(t)$ and the matrices X_{i0}, X_{ih}, and K_{di} satisfy the coupled DLMIs*

$$\begin{bmatrix} \dot{X}_i + A_i'X_i + X_iA_i + \sum_{i\in\mathbb{K}} \lambda_{ij}X_j & C_{zi}' \\ \bullet & -I \end{bmatrix} < 0,\ t \in [0, h) \tag{6.60}$$

subject to the LMI boundary condition

$$\begin{bmatrix} X_{ih} & X_{i0} + C_{yi}'K_{di}' \\ \bullet & X_{i0} \end{bmatrix} > 0 \tag{6.61}$$

for all $i \in \mathbb{K}$, then the solution of the convex programming problem with respect to X_i, K_{di} for all $i \in \mathbb{K}$

$$\varrho_2^2 = \inf \left\{ \sum_{i\in\mathbb{K}} \pi_{0i} \left(\mathbf{tr}(E_i'X_{ih}E_i) + \mathbf{tr}\left(E_{yi}'K_{di}'X_{i0}^{-1}K_{di}E_{yi}\right)\right) : (6.60)\text{–}(6.61) \right\} \tag{6.62}$$

provides the optimal time-invariant filter whose state space realization is given by

$$\hat{A}_{ci} = A_i,\ \hat{C}_{ci} = C_{zi} \tag{6.63}$$

$$\hat{A}_{di} = I + X_{i0}^{-1}K_{di}C_{yi},\ \hat{B}_{di} = -X_{i0}^{-1}K_{di} \tag{6.64}$$

for each mode $i \in \mathbb{K}$ of the Markov chain.

Proof Eliminating the matrix variables $K_{ci} = C_{zi}$ and $M_{ci} = -A_i'Z_i - X_iA_i$, the DLMI (6.51) is equivalent to (6.60) and $\dot{Z}_i + A_i'Z_i + Z_iA_i + \sum_{j\in\mathbb{K}} \lambda_{ij}Z_j < 0$, for all $i \in \mathbb{K}$. Moreover, eliminating the matrix variables $M_{di} = -X_{i0} - K_{di}C_{yi}$, it is seen that the LMI (6.52) is equivalent to (6.61) provided that $Z_{ih} > Z_{i0} > 0$ are arbitrarily close to the null matrix. The conclusion is that the matrix-valued function $Z_i(t)$ arbitrarily close to zero in the whole time interval $[0, h]$ is feasible. The fact that $0 < X_{i0} - Z_{i0} < X_{i0}$ makes clear that the objective function of problem (6.53) must be replaced by that of problem (6.62) by considering once again that $Z_{i0} > 0$, for all $i \in \mathbb{K}$ is arbitrarily close to zero. Taking this fact into account, from (6.54), we obtain $\hat{C}_{ci}(t) = C_{zi}$ and

$$\hat{A}_{ci}(t) = (Z_i - X_i)^{-1} \left(-X_iA_i - A_i'Z_i - \dot{Z}_i - \sum_{j\in\mathbb{K}} \lambda_{ij}Z_j \right)$$

$$= A_i \tag{6.65}$$

so as the formulas in (6.55) provide $\hat{A}_{di} = I + X_{i0}^{-1} K_{di} C_{yi}$ and $\hat{B}_{di} = -X_{i0}^{-1} K_{di}$, for all $i \in \mathbb{K}$, respectively. Finally, the fact that $X_{ih} > 0$ implies that all feasible solutions to the DLMI (6.60) are such that $X_i(t) > 0$, $\forall t \in [0, h]$, and then the constraint $X_i > Z_i > 0$, appearing in Theorem 6.4, is superfluous whenever $Z_i > 0$ is arbitrarily close to zero. The proof is concluded. □

The optimal filter is time-invariant and it is much simpler to be calculated and implemented when compared to the optimal filter provided by Theorem 6.4. Actually, the number of variables and the number of constraints have been significantly reduced, making the numerical determination of the optimal filter much more efficient. The fact that the optimal filter is time-invariant makes its practical implementation much simpler in terms of the filter matrix storage. As by-product, the optimal filter

$$\dot{\hat{x}}(t) = A_{\theta(t)} \hat{x}(t), \ \hat{x}(0) = 0 \tag{6.66}$$

$$\hat{x}(t_k) = \hat{x}(t_k^-) + \hat{B}_{d\theta[k]} \big(y[k] - C_{y\theta[k]} \hat{x}(t_k^-) \big) \tag{6.67}$$

in the time interval $t \in [t_k, t_{k+1})$, $k \in \mathbb{N}$, exhibits an observer-based structure which is mode dependent, making it necessary the online measurement of the modes $\theta(t) \in \mathbb{K}$ for all $t \in \mathbb{R}_+$. Notice that (6.65) makes it impossible the proposal of a suboptimal mode independent filter, unless $A_i = A$, $i \in \mathbb{K}$, is itself mode independent. The next example illustrates the theoretical results presented so far.

Example 6.2 This example is devoted to designing the optimal time-invariant filter from the results of Theorem 6.5. The sampled-data MJLS (6.43)–(6.45) has two modes ($N = 2$), initial probabilities $\pi_{01} = 0.75$, $\pi_{02} = 0.25$, and $h = 1.5$. The transition rate matrix and the mode dependent matrices are given by $(A_i, E_i) = (A(a_i), E(e_i))$, $i \in \mathbb{K}$, where $(a_1, e_1) = (1, 1)$, $(a_2, e_2) = (0, 2)$, and

$$\Lambda = \begin{bmatrix} -1 & 1 \\ 2 & -2 \end{bmatrix}, \ A(a) = \begin{bmatrix} 0 & 1 & 0 \\ 0 & 0 & 1 \\ a & -5 & -9 \end{bmatrix}, \ E(e) = \begin{bmatrix} 0 \\ 0 \\ e \end{bmatrix}$$

The remaining matrices are

$$C_z' = \begin{bmatrix} 0 \\ 1 \\ 0 \end{bmatrix}, \ C_{y1}' = \begin{bmatrix} 1 \\ 0 \\ 0 \end{bmatrix}, \ C_{y2}' = \begin{bmatrix} 0 \\ 0 \\ 1 \end{bmatrix}, \ E_{y1} = \begin{bmatrix} 5 \end{bmatrix}, \ E_{y2} = \begin{bmatrix} 1 \end{bmatrix}$$

which makes it clear that the choice $C_{z1} = C_{z2} = C_z$ expresses our intention to estimate the second component from the measurement of the first (mode "1") and the third (mode "2") components of the state vector. First of all, we have solved the

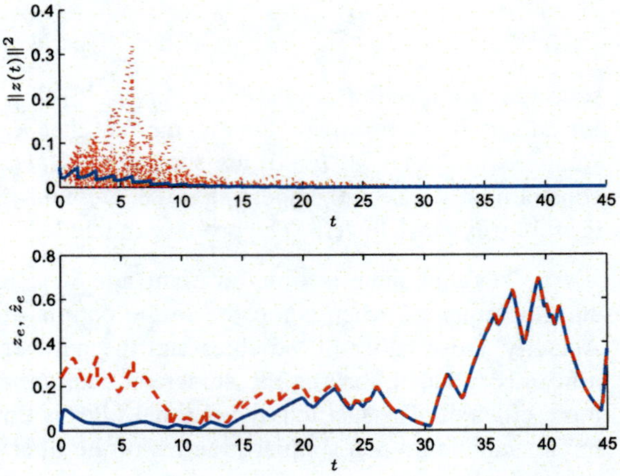

Fig. 6.1 Filter performance and estimation sample

optimal filter design problem (6.62) with $n_\phi = 64$ that provided $\varrho_2^2 = 0.3422$ and the associated optimal filter (6.46)–(6.48) with state space matrices $\hat{A}_{ci}(t) = A_i$, $\hat{C}_{ci}(t) = C_z$, $\hat{A}_{di} = I - \hat{B}_{di}C_{yi}$, for all $i \in \mathbb{K}$ and

$$\begin{bmatrix} \hat{B}_{d1} & \hat{B}_{d2} \end{bmatrix} = \begin{bmatrix} 0.4177 & -0.1869 \\ 0.0489 & -0.0196 \\ 0.0165 & 0.8019 \end{bmatrix}$$

In order to evaluate the performance of the optimal filter, we have implemented a Monte Carlo simulation with 100 simulations, considering simultaneous continuous- and discrete-time impulses applied in the w_c and w_d channels. On the top of Fig. 6.1, it is shown the trajectories (in dotted lines) of the quadratic errors for each simulation and the mean value (in solid line) for all simulations performed. It is clear that $||z(t)||$ goes to zero as $t > 0$ goes to infinity. The convergence of the error toward zero is fast, indicating the good performance of the designed filter. On the bottom of the same figure, we show a trajectory sample of the plant output (solid line) and the filter output (dashed line). The convergence puts in evidence the quality of the optimal filter, supporting thus the previous claim. □

The result of Theorem 6.5 must be viewed with the importance it deserves. The fact that the optimal filter is time-invariant and exhibits an observer-based structure puts the optimal filter for sampled-data MJLS at the same level of importance as that occupied by the celebrated Kalman filter in the context of LTI systems. Indeed, it can be designed from the solution of a convex programming problem whose global solution, whenever exists, can be calculated with no difficulty. Moreover, from Theorem 6.4, it is apparent that parameter uncertainty in convex bounded domains can also be incorporated in the filter design quite naturally, following the same path we have pursued in Chap. 4. We are now in a position to generalize the results obtained so far to deal with optimal control design.

6.3.3 Dynamic Output Feedback Design

The results obtained and the algebraic manipulations adopted so far are generalized to cope with sampled-data MJLS with the main task to design a full order dynamic output feedback controller. The plant and the controller structures are identical to those already designed in precedent chapters but with the only difference that they depend on $\theta(t) \in \mathbb{K}$, the mode of the Markov chain. The sampled-data MJLS of interest is as follows:

$$\dot{x}(t) = A_{\theta(t)}x(t) + B_{\theta(t)}u(t) + E_{\theta(t)}w_c(t), \;\; x(0) = 0 \tag{6.68}$$

$$y[k] = C_{y\theta[k]}x[k] + E_{y\theta[k]}w_d[k] \tag{6.69}$$

$$z(t) = C_{z\theta(t)}x(t) + D_{z\theta(t)}u(t) \tag{6.70}$$

where $x(\cdot) : \mathbb{R}_+ \to \mathbb{R}^{n_x}$ is the state, $y[\cdot] : \mathbb{N} \to \mathbb{R}^{n_y}$ is the measurement vector, and $z(\cdot) : \mathbb{R}_+ \to \mathbb{R}^{n_z}$ is the controlled variable used to build the performance index. The Markov chain has the same stochastic characteristics considered before in this chapter. The full order dynamic output feedback controller is mode dependent, which means that it has access to the mode $\theta(t)$ at each $t \in \mathbb{R}_+$. The state space realization is of the form

$$\dot{\hat{x}}(t) = \hat{A}_{c\theta(t)}(t)\hat{x}(t), \;\; \hat{x}(0) = 0 \tag{6.71}$$

$$\hat{x}(t_k) = \hat{A}_{d\theta[k]}\hat{x}(t_k^-) + \hat{B}_{d\theta[k]}y[k] \tag{6.72}$$

$$v(t) = \hat{C}_{c\theta(t)}(t)\hat{x}(t) \tag{6.73}$$

valid in the time interval $t \in [t_k, t_{k+1})$, $k \in \mathbb{N}$, and $\hat{x}(\cdot) : \mathbb{R}_+ \to \mathbb{R}^{n_x}$ indicates that the plant and the controller share the same dimension. According to the control structure introduced in the beginning of Chap. 3, we set $u(t) = v(t) - E_{u\theta(t)}w_c(t)$. Hence, the feedback connection of the plant and the controller can be written as the hybrid system (6.8)–(6.10) with the state variable $\psi(t)' = [x(t)' \; \hat{x}(t)']$ and matrices that are identical to those of (4.109)–(4.110) but being mode dependent $i \in \mathbb{K}$. The reader should keep in mind that, with no loss of generality, the effect of the external perturbation in the control channel can be incorporated in matrices $E_i \leftarrow E_i + B_iE_{ui}$, $i \in \mathbb{K}$. Finally, the relationships (4.111)–(4.113) and the calculations provided in Remarks 4.5 and 4.6, indexed by the Markov chain mode $i \in \mathbb{K}$, remain valid and are fully used in the sequel.

Theorem 6.6 *Let $h > 0$ be given and set $Z_{i0} = \bar{Z}_i > 0$, where $\bar{Z}_i$ is the stabilizing solution to the coupled algebraic Riccati equations*

$$A_i'\bar{Z}_i + \bar{Z}_iA_i - \big(D_{zi}'C_{zi} + B_i'\bar{Z}_i\big)'(D_{zi}'D_{zi})^{-1}\big(D_{zi}'C_{zi} + B_i'\bar{Z}_i\big) +$$

$$+C'_{zi}C_{zi} + \sum_{j\in\mathbb{K}} \lambda_{ij}\bar{Z}_j = 0 \tag{6.74}$$

for each $i \in \mathbb{K}$*. If the matrix-valued function* $X_i(t)$*, the symmetric matrices* X_{i0}*, and* X_{ih}*,* W_{di}*, and the matrix* K_{di} *satisfy the DLMI*

$$\begin{bmatrix} \dot{X}_i + A'_iX_i + X_iA_i + \sum_{j\in\mathbb{K}} \lambda_{ij}X_j & C'_{zi} \\ \bullet & -I \end{bmatrix} < 0, \; t \in [0, h) \tag{6.75}$$

subject to the LMI boundary condition

$$\begin{bmatrix} X_{ih} & X_{i0} + C'_{yi}K'_{di} & \bar{Z}_i \\ \bullet & X_{i0} & \bar{Z}_i \\ \bullet & \bullet & \bar{Z}_i \end{bmatrix} > 0 \tag{6.76}$$

then the solution of the convex programming problem with respect to X_i*,* K_{di}*,* W_{di} *for all* $i \in \mathbb{K}$

$$\varrho_2^2 = \inf \left\{ \sum_{i\in\mathbb{K}} \pi_{0i} \big(\mathbf{tr}(E'_iX_{ih}E_i) + \mathbf{tr}(W_{di})\big) \, : \, (6.75)\text{–}(6.76) \right\} \tag{6.77}$$

provides the optimal full order dynamic output feedback controller whose state space realization is given by

$$\hat{A}_{ci} = A_i + B_i\hat{C}_{ci} \tag{6.78}$$

$$\hat{C}_{ci} = -(D'_{zi}D_{zi})^{-1}\big(D'_{zi}C_{zi} + B'_i\bar{Z}_i\big) \tag{6.79}$$

$$\hat{A}_{di} = I - \hat{B}_{di}C_{yi} \tag{6.80}$$

$$\hat{B}_{di} = (\bar{Z}_i - X_{i0})^{-1}K_{di} \tag{6.81}$$

for each mode $i \in \mathbb{K}$ *of the Markov chain.*

Proof Once again, the result of Theorem 6.1 is used. In fact, our goal is to show that in the present case it is expressed by the solution of the convex programming problem (6.77). Taking into account the partitioned matrices involved, the fact that the choice $U_i = Y_i$, $\forall i \in \mathbb{K}$, can be done with no loss of generality, the DLMI (6.17), rewritten in the equivalent form (6.56), holds if and only if the coupled DLMIs

$$\dot{X}_i + A'_iX_i + X_iA_i + C'_{zi}C_{zi} + \sum_{j\in\mathbb{K}} \lambda_{ij}X_j < 0 \tag{6.82}$$

for all $i \in \mathbb{K}$ and the coupled differential nonlinear inequalities

$$\dot{Z}_i + (A_i + B_i K_{ci})' Z_i + Z_i (A_i + B_i K_{ci}) + \\ + (C_{zi} + D_{zi} K_{ci})'(C_{ci} + D_{zi} K_{ci}) + \sum_{j \in \mathbb{K}} \lambda_{ij} Z_j < 0 \tag{6.83}$$

where $Z_i = Y_i^{-1} > 0$ and $K_{ci} = \hat{C}_{ci}$ for all $i \in \mathbb{K}$ are satisfied. Simultaneously, the matrix-valued functions $M_{ci} = (Z_i - X_i)\hat{A}_{ci} + \dot{Z}_i$ must be determined from the decoupled equality conditions

$$A_i' Z_i + X_i (A_i + B_i K_{ci}) + M_{ci} + C_{zi}'(C_{zi} + D_{zi} K_{ci}) + \sum_{j \in \mathbb{K}} \lambda_{ij} Z_j = 0 \tag{6.84}$$

for all $i \in \mathbb{K}$, in the time interval $t \in [0, h]$. Now, it is important to observe that for each $i \in \mathbb{K}$, the boundary conditions and the objective function of the problems stated in Theorem 6.1 and Theorem 4.1 are identical. Hence, the same arguments used in the proof of Theorem 4.7 are still valid. Furthermore, similarly to Remark 2.1, from the perspective of the objective function and the boundary conditions, the less constrained inequality follows from the choice of $Z_{ih} > 0$ and $Z_{i0} > 0$ arbitrarily close to the minimal matrix $\bar{Z}_i > 0$, which solves the coupled algebraic Lyapunov equation

$$(A_i + B_i K_{ci})' Z_i + Z_i (A_i + B_i K_{ci}) + (C_{zi} + D_{zi} K_{ci})'(C_{ci} + D_{zi} K_{ci}) + \\ + \sum_{j \in \mathbb{K}} \lambda_{ij} Z_j = 0 \tag{6.85}$$

for each $i \in \mathbb{K}$ which, by its turn, admits a minimal feasible solution. Indeed, the pair $(\bar{K}_i, \bar{Z}_i)$ with $\bar{Z}_i > 0$ being the stabilizing solution of the Riccati equation (6.74) and $\bar{K}_i = -(D_{zi}' D_{zi})^{-1}(D_{zi}' C_{zi} + B_i' \bar{Z}_i)$ the corresponding stabilizing gain is such that $Z_i \geq \bar{Z}_i$ for all feasible pairs (K_{ci}, Z_i). In addition, defining $\Xi_i = X_i - \bar{Z}_i$ and subtracting the Schur Complement of (6.75) to the Riccati equation, (6.74) provides

$$\dot{\Xi}_i + A_i' \Xi_i + \Xi_i A_i + \bar{K}_i' (D_{zi}' D_{zi}) \bar{K}_i + \sum_{j \in \mathbb{K}} \lambda_{ij} \Xi_j < 0 \tag{6.86}$$

valid for all $t \in [0, h)$ and subject to the final boundary condition $\Xi_i(h) = X_{ih} - \bar{Z}_i > 0$ imposed by (6.76). Consequently, the inequality $X_i(t) > \bar{Z}_i$ holds in the whole time interval $[0, h]$. Setting $Z_{i0} > 0$ arbitrarily close to $\bar{Z}_i$, the optimal controller design problem is formulated as (6.77).

Let us now proceed by calculating the controller state space matrices from the one-to-one change of variables adopted in the output feedback control design. First, $K_{ci} = \bar{K}_i$ gives $\hat{C}_{ci} = \bar{K}_i$ as in (6.79). In addition, setting $Z_i(t) = Z_{i0} = Z_{ih} = \bar{Z}_i$ and $K_{ci} = \bar{K}_i$ for all $t \in [0, h]$, using (6.84) and (6.74), we have

$$
\begin{aligned}
M_{ci} &= -A_i' Z_i - X_i(A_i + B_i K_{ci}) - C_{zi}'(C_{zi} + D_{zi} K_{ci}) - \sum_{j \in \mathbb{K}} \lambda_{ij} Z_j \\
&= (Z_i - X_i)(A_i + B_i K_{ci}) + K_{ci}' \big(D_{zi}'(C_{zi} + D_{zi} K_{ci}) + B_i' Z_i \big) \\
&= (Z_i - X_i)(A_i + B_i K_{ci}) \qquad (6.87)
\end{aligned}
$$

and, consequently, taking into account that $\dot{Z}_i = 0$, the inverse transformation formula $\hat{A}_{ci} = (Z_i - X_i)^{-1} M_{ci}$ reproduces (6.78). Finally, the same transformation provides $\hat{B}_{di} = (Z_{i0} - X_{i0})^{-1} K_{di}$ given in (6.81) and

$$
\begin{aligned}
M_{di} &= Z_{i0} - X_{i0} - K_{di} C_{yi} \\
&= (Z_{i0} - X_{i0}) \big(I - (Z_{i0} - X_{i0})^{-1} K_{di} C_{yi} \big) \qquad (6.88)
\end{aligned}
$$

together with $\hat{A}_{di} = (Z_{i0} - X_{i0})^{-1} M_{di}$ yield (6.80), proving thus the claim. □

The aspect of main importance to point out is the time-invariant characteristic of the optimal output feedback controller. It was already known, see Chap. 4, that this property is true for LTI systems but, from the result of Theorem 6.6, it is generalized to sampled-data MJLS. This is a major result in both theoretical and practical viewpoints. Notice that, as far as computational efficiency is concerned, the control design problem (6.77) is convex and with relatively small dimension since the determination of the stabilizing solution $\bar{Z}_i, \forall i \in \mathbb{K}$, can be done *a priori*. These aspects are discussed in the next illustrative example.

Example 6.3 Let us reconsider the sampled-data MJLS already treated in Example 6.1, but with some adaptations to cope with dynamic output feedback control design. The state space realization is given in (6.68)–(6.70) with $h = 1.5$ and a Markov chain with $N = 2$ modes and initial probabilities $\mathcal{P}(\theta_0 = 1) = \pi_{01} = 0.25$, $\mathcal{P}(\theta_0 = 2) = \pi_{02} = 0.75$. The transition rate matrix and the mode dependent matrices are given by $(A_i, B_i) = (A(a_i), B(b_i)), i \in \mathbb{K}$, where $(a_1, b_1) = (0, 1)$, $(a_2, b_2) = (9, 2)$, and

$$
\Lambda = \begin{bmatrix} -2 & 2 \\ 1 & -1 \end{bmatrix}, \; A(a) = \begin{bmatrix} 0 & 1 & 0 \\ 0 & 0 & 1 \\ a & -5 & -9 \end{bmatrix}, \; B(b) = \begin{bmatrix} 0 \\ 0 \\ b \end{bmatrix}
$$

The controlled output matrices that define the performance index are mode independent

$$
C_z = \begin{bmatrix} 1 & 1 & 1 \\ 0 & 0 & 0 \end{bmatrix}, \; D_z = \begin{bmatrix} 0 \\ 1 \end{bmatrix}
$$

as well as the remaining matrices that are given by

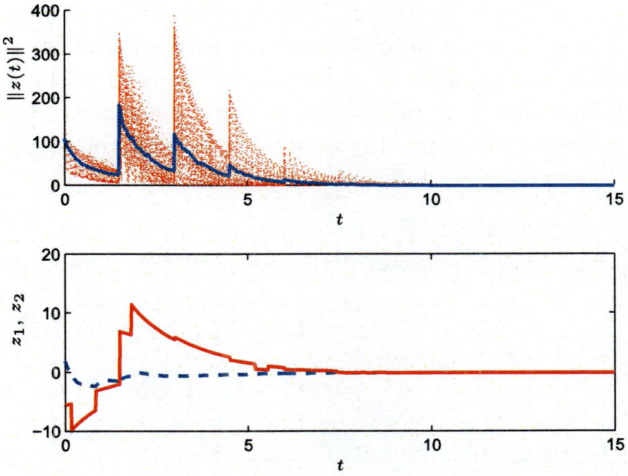

Fig. 6.2 Controller performance and a trajectory sample

$$E = \begin{bmatrix} 0 \\ 0 \\ 1 \end{bmatrix}, \; E_u = \begin{bmatrix} 1 \end{bmatrix}, \; C_y' = \begin{bmatrix} 1 \\ 0 \\ 0 \end{bmatrix}, \; E_y = \begin{bmatrix} 1 \end{bmatrix}$$

Notice that, in this case, we have to redefine matrices $E_i \leftarrow E + B_i E_u$, for $i \in \mathbb{K}$. Furthermore, both matrices $A_i, \forall i \in \mathbb{K}$, are not Hurwitz stable, which makes it clear the necessity of a control action such that the closed-loop system evolves with minimum $\mathcal{H}_2$ performance index.

At a first step, we have solved the control design problem (6.77) with the piecewise linear approximation algorithm and $n_\phi = 32$ that yielded $\varrho_2^2 = 579.3464$ and the corresponding optimal time-invariant controller realization (6.78)–(6.81) with continuous-time gains

$$\begin{bmatrix} \hat{C}_{c1} \\ \hat{C}_{c2} \end{bmatrix} = \begin{bmatrix} -4.1247 & -4.3988 & -0.5180 \\ -8.1392 & -7.0731 & -0.8315 \end{bmatrix}$$

and the discrete-time ones

$$\begin{bmatrix} \hat{B}_{d1}' \\ \hat{B}_{d2}' \end{bmatrix} = \begin{bmatrix} 0.8827 & 0.4343 & -0.0955 \\ 0.8502 & 0.5545 & 0.4483 \end{bmatrix}$$

We have performed 100 runs of a Monte Carlo simulation with impulses applied in the continuous- and discrete-time channels of the closed-loop system, simultaneously. On the top of Fig. 6.2, it is shown, against time, the value of $||z(t)||^2$ in dotted lines and the value of its mean in solid line, for all runs. It is seen that the square mean tends to zero as time increases. On the bottom of the same figure, a sample of the controlled output is shown, $z_1(t)$ in dashed line and $z_2(t)$ in solid line. The designed sample-data controller is very effective and optimal. □

As we have done in Chap. 4, our intention was to design a pure discrete-time full order output feedback controller for sampled-data MJLS. Unfortunately, following the same reasoning, it was not possible to obtain similar design conditions based on the solution of a convex programming problem expressed by DLMIs and LMIs. We want to make this clear by stating the following open problem, needing further attention, within this research subject.

Open Problem 6.1 *In the framework of $\mathcal{H}_2$ performance, design for a sampled-data MJLS a pure discrete-time full order output feedback controller of the form*

$$\hat{x}[k] = \hat{A}_{d\theta[k]}\hat{x}[k-1] + \hat{B}_{d\theta[k]}y[k] \tag{6.89}$$

$$v[k] = \hat{C}_{d\theta[k]}\hat{x}[k-1] + \hat{D}_{d\theta[k]}y[k] \tag{6.90}$$

evolving from the initial condition $\hat{x}[-1] = 0$ and with the feedback connection to the plant (6.68)–(6.70) given by $u[k] = v[k] + E_u w_d[k]$, for all $k \in \mathbb{N}$.

It is worth noticing the main difficulty in facing this problem. First, the hybrid system associated with the closed-loop sampled-data MJLS is readily calculated with no major difficulty and it is seen that only the matrices that define the jump equation depend on the controller matrix variables. As a consequence, between successive sampling times, the hybrid system evolves freely. These observations enable us to say that (apparently) we do not have enough independent matrix variables to cope successfully with the sums $\sum_{j\in\mathbb{K}} \lambda_{ij} P_j, \forall i \in \mathbb{K}$ preserving convexity. For that, the strategy was to set $U_i = Y_i, \forall i \in \mathbb{K}$, with no loss of generality due to the matrix variables and constraints involved, but, unfortunately, it cannot be adopted in the present case for the design of controllers of the class (6.89)–(6.90). Of course, this discussion intends to preserve optimality as well, but, taking into account the importance of the design problem we are dealing with, suboptimal solutions are welcome.

6.4 $\mathcal{H}_\infty$ Performance Analysis and Design

Naturally, we now move our attention to the $\mathcal{H}_\infty$ performance index. The purpose is again to follow the same path adopted in Chap. 5 concerning analysis, filtering, and control design. Of course, the algebraic manipulations pointed out so far, in this chapter, are useful and are on the basis of the generalizations to be presented in the sequel. The sampled-data MJLS to be dealt with is the one whose state space realization is given in (6.4)–(6.6) and satisfies the assumptions discussed in that opportunity. The $\mathcal{H}_\infty$ performance is stated as follows.

Definition 6.2 The $\mathcal{H}_\infty$ performance index associated with the sampled-data Markov jump linear system (6.4)–(6.6) is given by

$$\varrho_\infty^2 = \sup_{0 \neq (w_c, w_d) \in \mathcal{L}_2 \times \ell_2} \frac{||z||_2^2}{||w_c||_2^2 + ||w_d||_2^2} \tag{6.91}$$

where $z(t)$ is the output due to a finite norm continuous-time input $w_c(t)$ and a finite norm discrete-time input $w_d[k]$ applied in each perturbation channel. The mathematical expectation $||z||_2^2 = \mathcal{E}\{||z||^2\}$ is also performed with respect to the initial Markov state $\theta_0 \in \mathbb{K}$ with probability $\mathcal{P}(\theta_0 = i) = \pi_{0i} \geq 0,\ i \in \mathbb{K}$.

It is apparent that this definition about $\mathcal{H}_\infty$ performance index is commonly adopted in the framework of MJLS and it is similar to that associated with LTI systems. The only but remarkable difference is the presence of the mathematical expectation operator, necessary for evaluating the norms of the input (w_c, w_d) and output z stochastic signals, respectively. Furthermore, it is important to stress that the hybrid MJLS (6.8)–(6.10) and the content of Remark 6.1 remain valid in the present context.

Theorem 6.7 *Let $h > 0$ be given. The hybrid Markov jump linear system defined by (6.8)–(6.10) is mean square stable with ϱ_∞^2 performance index if and only if the coupled DLMIs*

$$\begin{bmatrix} \begin{pmatrix} \dot{P}_i(t) + F_i' P_i(t) + P_i(t) F_i + G_i' G_i \\ + \sum_{j \in \mathbb{K}} \lambda_{ij} P_j(t) \end{pmatrix} & P_i(t) J_{ci} \\ \bullet & -\gamma^2 I \end{bmatrix},\ t \in [0, h) \tag{6.92}$$

subject to the LMI boundary conditions

$$P_{ih} > 0,\ \gamma^2 I > J_{di}' P_{i0} J_{di},\ P_{ih} > H_i' \left(P_{i0}^{-1} - \gamma^{-2} J_{di} J_{di}' \right)^{-1} H_i \tag{6.93}$$

for all $i \in \mathbb{K}$ are feasible. In the affirmative case, the equality

$$\varrho_\infty^2 = \inf_{\gamma > 0, P_i(\cdot)} \left\{ \gamma^2 : (6.92)\text{–}(6.93) \right\} \tag{6.94}$$

holds.

Proof For the sufficient part of the proof, assume that the HMJLS (6.8)–(6.10) evolves from $\psi(0) \in \mathbb{R}^{n_\psi}$ arbitrary and the coupled DLMIs (6.92) admit a solution, which means that the quadratic cost function $V_\theta(\psi, t) = \psi' P_\theta(t) \psi$ satisfies the Hamilton–Jacobi–Bellman inequality

$$\sup_{w_c \in \mathcal{L}_2} \left\{ \frac{\partial V_i'}{\partial \psi} (F_i \psi + J_{ci} w_c) + \psi' G_i' G_i \psi - \gamma^2 w_c' w_c \right\} + \frac{\partial V_i}{\partial t} < - \sum_{j \in \mathbb{K}} \lambda_{ij} V_j \tag{6.95}$$

for $t \in [0, h)$, and for each $i \in \mathbb{K}$, which by integration yields

$$V_i(\xi, t_k) > \sup_{w_c \in \mathcal{L}_2} \mathcal{E}^{\nu(t_k)} \left\{ \int_{t_k}^{t_{k+1}} \left(z(t)'z(t) - \gamma^2 w_c(t)'w_c(t) \right) dt + \right.$$

$$\left. + V_{\theta(t_{k+1})}(\psi(t_{k+1}^-), t_{k+1}) \right\} \tag{6.96}$$

Now, based on the fact that $P_{ih} > 0$ implies $P_{i0} > 0, \forall i \in \mathbb{K}$, let us define the positive definite quadratic function $v_\theta(\psi) = \psi' P_\theta(0)\psi$. Due to Remark 6.1 and the time-invariant nature of the hybrid system, the functions $V_i(\psi, t) = \psi' P_i(t - t_k)\psi, i \in \mathbb{K}$, remain feasible in any subsequent time interval, $k \in \mathbb{N}$, provided that the boundary conditions (6.93) are imposed. Hence, by construction, we have $V_i(\psi, t_k) = v_i(\psi)$, for all $i \in \mathbb{K}$. Applying the Schur complement, the boundary conditions (6.93) can be rewritten equivalently as

$$\begin{bmatrix} P_{jh} & 0 \\ 0 & \gamma^2 I \end{bmatrix} > \begin{bmatrix} H_j' \\ J_{dj}' \end{bmatrix} P_{j0} \begin{bmatrix} H_j & J_{dj} \end{bmatrix}, \ \forall j \in \mathbb{K} \tag{6.97}$$

which when multiplied to the left by $[\psi(t_{k+1}^-)' \ w_d[k+1]']$ and to the right by its transpose, together with the jump equation written as $\psi(t_{k+1}) = H_j\psi(t_{k+1}^-) + J_{dj} w_d[k+1]$, with $j = \theta(t_{k+1}) \in \mathbb{K}$, we obtain

$$\begin{aligned} V_j(\psi(t_{k+1}^-), t_{k+1}) &= \psi(t_{k+1}^-)' P_{jh} \psi(t_{k+1}^-) \\ &> \psi(t_{k+1})' P_{j0} \psi(t_{k+1}) - \gamma^2 w_d[k+1]' w_d[k+1] \\ &= v_{\theta(t_{k+1})}(\psi(t_{k+1})) - \gamma^2 w_d[k+1]' w_d[k+1] \end{aligned} \tag{6.98}$$

which is valid for all $w_d \in \ell_2$. Consequently, from the previous calculations and inequalities (6.96) and (6.98), we have

$$v_i(\xi) > \sup_{(w_c, w_d) \in \mathcal{L}_2 \times \ell_2} \mathcal{E}^{\nu(t_k)} \left\{ \int_{t_k}^{t_{k+1}} \left(z(t)'z(t) - \gamma^2 w_c(t)'w_c(t) \right) dt - \right.$$

$$\left. - \gamma^2 w_d[k+1]' w_d[k+1] + v_{\theta(t_{k+1})}(\psi(t_{k+1})) \right\} \tag{6.99}$$

Since this inequality holds, in particular, for any initial condition $\psi(0) \neq 0$ and for $(w_c, w_d) = (0, 0)$, then it implies that there exists $\varepsilon > 0$ sufficiently small such that $\mathcal{E}^{\nu(t_k)}\{v_{\theta(t_{k+1})}(\psi(t_{k+1}))\} \leq (1 - \varepsilon) v_{\theta(t_k)}(\psi(t_k))$ meaning that the sequence $\mathcal{E}\{v_{\theta(t_k)}(\psi(t_k))\}$ converges to zero as $k \in \mathbb{N}$ goes to infinity indicating that the HMJLS under consideration is mean square stable. Moreover, for any initial condition $\psi(0)$ and any $0 \neq (w_c, w_d) \in \mathcal{L}_2 \times \ell_2$, the telescoping sum of (6.99) yields

$$\begin{aligned}\mathcal{E}\{v_{\theta_0}(\psi(0))\} &= \mathcal{E}\left\{\sum_{k\in\mathbb{N}} \left(v_{\theta(t_k)}(\psi(t_k)) - v_{\theta(t_{k+1})}(\psi(t_{k+1}))\right)\right\} \\ &> \mathcal{E}\left\{\sum_{k\in\mathbb{N}} \mathcal{E}^{\nu(t_k)}\left\{\int_{t_k}^{t_{k+1}} \left(z(t)'z(t) - \gamma^2 w_c(t)'w_c(t)\right) dt - \right.\right. \\ &\quad \left.\left. - \gamma^2 w_d[k+1]'w_d[k+1]\right\}\right\} \\ &= \|z\|_2^2 - \gamma^2\left(\|w_c\|_2^2 + \|w_d\|_2^2\right) \end{aligned} \tag{6.100}$$

where it has been used the fact that the HMJLS is mean square stable. Hence, setting $\psi(t_0^-) = 0$ and $w_d[0] = 0$ to impose $\psi(0) = \psi(t_0) = H_{\theta_0}\psi(t_0^-) + J_{d\theta_0} w_d[0] = 0$, it can be readily concluded that the upper bound $\varrho_\infty^2 < \gamma^2$ holds. By Definition 6.2, the smallest upper bound provided by the solution of problem (6.94) equals the $\mathcal{H}_\infty$ performance index ϱ_∞^2 as it is shown in the necessity part of the proof.

Necessity follows from Bellman's Principle of Optimality applied to the hybrid linear system. To this end, let us define the cost-to-go function

$$\begin{aligned} V_c(\xi, t_k) = \sup_{(w_c, w_d)\in\mathcal{L}_2\times\ell_2} & \mathcal{E}^{\nu(t_k)}\left\{\int_{t_k}^{\infty} \left(\psi(t)'G_{\theta(t)}'G_{\theta(t)}\psi(t) - \gamma^2 w_c(t)'w_c(t)\right) dt - \right. \\ & \left. - \gamma^2 \sum_{n=k}^{\infty} w_d[n+1]'w_d[n+1]\right\} \end{aligned} \tag{6.101}$$

where for $t \geq t_k$, the trajectory $\psi(t)$ is provided by the linear differential equation (6.8) and the jump equation (6.10), subject to the initial condition $\psi(t_k) = \xi$. It is positive definite because $(w_c, w_d) = (0, 0) \in \mathcal{L}_2 \times \ell_2$.

Assuming that the hybrid system is globally asymptotically stable and the $\mathcal{H}_\infty$ performance index ϱ_∞^2 is well defined and finite, our goal is to prove that it is equal to the smallest value of γ^2 such that the stationary version of the cost-to-go function $v_\theta(\psi)$ satisfies the inequality (6.99) arbitrarily close to equality and is a quadratic function of the form $v_\theta(\xi) = \xi' S_\theta \xi$ for some $S_i > 0, i \in \mathbb{K}$. To prove this claim, once again, let us consider that this actually occurs at sampling time t_{k+1} and split the whole time interval $t_k \mapsto t_{k+1}$ into two successive events, namely $t_k \mapsto t_{k+1}^-$ and $t_{k+1}^- \mapsto t_{k+1}$. For the second part of the time interval, we first point out that, for each $i \in \mathbb{K}$, the quadratic programming problem

$$\sup_{w_d}\{v_i(H_i\psi + J_{di}w_d) - \gamma^2 w_d' w_d\} = \psi' H_i'\left(S_i^{-1} - \gamma^{-2} J_{di} J_{di}'\right)^{-1} H_i \psi \tag{6.102}$$

admits a finite optimal solution with respect to w_d if and only if the objective is strictly concave, which means that $\gamma^2 I > J_{di}' S_i J_{di}$, and the equality (6.102) holds.

Now consider $P_{ih} > 0$ satisfying

$$P_{ih} > H_i'\Big(S_i^{-1} - \gamma^{-2} J_{di} J_{di}'\Big)^{-1} H_i \tag{6.103}$$

but arbitrarily close to the matrix on the right hand side of (6.103). Moving our attention to the first part of the time interval, particularizing (6.102) for $\psi = \psi(t_{k+1}^-)$, $w_d = w_d[k+1]$, and plugging the result into the right hand side of inequality (6.99), we have

$$\begin{aligned} v_i(\xi) > \sup_{w_c \in \mathcal{L}_2} \mathcal{E}^{\nu(t_k)} \Bigg\{ & \int_{t_k}^{t_{k+1}} \Big(z(t)'z(t) - \gamma^2 w_c(t)' w_c(t) \Big) dt + \\ & + \psi(t_{k+1}^-)' P_{\theta(t_k+1)}(h) \psi(t_{k+1}^-) \Bigg\} \\ = \xi' P_i(0) \xi \end{aligned} \tag{6.104}$$

for all $i \in \mathbb{K}$, where the equality follows from the well known calculations of pure MJLS $\mathcal{H}_\infty$ Theory. This states that the supremum indicated in (6.104) is given by the quadratic function $V_{\theta(t)}(\psi(t), t) = \psi(t)' P_{\theta(t)}(t) \psi(t)$ evaluated at $t = 0$, with $P_i(t), i \in \mathbb{K}$, being the positive definite solution to the coupled differential Riccati equations, which follow from the Schur Complement of (6.92), and taking the result arbitrarily close to equality, subject to the final boundary conditions $P_i(h) = P_{ih}, \forall i \in \mathbb{K}$. As a consequence of (6.104), the smallest cost is quadratic and is reached by setting $S_i = P_i(0) = P_{i0}, \forall i \in \mathbb{K}$, which together with (6.103) implies that the boundary conditions (6.93) hold. Finally, since $\psi(0) = 0$ requires $w_d[0] = 0$ and the worst signals w_c and w_d depend linearly on $\psi(\cdot)$, the mean square stability assures that $(w_c, w_d) \in \mathcal{L}_2 \times \ell_2$, concluding thus the proof. □

It is clear that the result of Theorem 6.7 is valid for any HMJLS of the form (6.8)–(6.10) which, in particular, may represent a sampled-data MJLS with state space realization (6.4)–(6.6). For that, it suffices to set the involved matrices according to (6.28). Moreover, the DLMI (6.92) rewritten in the equivalent form

$$\begin{bmatrix} \begin{pmatrix} \dot{P}_i(t) + F_i' P_i(t) + P_i(t) F_i \\ + \sum_{j \in \mathbb{K}} \lambda_{ij} P_j(t) \end{pmatrix} & P_i(t) J_{ci} & G_i' \\ \bullet & -\gamma^2 I & 0 \\ \bullet & \bullet & -I \end{bmatrix} < 0, \ t \in [0, h) \tag{6.105}$$

for each $i \in \mathbb{K}$ puts in evidence that it is of the same type as that given in (6.29) which, to be handled, needs the inequality (6.30) that induces a useful decomposition property. For this reason, the control design results to be presented in the sequel follow from similar algebraic manipulations already adopted when we have dealt with the $\mathcal{H}_2$ performance index.

6.4.1 State Feedback Design

Consider the sampled-data MJLS with state space realization given in (6.32)–(6.34) governed by a Markov chain with the same stochastic characteristics. The matrices of the HMJLS are the same given in (6.28) and (6.35). The next theorem summarizes the first result concerning state feedback control synthesis.

Theorem 6.8 *Let $h > 0$ be given. Consider that together with the coupled DLMIs*

$$\begin{bmatrix} \begin{pmatrix} -\dot{Q}_i + Q_i F_i' + F_i Q_i \\ +\sum_{j\in\mathbb{K}} \lambda_{ij} S_{ij} \end{pmatrix} & Q_i G_i' & J_{ci} \\ \bullet & -I & 0 \\ \bullet & \bullet & -\mu I \end{bmatrix} < 0, \ t \in [0, h) \tag{6.106}$$

subject to the LMI boundary conditions

$$\begin{bmatrix} Q_{ih} & Q_{ih} \begin{bmatrix} I \\ 0 \end{bmatrix} \\ \bullet & W_i \end{bmatrix} > 0, \quad \begin{bmatrix} W_i & [W_i \ K_i'] & 0 \\ \bullet & Q_{i0} & J_{di} \\ \bullet & \bullet & \mu I \end{bmatrix} > 0 \tag{6.107}$$

the optimal solution of the convex programming problem with respect to $\mu > 0$ and the matrix variables $Q_i(t)$, W_i, K_i for all $i \in \mathbb{K}$ and $S_{ij}(t)$, for all $\forall i \neq j \in \mathbb{K} \times \mathbb{K}$

$$\varrho_\infty^2 = \inf\{\mu : (6.31), (6.106)–(6.107)\} \tag{6.108}$$

provides the matrix gains $L_i = K_i W_i^{-1}$, $i \in \mathbb{K}$. Then, the closed-loop sampled-data control system (6.32)–(6.34) is mean square stable and operates with minimum ϱ_∞^2 performance index.

Proof The proof is almost immediate in the sense that it is based on arguments already used. The coupled DLMIs (6.106) are equivalent to (6.105) whenever $Q_i(t) = P_i(t)^{-1} > 0$, which is due to $Q_{i0} > 0$, for all $i \in \mathbb{K}$. In addition, the constraints (6.107) are equivalent to (6.93). Indeed, it has been proven in several opportunities that the LMIs (6.107) can be equivalently written as (6.93), by taking into account the presence of the state feedback gain matrix variables L_i, $\forall i \in \mathbb{K}$. Hence, the convex programming problem (6.108) is a joint statement of problem (6.94). The proof is concluded. □

Comparing the coupled DLMIs and the boundary conditions of Theorem 6.2 with those of Theorem 6.8, it is seen that the last reduces to the former whenever $\mu \to +\infty$. In both cases, the number of DLMIs and the number of LMIs are the same. The dimensions of the DLMIs and LMIs are bigger in the $\mathcal{H}_\infty$ case, but that is not what determines the computational burden involved in the solution of the respective control design problem. In general, the number of matrix variables is very high if one wants to keep the decomposition induced by variables S_{ij} satisfying (6.30) and (6.31). The next theorem provides equivalent state feedback design conditions, but expressed through a smaller number of matrix variables.

Theorem 6.9 *Let $h > 0$ be given. Consider that together with the coupled DLMIs*

$$\begin{bmatrix} \begin{pmatrix} -\dot{Q}_i + Q_i F_i' + F_i Q_i \\ +\sum_{j\in\mathbb{K}} \lambda_{ij} S_{ij} \end{pmatrix} & Q_i G_i' & J_{ci} \\ \bullet & -I & 0 \\ \bullet & \bullet & -\mu I \end{bmatrix} < 0, \; t \in [0, h) \tag{6.109}$$

subject to the LMI boundary conditions

$$\begin{bmatrix} Q_{i0} & J_{di} \\ \bullet & \mu I \end{bmatrix} > 0, \; I_x'(Q_{i0} - Q_{ih})I_x > 0 \tag{6.110}$$

the optimal solution of the convex programming problem with respect to $\mu > 0$ and the matrix variables $Q_i(t)$, for all $i \in \mathbb{K}$ and $S_{ij}(t)$, for all $\forall i \neq j \in \mathbb{K} \times \mathbb{K}$

$$\varrho_\infty^2 = \inf\{\mu : (6.31), (6.109)\text{–}(6.110)\} \tag{6.111}$$

provides the matrix gains $L_i = U_{i0}' Y_{i0}^{-1}$, constructed with the partitions of matrices $Q_{i0} = P_{i0}^{-1}, \forall i \in \mathbb{K}$. Then, the closed-loop sampled-data control system (6.32)–(6.34) is mean square stable and operates with minimum ϱ_∞^2 performance index.

Proof It follows from the fact that the coupled DLMIs (6.109) and (6.105) are identical. Moreover, the boundary conditions (6.110), for all $i \in \mathbb{K}$, are identical to those of Theorem 5.3 enabling us to apply the same arguments in order to obtain the optimal state feedback gains, completing thus the proof. □

The reduction on the computational effort needed to solve problem (6.111) compared to the one of problem (6.108) is, in general, small and depends strongly on the number of Markov modes. However, from the theoretical viewpoint, it is interesting to observe that the same reasoning adopted to face deterministic and stochastic sampled-data state feedback control system design are similar and in many instances identical. Indeed, whenever $N = 1$, which models an MJLS with only one mode, all results of this section reduce to the corresponding ones of the previous chapters.

6.4.2 Filter Design

The sampled-data MJLS of interest is the one with state space realization (6.43)–(6.45), while the filter has the structure given in (6.46)–(6.48). The stochastic characteristics of the Markov chain are the same as already treated in this chapter. As a consequence, the associated hybrid linear system has the form (6.8)–(6.10) with matrices given in (6.49)–(6.50). The algebraic manipulations provided in Chap. 5 are essential to develop the filter design conditions to be presented in the sequel.

Theorem 6.10 *Let $h > 0$ be given. If the matrix-valued functions $Z_i(t)$, $X_i(t)$, $M_{ci}(t)$, $K_{ci}(t)$, the matrices M_{di}, K_{di}, and the scalar $\mu > 0$ satisfy the coupled DLMIs*

$$\begin{bmatrix} \begin{pmatrix} \dot{X}_i + A_i'X_i + X_iA_i \\ +\sum_{j\in\mathbb{K}} \lambda_{ij}X_j \end{pmatrix} & X_iA_i + A_i'Z_i + M_{ci} & X_iE_i & C_{zi}' \\ \bullet & \begin{pmatrix} \dot{Z}_i + A_i'Z_i + Z_iA_i \\ +\sum_{j\in\mathbb{K}} \lambda_{ij}Z_j \end{pmatrix} & Z_iE_i & C_{zi}' - K_{ci}' \\ \bullet & \bullet & -\mu I & 0 \\ \bullet & \bullet & \bullet & -I \end{bmatrix} < 0 \tag{6.112}$$

for all $t \in [0, h)$, subject to the LMI boundary conditions

$$\begin{bmatrix} X_{ih} & Z_{ih} & X_{i0} + C_{yi}'K_{di}' & Z_{i0} & 0 \\ \bullet & Z_{ih} & X_{i0} + C_{yi}'K_{di}' + M_{di}' & Z_{i0} & 0 \\ \bullet & \bullet & X_{i0} & Z_{i0} & K_{di}E_{yi} \\ \bullet & \bullet & \bullet & Z_{i0} & 0 \\ \bullet & \bullet & \bullet & \bullet & \mu I \end{bmatrix} > 0 \tag{6.113}$$

for all $i \in \mathbb{K}$, then the solution of the convex programming problem with respect to $X_i > Z_i > 0$, M_{ci}, K_{ci}, M_{di}, K_{di} for all $i \in \mathbb{K}$ and the scalar $\mu > 0$

$$\varrho_\infty^2 = \inf\{\mu : (6.112)\text{–}(6.113)\} \tag{6.114}$$

provides the optimal filter whose state space realization is given by

$$\hat{A}_{ci} = (Z_i - X_i)^{-1}\left(M_{ci} - \dot{Z}_i - \sum_{j\in\mathbb{K}} \lambda_{ij}Z_j\right), \quad \hat{C}_{ci} = K_{ci} \tag{6.115}$$

$$\hat{A}_{di} = (Z_{i0} - X_{i0})^{-1}M_{di}, \quad \hat{B}_{di} = (Z_{10} - X_{i0})^{-1}K_{di} \tag{6.116}$$

Proof The proof follows from Theorem 6.7. Multiplying, for each $i \in \mathbb{K}$, the DLMI (6.92) to the left by $\text{diag}(\Gamma_i', I)$ and to the right by its transpose, calculating the Schur Complement of the result, and setting $\mu = \gamma^2$, the one-to-one change of variables $M_{ci} = V_i\hat{A}_{ci}U_i'Z_i + \dot{V}_iU_i'Z_i + \dot{X}_i + \sum_{j\in\mathbb{K}} \lambda_{ij}(X_j + V_jU_i'Z_i)$ and $K_{ci} = \hat{C}_{ci}U_i'Z_i$ yields the DLMI (6.112), where the second main diagonal block is the minimal value attained by the particular choice $U_i = Y_i = Z_i^{-1}$, $i \in \mathbb{K}$, which, as discussed before, can be done preserving feasibility and optimality. On the other hand, the Schur Complement of the boundary conditions (6.93) provides, for each $i \in \mathbb{K}$, the LMI

$$\begin{bmatrix} P_{ih} & H_i' P_{i0} & 0 \\ \bullet & P_{i0} & P_{i0} J_{di} \\ \bullet & \bullet & \mu I \end{bmatrix} > 0, \ i \in \mathbb{K} \tag{6.117}$$

which, when multiplied to the left by $\text{diag}(\Gamma_{ih}', \Gamma_{i0}', I)$ and to the right by its transpose, reproduces the LMI (6.113) by applying the one-to-one change of variables $M_{di} = V_{i0}\hat{A}_{di}U_{ih}'Z_{ih}$ and $K_{di} = V_{i0}\hat{B}_{di}$. As a conclusion, problems (6.94) and (6.114) are identical, in which case, it remains to extract the optimal filter state space realization from its optimal solution. This task is accomplished by adopting the only possible choice $U_i = Y_i = Z_i^{-1}$, which imposes $V_i = Z_i - X_i < 0$, $i \in \mathbb{K}$, leading to the matrix-valued functions $\hat{C}_{ci} = K_{ci}(U_i'Z_i)^{-1} = K_{ci}$ and

$$\begin{aligned} \hat{A}_{ci} &= V_i^{-1}\left(M_{ci} - \dot{V}_i U_i' Z_i - \dot{X}_i - \sum_{j \in \mathbb{K}} \lambda_{ij} Z_j \right)(U_i'Z_i)^{-1} \\ &= (Z_i - X_i)^{-1}\left(M_{ci} - \dot{Z}_i - \sum_{j \in \mathbb{K}} \lambda_{ij} Z_j \right) \end{aligned} \tag{6.118}$$

as well as the matrices $\hat{A}_{di} = V_{i0}^{-1}M_{di}(U_{ih}'Z_{ih})^{-1} = (Z_{i0} - X_{i0})^{-1}M_{di}$ and $\hat{B}_{di} = V_{i0}^{-1}K_{di} = (Z_{i0} - X_{i0})^{-1}K_{di}$ that all together reproduce the filter state space matrices (6.115) and (6.116), respectively. The proof is concluded. □

This theorem shows that the design conditions are linear with respect to the plant state space realization matrices. As a consequence, convex bounded uncertainty can be handled naturally as we have done in the previous chapter. However, in the present design framework, the filter state space realization (6.115) depends on the elements of the transition rate matrix, implying that they must be free of uncertainty. Observe that the optimal filter is time-varying but a time-invariant version keeping optimality is presented in the next theorem.

Theorem 6.11 *Let $h > 0$ be given. If the matrix-valued functions $X_i(t)$, the matrices X_{i0}, X_{ih}, K_{di}, and the scalar $\mu > 0$ satisfy the coupled DLMIs*

$$\begin{bmatrix} \dot{X}_i + A_i'X_i + X_iA_i + \sum_{i \in \mathbb{K}} \lambda_{ij} X_j & X_iE_i & C_{zi}' \\ \bullet & -\mu I & 0 \\ \bullet & \bullet & -I \end{bmatrix} < 0, \ t \in [0, h) \tag{6.119}$$

subject to the LMI boundary condition

$$\begin{bmatrix} X_{ih} & X_{i0} + C'_{yi}K'_{di} & 0 \\ \bullet & X_{i0} & K_{di}E_{yi} \\ \bullet & \bullet & \mu I \end{bmatrix} > 0 \tag{6.120}$$

for all $i \in \mathbb{K}$, then the solution of the convex programming problem with respect to X_i, K_{di}, for all $i \in \mathbb{K}$, and the scalar $\mu > 0$

$$\varrho_\infty^2 = \inf\{\mu : (6.119)\text{–}(6.120)\} \tag{6.121}$$

provides the optimal time-invariant filter whose state space realization is given by

$$\hat{A}_{ci} = A_i, \ \hat{C}_{ci} = C_{zi} \tag{6.122}$$

$$\hat{A}_{di} = I + X_{i0}^{-1}K_{di}C_{yi}, \ \hat{B}_{di} = -X_{i0}^{-1}K_{di} \tag{6.123}$$

for each mode $i \in \mathbb{K}$ of the Markov chain.

Proof For each $i \in \mathbb{K}$, eliminating the matrix variables $K_{ci} = C_{zi}$ and $M_{ci} = -A_i'Z_i - \mu^{-1}X_iE_iE_i'Z_i - X_iA_i$, the DLMI (6.112) is equivalent to (6.119) and $\dot{Z}_i + A_i'Z_i + Z_iA_i + \mu^{-1}Z_iE_iE_i'Z_i + \sum_{j\in\mathbb{K}}\lambda_{ij}Z_j < 0$, for all $i \in \mathbb{K}$. Moreover, eliminating the matrix variables $M_{di} = -X_{i0} - K_{di}C_{yi}$, it is seen that the LMI (6.113) is equivalent to (6.120) provided that $Z_{ih} > Z_{i0} > 0$ are arbitrarily close to the null matrix. As a result, the matrix-valued function $Z_i(t)$ arbitrarily close to zero in the whole time interval $[0, h]$ is feasible. Taking this fact into account, (6.115) yields $\hat{C}_{ci}(t) = C_{zi}$ and

$$\begin{aligned} \hat{A}_{ci}(t) &= (Z_i - X_i)^{-1}\left(-X_iA_i - A_i'Z_i - \mu^{-1}X_iE_iE_i'Z_i - \dot{Z}_i - \sum_{j\in\mathbb{K}}\lambda_{ij}Z_j\right) \\ &= A_i \end{aligned} \tag{6.124}$$

Moreover, (6.116) provides $\hat{A}_{di} = I + X_{i0}^{-1}K_{di}C_{yi}$ and $\hat{B}_{di} = -X_{i0}^{-1}K_{di}$, for all $i \in \mathbb{K}$, respectively. Finally, because $X_{ih} > 0$ implies that all feasible solutions to the DLMI (6.119) are such that $X_i(t) > 0$, $\forall t \in [0, h]$, then the constraint $X_i > Z_i > 0$, appearing in Theorem 6.10, is superfluous whenever $Z_i > 0$ is arbitrarily close to zero. The proof is concluded. □

Putting aside the importance of this result whenever one is concerned with practical implementation, from the theoretical viewpoint, it is remarkable since a robust filter design facing norm bounded parameter uncertainty is perfectly approachable. Indeed, the constant and the quadratic terms present in the coupled DLMIs (6.119) can be used to establish an upper bound to the effect of all feasible uncertainty. Of course, by doing this, the robust filter preserves mean square stability (the estimation error converges to zero), but, in general, optimality with respect to a desired criterion must be further imposed. These aspects are deeply analyzed and are illustrated by means of an example, presented in the next section.

6.4.3 Dynamic Output Feedback Design

The plant to be controlled and the dynamic output feedback controller structure are given in (6.68)–(6.70) and (6.71)–(6.73), respectively. As usual, the stochastic characteristics of the Markov chain remain unchanged so as to keep valid the result of Theorem 6.7. The reader is requested to see the discussion about $\mathcal{H}_2$ performance optimization of dynamic output feedback controllers in this chapter because it is useful to follow and understand the material to be presented in the sequel.

Theorem 6.12 *Let $h > 0$ be given and set $Z_{i0}(\mu) = \bar{Z}_i > 0$, where $\bar{Z}_i$ is the stabilizing solution to the coupled algebraic Riccati equations*

$$A_i'\bar{Z}_i + \bar{Z}_i A_i - \big(D_{zi}'C_{zi} + B_i'\bar{Z}_i\big)'(D_{zi}'D_{zi})^{-1}\big(D_{zi}'C_{zi} + B_i'\bar{Z}_i\big) + \\ + \mu^{-1}\bar{Z}_i E_i E_i'\bar{Z}_i + C_{zi}'C_{zi} + \sum_{j\in\mathbb{K}} \lambda_{ij}\bar{Z}_j = 0 \tag{6.125}$$

for all $i \in \mathbb{K}$, for each feasible value of $\mu > 0$. If the matrix-valued function $X_i(t)$, the symmetric matrices X_{i0}, X_{ih}, and the matrix K_{di} satisfy the DLMI

$$\begin{bmatrix} \dot{X}_i + A_i'X_i + X_iA_i + \sum_{i\in\mathbb{K}}\lambda_{ij}X_j & X_iE_i & C_{zi}' \\ \bullet & -\mu I & 0 \\ \bullet & \bullet & -I \end{bmatrix} < 0, \ t \in [0, h) \tag{6.126}$$

subject to the LMI boundary condition

$$\begin{bmatrix} X_{ih} & X_{i0} + C_{yi}'K_{di}' & Z_{i0}(\mu) & 0 \\ \bullet & X_{i0} & Z_{i0}(\mu) & K_{di}E_{yi} \\ \bullet & \bullet & Z_{i0}(\mu) & 0 \\ \bullet & \bullet & \bullet & \mu I \end{bmatrix} > 0 \tag{6.127}$$

then the solution of the non-convex programming problem with respect to the variables X_i, K_{di}, for all $i \in \mathbb{K}$, and the scalar $\mu > 0$

$$\varrho_\infty^2 = \inf\{\mu : (6.126)\text{–}(6.127)\} \tag{6.128}$$

provides the optimal full order dynamic output feedback controller whose state space realization is given by

$$\hat{A}_{ci} = A_i + \mu^{-1}E_iE_i'\bar{Z}_i + B_i\hat{C}_{ci} \tag{6.129}$$

$$\hat{C}_{ci} = -(D_{zi}' D_{zi})^{-1}\big(D_{zi}' C_{zi} + B_i' \bar{Z}_i\big) \tag{6.130}$$

$$\hat{A}_{di} = I - \hat{B}_{di} C_{yi} \tag{6.131}$$

$$\hat{B}_{di} = (\bar{Z}_i - X_{i0})^{-1} K_{di} \tag{6.132}$$

for each mode $i \in \mathbb{K}$ of the Markov chain.

Proof Our purpose is to show that the convex programming problems (6.94) and (6.128) share the same optimal solution. Taking into account the partitioned matrices involved, the fact that the choice $U_i = Y_i, \forall i \in \mathbb{K}$, can be done with no loss of generality, the DLMI (6.92) multiplied to the left by $\text{diag}(\Gamma_i', I)$ and to the right by its transpose, holds if and only if (6.126) holds. Moreover, the coupled differential nonlinear inequalities

$$\begin{aligned} &\dot{Z}_i + (A_i + B_i K_{ci})' Z_i + Z_i (A_i + B_i K_{ci}) + \mu^{-1} Z_i E_i E_i' Z_i + \\ &+ (C_{zi} + D_{zi} K_{ci})'(C_{ci} + D_{zi} K_{ci}) + \sum_{j \in \mathbb{K}} \lambda_{ij} Z_j < 0 \end{aligned} \tag{6.133}$$

where $Z_i = Y_i^{-1} > 0$, $K_{ci} = \hat{C}_{ci}$ for all $i \in \mathbb{K}$, are satisfied and, simultaneously, the matrix-valued functions $M_{ci} = (Z_i - X_i)\hat{A}_{ci} + \dot{Z}_i$ must be determined from the decoupled equality conditions

$$\begin{aligned} &A_i' Z_i + X_i (A_i + B_i K_{ci}) + M_{ci} + \mu^{-1} X_i E_i E_i' Z_i + \\ &+ C_{zi}'(C_{zi} + D_{zi} K_{ci}) + \sum_{j \in \mathbb{K}} \lambda_{ij} Z_j = 0 \end{aligned} \tag{6.134}$$

for all $i \in \mathbb{K}$, in the time interval $t \in [0, h]$. Now, observe that for each $i \in \mathbb{K}$, the boundary conditions stated in Theorem 6.2 and Theorem 5.1 are identical. Hence, the same arguments used in the proof of Theorem 5.6 are still valid. From the perspective of the boundary conditions, the less constrained inequality follows from the choice of $Z_{ih} > 0$ and $Z_{i0} > 0$ arbitrarily close to $\bar{Z}_i > 0$, which similarly to Remark 2.1 must solve the coupled algebraic Riccati equation

$$\begin{aligned} &(A_i + B_i K_{ci})' Z_i + Z_i (A_i + B_i K_{ci}) + \mu^{-1} Z_i E_i E_i' Z_i + \\ &+ (C_{zi} + D_{zi} K_{ci})'(C_{ci} + D_{zi} K_{ci}) + \sum_{j \in \mathbb{K}} \lambda_{ij} Z_j = 0 \end{aligned} \tag{6.135}$$

for each $i \in \mathbb{K}$ which, by its turn, admits a minimal feasible solution. Indeed, the pair $(\bar{K}_i, \bar{Z}_i)$ with $\bar{Z}_i > 0$ being the stabilizing solution of the Riccati equation (6.125) and the corresponding stabilizing gain $\bar{K}_i = -(D_{zi}' D_{zi})^{-1}(D_{zi}' C_{zi} + B_i' \bar{Z}_i)$ is such that $Z_i \geq \bar{Z}_i$ for all feasible pairs (K_{ci}, Z_i). In addition, defining $\Xi_i = X_i - \bar{Z}_i$, subtracting the Schur Complement of (6.126) to the Riccati equation (6.125) yields

$$\dot{\Xi}_i + \bar{A}_i'\Xi_i + \Xi_i\bar{A}_i + \mu^{-1}\Xi_i E_i E_i'\Xi_i + \bar{K}_i'(D_{zi}'D_{zi})\bar{K}_i + \sum_{j\in\mathbb{K}} \lambda_{ij}\Xi_j < 0 \tag{6.136}$$

valid for all $t \in [0, h)$ with $\bar{A}_i = A_i + \mu^{-1}E_iE_i'\bar{Z}_i$, $i \in \mathbb{K}$, and subject to the final boundary condition $\Xi_i(h) = X_{ih} - \bar{Z}_i > 0$ imposed by (6.127). Consequently, the inequality $X_i(t) > \bar{Z}_i$ holds in the whole time interval $[0, h]$. Setting $Z_{i0} > 0$ arbitrarily close to $\bar{Z}_i$, the optimal controller design problem reduces to (6.128).

Let us now proceed by calculating the controller state space matrices from the one-to-one change of variables adopted in the output feedback control design. First, $K_{ci} = \bar{K}_i$ gives $\hat{C}_{ci} = \bar{K}_i$ as in (6.130). In addition, setting $Z_i(t) = Z_{i0} = Z_{ih} = \bar{Z}_i$ and $K_{ci} = \bar{K}_i$ for all $t \in [0, h]$, using (6.134) and (6.125), we have

$$\begin{aligned}
M_{ci} &= -A_i'Z_i - X_i(A_i + B_iK_{ci}) - \mu^{-1}X_iE_iE_i'Z_i - \\
&\quad - C_{zi}'(C_{zi} + D_{zi}K_{ci}) - \sum_{j\in\mathbb{K}} \lambda_{ij}Z_j \\
&= (Z_i - X_i)(A_i + \mu^{-1}E_iE_i'Z_i + B_iK_{ci}) + K_{ci}'\big(D_{zi}'(C_{zi} + D_{zi}K_{ci}) + B_i'Z_i\big) \\
&= (Z_i - X_i)(A_i + \mu^{-1}E_iE_i'Z_i + B_iK_{ci})
\end{aligned} \tag{6.137}$$

and, consequently, taking into account that $\dot{Z}_i = 0$, the inverse transformation formula $\hat{A}_{ci} = (Z_i - X_i)^{-1}M_{ci}$ reproduces (6.129). Finally, the same transformation provides $\hat{B}_{di} = (Z_{i0} - X_{i0})^{-1}K_{di}$ given in (6.132) and

$$\begin{aligned}
M_{di} &= Z_{i0} - X_{i0} - K_{di}C_{yi} \\
&= (Z_{i0} - X_{i0})\big(I - (Z_{i0} - X_{i0})^{-1}K_{di}C_{yi}\big)
\end{aligned} \tag{6.138}$$

together with $\hat{A}_{di} = (Z_{i0} - X_{i0})^{-1}M_{di}$ yields (6.131), proving thus the claim. □

At this stage, we could convert the non-convex problem (6.128) into a convex one as we have done in Corollary 5.3 of Theorem 5.6. However, we want to investigate another possibility that takes advantage of the fact that $\mu > 0$ is scalar. It consists of the determination of the following sequence:

$$\mu_{n+1} = \inf\big\{\mu : \text{(6.126)–(6.127)}, Z_{i0}(\mu_n) = \bar{Z}_i,\ i \in \mathbb{K}\big\} \tag{6.139}$$

which starts with $\mu_0 > 0$ such that the coupled Riccati equation (6.125) admits a positive definite stabilizing solution. At each iteration, we have to solve a convex programming problem. We conjecture that the sequence generated by (6.139) converges (if one exists) to the equilibrium point $\mu_{n+1} = \mu_n = \varrho_\infty^2$. The next example illustrates some aspects of the theory presented so far.

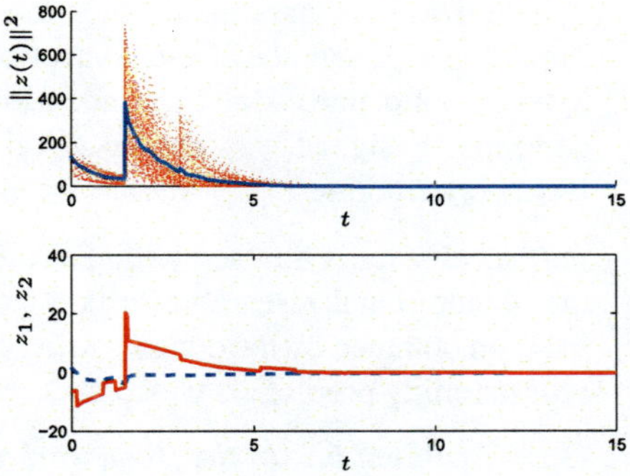

Fig. 6.3 Controller performance and a trajectory sample

Example 6.4 Let us consider the sampled-data MJLS already treated in Example 6.3 in the context of $\mathcal{H}_2$ performance design. The state space realization of the plant is given in (6.68)–(6.70) with $h = 1.5$ and a Markov chain with $N = 2$ modes. For more details, the reader is requested to see Example 6.3.

The problem on the right hand side of (6.139) has been solved by the piecewise linear approximation algorithm with $n_\phi = 32$. We have obtained the convergent sequence

$$\{\mu_n\}_{n=0}^{4} = \{0.1000,\ 1.0508,\ 1.0283,\ 1.0284,\ 1.0284\} \times 10^3$$

indicating that the equilibrium point has been reached within the adopted precision, providing thus $\varrho_\infty^2 = 1.0284e{+}03$ and the corresponding optimal time-invariant controller state space realization (6.129)–(6.132) with continuous-time gains

$$\begin{bmatrix} \hat{C}_{c1} \\ \hat{C}_{c2} \end{bmatrix} = \begin{bmatrix} -4.1351 & -4.4103 & -0.5193 \\ -8.1578 & -7.0897 & -0.8333 \end{bmatrix}$$

and the discrete-time ones

$$\begin{bmatrix} \hat{B}'_{d1} \\ \hat{B}'_{d2} \end{bmatrix} = \begin{bmatrix} 1.0000 & 0.5061 & -0.1059 \\ 1.0000 & 0.6344 & 0.5032 \end{bmatrix}$$

To confirm that the sequence (6.139) converged to the unique equilibrium point, we have calculated another sequence

$$\{\mu_n\}_{n=0}^{3} = \{10.0000,\ 1.0262,\ 1.0284,\ 1.0284\} \times 10^3$$

with different initialization and, as it can be seen, it converged faster to the same optimal solution. We have performed 100 runs of a Monte Carlo simulation with impulses applied to the continuous- and discrete-time channels of the closed-loop system, simultaneously. On the top of Fig. 6.3, it is shown, against time, the value of $||z(t)||^2$ in dotted lines and the value of its mean in solid line, over all runs. It is seen that the square mean tends to zero as time increases. On the bottom of the same figure, a sample of the controlled output is shown, $z_1(t)$ in dashed line and $z_2(t)$ in solid line. Albeit the open-loop system is unstable, the output feedback controller designed was able to stabilize the plant and to improve its robustness facing uncertainties. □

This chapter is finished with the formulation of the following open problem that merits additional research efforts toward its solution. Even a partial (suboptimal) solution obtained by introducing some conservatism but calculated through a convex programming problem is welcome.

Open Problem 6.2 *In the framework of $\mathcal{H}_\infty$ performance, design, for a sampled-data MJLS, a pure discrete-time full order output feedback controller of the form*

$$\hat{x}[k] = \hat{A}_{d\theta[k]}\hat{x}[k-1] + \hat{B}_{d\theta[k]}y[k] \tag{6.140}$$

$$v[k] = \hat{C}_{d\theta[k]}\hat{x}[k-1] + \hat{D}_{d\theta[k]}y[k] \tag{6.141}$$

evolving from the initial condition $\hat{x}[-1] = 0$ and with the feedback connection to the plant (6.68)–(6.70) given by $u[k] = v[k] + E_u w_d[k]$, for all $k \in \mathbb{N}$.

This open problem formulation is a natural generalization of the previous one to cope with $\mathcal{H}_\infty$ instead of $\mathcal{H}_2$ performance. As we have seen just before, both are strongly related, meaning that the solution of one probably contributes decisively to the solution of the other. The importance of the controller structure (6.140)–(6.141) is its pure discrete-time nature, making its practical implementation possible by using digital facilities, exclusively. It is important to stress that, unfortunately, the one-to-one change of variables that works well with sampled-data LTI systems does not work whenever sampled-data MJLS control design is concerned.

6.5 Bibliography Notes

The classical Markov jump linear system, identified by the acronym MJLS, can be viewed as a set of linear time-invariant systems, evolving in time, following a Markov chain that selects in each instant of time, the one to be connected. Between successive jumps, naturally, it behaves as an LTI system, a fact that is decisive for numerical implementation of simple and precise Monte Carlo simulation. The importance of this class of systems from the theoretical and practical viewpoints is well presented, including historical remarks and numerical examples in book [15].

There are countless papers and books dealing with MJLS analysis and control design, as it can be viewed in the References section of [15]. In general, the problems are tackled by applying classical tools based on Lyapunov and Riccati equations. Another possibility is to put in evidence convexity by expressing analysis or synthesis conditions through LMIs. This is the case of reference [17] where, for the first time, full order dynamic output feedback controllers have been designed from LMI conditions. However, the same cannot be said for sampled-data MJLS analysis, filtering, and control synthesis. Exceptions are [18], a text entirely devoted to sampled-data MJLS (see also the references therein), and the paper [20].

Finally, it is worth mentioning some relevant aspects related to time-simulation of sampled-data MJLS. In this context, book [40] contains valuable information and discussions. The crucial point is how to integrate the differential equations subject to stochastic jumps. This is numerically resolved by noticing that the Markov state spends a time interval (waiting time) in a certain mode, defined by an exponential distribution with known mean, and jumps to another mode according to some known probability, both depending on the stochastic characteristics of the Markov chain. During the time interval between successive jumps, the MJLS evolves as a mere LTI system whose equation can be precisely solved without difficulty.

Chapter 7
Nonlinear Systems Control

7.1 Introduction

The main goal of this chapter is to generalize the previous sampled-data control design conditions to nonlinear dynamic systems. In fact, we focus on the particular but important class of nonlinear systems known as Lur'e systems, which is fully characterized in the literature and in the very beginning of this chapter. The results reported afterward may have a useful impact in sampled-data control systems due to the use of the ideas that stemmed from the celebrated Popov stability criterion. Naturally, only sufficient design conditions, carrying thus some conservatism, are provided. As a by-product of the theoretical results specially developed to cope with sampled-data control systems, it is worth mentioning the ones related to stability and performance optimization analysis of nonlinear hybrid systems. A simple design problem is tackled, in order to put in evidence the mathematical manipulations involved, but we are aware that some identified open problems are left to be faced in the near future.

7.2 Sampled-Data Lur'e Systems

In this chapter, we restrict our attention to the case of sampled-data feedback control design of nonlinear systems with the following state space realization:

$$\dot{x}(t) = Ax(t) + Bu(t) + Eq(t) \tag{7.1}$$

$$p(t) = C_p x(t) \tag{7.2}$$

$$z(t) = C_z x(t) + D_z u(t) \tag{7.3}$$

$$u(t) = L_x x[k] + L_u u[k-1],\ t \in [t_k, t_{k+1}),\ \forall k \in \mathbb{N} \tag{7.4}$$

J. C. Geromel, *Differential Linear Matrix Inequalities*,
https://doi.org/10.1007/978-3-031-29754-0_7

that evolves from the initial conditions $x(0) = x_0 \in \mathbb{R}^{n_x}$ and $u[-1] = 0 \in \mathbb{R}^{n_u}$. The state, control, and output controlled variables are, as usual, denoted by $x(\cdot) : \mathbb{R}_+ \to \mathbb{R}^{n_x}$, $u(\cdot) : \mathbb{R}_+ \to \mathbb{R}^{n_u}$, and $z(\cdot) : \mathbb{R}_+ \to \mathbb{R}^{n_z}$, respectively. The input $q(\cdot) : \mathbb{R}_+ \to \mathbb{R}^N$ and the output $p(\cdot) : \mathbb{R}_+ \to \mathbb{R}^N$ with the same dimensions are used to define the nonlinear part of the model, that is,

$$q_i(t) = -\phi_i(p_i(t)), \ i \in \mathbb{K} \tag{7.5}$$

where $\phi_i(\cdot) : \mathbb{R} \to \mathbb{R}$ is a continuous function whose graph belongs to the sector $[0, \kappa_i]$, with $\kappa_i \geq 0$ given. Clearly, the sector condition requires that $\phi_i(0) = 0$, for all $i \in \mathbb{K}$. All functions $\phi(\cdot) : \mathbb{R}^N \to \mathbb{R}^N$ with components $\phi_i(\cdot), i \in \mathbb{K}$ constitute the set Φ. In general, a specific function $\phi \in \Phi$ is not known, but the set Φ is, through the sector upper bound matrix $K = \text{diag}(\kappa_1, \cdots, \kappa_N) \geq 0$. Notice that system (7.1)–(7.4) has two parts, namely: one constituted by an LTI system coupled to another one characterized by the unknown nonlinear input–output $p \mapsto q$ relationship (7.5). This nonlinear system, known generically as Lur'e system, has been largely studied in the literature. On this matter, the reader is requested to see the Bibliography notes of this chapter. Of course, the main challenge is to design a sampled-data control assuring global asymptotic stability and a good performance with respect to a chosen criterion.

The state feedback control structure (7.4) depends on the matrix gain $L = [L_x \ L_u] \in \mathbb{R}^{n_u \times (n_x + n_u)}$ partitioned into two blocks. The first one implements the feedback of the state variable, while the second one is used to include a memory effect in the control action. This control structure can be designed with no additional difficulty. Clearly, it represents an additional control variable that, whenever adequately determined, may improve the closed-loop sampled-data control system performance. Actually, our main goal is to analyze and solve the optimal control problem

$$\inf_L \sup_{\phi \in \Phi} \left\{ \int_0^\infty z(t)' z(t) dt : (7.1)\text{–}(7.5) \right\} \tag{7.6}$$

whose solution prevents the closed-loop system to be unstable. In general, the optimal solution of this problem cannot be exactly calculated. For this reason, we search for a suboptimal solution with minimum guaranteed cost that results from the adoption of the ideas related to the celebrated Popov stability criterion. The first step is the definition of the associated hybrid Lur'e system as follows:

$$\dot{\psi}(t) = F\psi(t) + J_q q(t) \tag{7.7}$$

$$p(t) = G_p \psi(t) \tag{7.8}$$

$$z(t) = G\psi(t) \tag{7.9}$$

$$\psi(t_k) = H\psi(t_k^-) \tag{7.10}$$

where $t \in [t_k, t_{k+1})$, $k \in \mathbb{N}$, and the state space vector is $\psi(t) = [x(t)' \ u(t)']' \in \mathbb{R}^{n_x+n_u}$. The nonlinear relationship (7.5), valid for each $i \in \mathbb{K}$, imposes $q(t) = -\phi(p(t))$ for some $\phi \in \Phi$. The matrices present in the hybrid system model, of appropriate dimensions, are given by

$$F = \begin{bmatrix} A & B \\ 0 & 0 \end{bmatrix}, \ G = \begin{bmatrix} C_z & D_z \end{bmatrix}, \ H = \begin{bmatrix} I & 0 \\ L_x & L_u \end{bmatrix} \tag{7.11}$$

and

$$J_q = \begin{bmatrix} E \\ 0 \end{bmatrix}, \ G_p = \begin{bmatrix} C_p & 0 \end{bmatrix} \tag{7.12}$$

which readily follow from the Lur'e system under consideration. The next results are valid for a general hybrid Lur'e system of the form (7.7)–(7.10) with matrices not necessarily given as in (7.11) and (7.12), at one important exception: the equality $G_pH = G_p$ cannot be dropped. Finally, it is important to mention that $\phi \in \Phi$ if and only if

$$\phi_i(p_i)\big(\phi_i(p_i) - \kappa_i p_i\big) \le 0, \ i \in \mathbb{K} \tag{7.13}$$

or equivalently, from (7.5), that $q_i(q_i + \kappa_i p_i) \le 0, i \in \mathbb{K}$. As a consequence, for any matrix $\Lambda = \text{diag}(\lambda_i, \cdots, \lambda_N) \ge 0$, it follows that

$$\begin{aligned} q'\Lambda(q + Kp) &= \sum_{i\in\mathbb{K}} \lambda_i q_i\big(q_i + \kappa_i p_i\big) \\ &\le 0 \end{aligned} \tag{7.14}$$

whenever $q = -\phi(p)$ and $\phi \in \Phi$. These algebraic manipulations are useful to establish one of the main results of this chapter synthesized in the next theorem. They are also discussed in more detail in the Bibliography notes.

Theorem 7.1 *Let $h > 0$, $\varepsilon > 0$ be given and assume that $G_pH = G_p$. If there exist a positive function $v_d(\psi)$ and a function $V_c(\psi, t)$ satisfying*

$$\sup_{\phi\in\Phi} \left\{ \frac{\partial V_c'}{\partial \psi}\big(F\psi - J_q\phi(G_p\psi)\big) + \psi'G'G\psi \right\} + \frac{\partial V_c}{\partial t} \le -\varepsilon\psi'\psi, \ t \in [0, h) \tag{7.15}$$

for all $\phi \in \Phi$, subject to the boundary conditions

$$v_d(\psi) \ge V_c(\psi, 0), \ V_c(\psi, h) \ge v_d(H\psi), \ \forall \psi \in \mathbb{R}^{n_x+n_u} \tag{7.16}$$

then the hybrid Lur'e system (7.7)–(7.10) is globally asymptotically stable and evolving from the initial condition $\psi(0^-) = \psi_0$ satisfies

$$\int_0^\infty z(t)'z(t)dt < v_d(H\psi_0) \tag{7.17}$$

for all $\phi \in \Phi$.

Proof Assume that the hybrid Lur'e system (7.7)–(7.10) evolves from the initial condition $\phi(0) = H\psi_0 \neq 0$. The fact that the Hamilton–Jacobi–Bellman inequality (7.15) holds, invoking the time-invariant nature of the system under consideration, we have

$$V_c(\psi(t_k), t_k) \geq \int_{t_k}^{t_{k+1}} \big(z(t)'z(t) + \varepsilon\psi(t)'\psi(t)\big)dt + V_c(\psi(t_{k+1}^-), t_{k+1}) \tag{7.18}$$

for all $k \in \mathbb{N}$ and all $\phi \in \Phi$. Now, the boundary conditions (7.16) yield $v_d(\psi(t_k)) \geq V_c(\psi(t_k), t_k)$ and $V_c(\psi(t_{k+1}^-), t_{k+1}) \geq v_d(H\psi(t_{k+1}^-)) = v_d(\psi(t_{k+1}))$, where the last equality is due to the jump equation (7.10). Hence, these relationships allow us to rewrite (7.18) as

$$v_d(\psi(t_k)) \geq \int_{t_k}^{t_{k+1}} \big(z(t)'z(t) + \varepsilon\psi(t)'\psi(t)\big)dt + v_d(\psi(t_{k+1})) \tag{7.19}$$

valid for all $k \in \mathbb{N}$ and all $\phi \in \Phi$. Taking into account that $\varepsilon > 0$ and $v_d(\cdot)$ is a positive function, then $v_d(\psi(t_{k+1})) < v_d(\psi(t_k)), \forall k \in \mathbb{N}$, which implies that the discrete-time sequence $\{\psi(t_k)\}_{k\in\mathbb{N}}$ converges asymptotically to zero. Hence, the telescoping sum of (7.19) provides the upper bound

$$\begin{aligned}
\int_0^\infty \big(z(t)'z(t) + \varepsilon\psi(t)'\psi(t)\big)dt &\leq \sum_{k\in\mathbb{N}} \Big(v_d(\psi(t_k)) - v_d(\psi(t_{k+1}))\Big) \\
&= v_d(\psi(0)) \\
&= v_d(H\psi_0)
\end{aligned} \tag{7.20}$$

from which two conclusions can be drawn. First, since $\varepsilon > 0$, then (7.20) yields

$$\int_0^\infty \psi(t)'\psi(t)dt \leq \varepsilon^{-1} v_d(H\psi_0) < \infty \tag{7.21}$$

and consequently $\psi(t) \to 0$ as $t \to \infty$, proving global asymptotic stability. Second, for the same reason, (7.20) implies that (7.17) holds as well, concluding thus the proof. □

As far as the performance is concerned, it is clear that we have all interest to set $\varepsilon > 0$ arbitrarily small. On the other hand, Theorem 6.1 generalizes the results reported in Chap. 4 to nonlinear hybrid systems. The cost functions $v_d(\psi)$ and $V_c(\psi, t)$ are not restricted to be quadratic. This point is important because it makes it possible to search for a feasible solution to (7.15) based on the celebrated Lyapunov–Popov function, namely

$$V_c(\psi, t) = \psi' P(t)\psi + 2\sum_{i\in\mathbb{K}} \theta_i \int_0^{G_{pi}\psi} \phi_i(\eta)d\eta \tag{7.22}$$

where $G_{pi} = e_i' G_p$ with $e_i \in \mathbb{R}^N$ is the i-th column of the identity matrix of appropriate dimension, which selects $p_i = e_i' p = e_i' G_p \psi$ for each $i \in \mathbb{K}$. It depends on the matrix-valued function $P(t)$ with domain, the time interval $[0, h]$, and the scalars $\theta_i \in \mathbb{R}, i \in \mathbb{K}$. Simple but tedious calculations provide the partial derivatives

$$\begin{aligned}
\frac{\partial V_c}{\partial \psi} &= 2P(t)\psi + 2\sum_{i\in\mathbb{K}} \theta_i \phi_i(p_i) G_{pi}' \\
&= 2\Big(P(t)\psi - \sum_{i\in\mathbb{K}} G_{pi}' \theta_i q_i\Big) \\
&= 2\Big(P(t)\psi - G_p' \Theta q\Big)
\end{aligned} \tag{7.23}$$

where $\Theta = \mathrm{diag}(\theta_1, \cdots, \theta_N)$ and

$$\frac{\partial V_c}{\partial t} = \psi' \dot{P}(t)\psi \tag{7.24}$$

used to construct a feasible solution to the inequality (7.15) with $\varepsilon > 0$ arbitrarily small. Notice that the exact value of the sup operator in (7.15) is virtually impossible to calculate. What is possible is to impose a condition that forces that inequality to hold. This aspect is treated in the next lemma.

Lemma 7.1 *Consider the function $V_c(\psi, t)$ defined in (7.22) and the function*

$$v_d(\psi) = \psi' S\psi + 2\sum_{i\in\mathbb{K}} \theta_i \int_0^{G_{pi}\psi} \phi_i(\eta)d\eta \tag{7.25}$$

with $S > 0$ and $\theta_i \geq 0, i \in \mathbb{K}$. The following statements hold:

1. *If $\phi \in \Phi$, then $v_d(\cdot)$ is positive definite.*
2. *If $S \geq P_0$, $P_h \geq H' S H$, and $G_p H = G_p$, then the boundary conditions (7.16) are satisfied.*

Proof The first part is immediate, because $v_d(0) = 0$ and $v_d(\psi) \geq \psi' S\psi > 0$ for all $\psi \neq 0$, since $\phi \in \Phi$ makes the integral in (7.25) non-negative. To prove the second part, let us calculate each boundary condition separately by taking $\psi \in \mathbb{R}^{n_x+n_u}$ arbitrary. The first one yields

$$\begin{aligned} v_d(\psi) &\geq \psi' P_0 \psi + 2\sum_{i\in\mathbb{K}} \theta_i \int_0^{G_{pi}\psi} \phi_i(\eta)d\eta \\ &= V_c(\psi, 0) \end{aligned} \tag{7.26}$$

while the second, taking into account the equality $G_pH = G_p$, is such that

$$\begin{aligned} v_d(H\psi) &= \psi' H' S H \psi + 2\sum_{i\in\mathbb{K}} \theta_i \int_0^{G_{pi}H\psi} \phi_i(\eta)d\eta \\ &\leq \psi' P_h \psi + 2\sum_{i\in\mathbb{K}} \theta_i \int_0^{G_{pi}\psi} \phi_i(\eta)d\eta \\ &= V_c(\psi, h) \end{aligned} \tag{7.27}$$

which reproduces the boundary conditions (7.16), concluding thus the proof. □

This lemma brings to light a good surprise as far as the boundary conditions (7.16) are concerned. They are enforced by means of simple LMI conditions, similar to those handled in Chap. 4 albeit the involved functions are not quadratic. The proof of this lemma makes it clear the need of the condition $G_pH = G_p$ to obtain the stated result. Notice that the second term of $v_d(\psi)$, corresponding to the nonlinear part of the system, is identical to that present in $V_c(\psi, t)$. This is a tipping point that makes the result of the next theorem possible.

Lemma 7.2 *Let $h > 0$ be given. Define the multiplier dependent matrices $N_{\lambda\theta} = \Lambda K G_p + \Theta G_p F$ and $M_{\lambda\theta} = \Lambda + \Theta G_p J_q$. If there exist diagonal matrices $\Lambda \geq 0$ and $\Theta \geq 0$ and a matrix-valued function $P(t)$ satisfying the DLMI*

$$\begin{bmatrix} \dot{P} + F'P + PF & PJ_q - N'_{\lambda\theta} & G' \\ \bullet & -M_{\lambda\theta} - M'_{\lambda\theta} & 0 \\ \bullet & \bullet & -I \end{bmatrix} < 0, t \in [0, h) \tag{7.28}$$

then the function $V_c(\psi, t)$ given in (7.22) solves the differential inequality (7.15) with $\varepsilon > 0$ small enough for all $\phi \in \Phi$.

Proof Assuming that the DLMI (7.28) holds, perform the Schur complement with respect to the third main diagonal element and multiply the result from the left by $[\psi' \ q']$ and from the right by its transpose, yielding

$$\psi'\dot{P}\psi + 2\psi'PF\psi + 2\psi'(PJ_q - N'_{\lambda\theta})q + \psi'G'G\psi - 2q'M_{\lambda\theta}q < 0 \tag{7.29}$$

and, consequently, replacing the matrices that depend on the diagonal multipliers Λ and Θ there exists $\varepsilon > 0$ sufficiently small such that

$$\psi'\dot{P}\psi + 2(P\psi - G_p'\Theta q)'(F\psi + J_q q) + \psi'G'G\psi \leq -\varepsilon\psi'\psi + 2q'\Lambda(q + KG_p\psi) \tag{7.30}$$

This inequality, together with (7.23) and (7.24), can be rewritten as

$$\begin{aligned} \frac{\partial V_c}{\partial t} + \frac{\partial V_c}{\partial t}'(F\psi - J_q\phi(G_p\psi)) + \psi'G'G\psi &\leq -\varepsilon\psi'\psi + 2q'\Lambda(q + Kp) \\ &\leq -\varepsilon\psi'\psi, \ \forall\phi \in \Phi \end{aligned} \tag{7.31}$$

where the last inequality is due to (7.14) that holds for all $\phi \in \Phi$. As a consequence, the function $V_c(\psi, t)$ given in (7.22) satisfies (7.15) for all $t \in [0, h)$ and all $\phi \in \Phi$. The proof is complete. □

This result deserves many considerations in order to make its importance clear in the present context. The DLMI (7.28) admits a feasible constant solution $P(t) = P > 0, \forall t \in [0, h)$, whenever the celebrated Popov stability criterion holds. However, this is not possible, because F is not Hurwitz stable and also due to the necessity to satisfy the boundary conditions. In other words, in the context of sampled-data Lur'e systems, the existence of a non-constant solution is essential.

From Lemma 7.1, a simple way to impose the boundary conditions (7.16) is by setting $P_h > 0$ and $P_h > H'P_0H$ and choose $S = P_0$. This most favorable choice is possible because, due to the first main diagonal element, any feasible solution to the DLMI (7.28) satisfies

$$P(t) > e^{F'(h-t)}P_h e^{F(h-t)}, \ t \in [0, h] \tag{7.32}$$

which means that $P_h > 0$ implies $P(t) > 0$ in the whole time interval $t \in [0, h]$ and specially $P_0 > 0$. Interesting enough is the fact that the boundary conditions are exactly those of Theorem 4.1, where the $\mathcal{H}_2$ performance of linear sampled-data system is concerned. We are in position to state and prove one of the main results of this chapter.

Theorem 7.2 *Let $h > 0$ be given and assume that $G_pH = G_p$. If the DLMI (7.28) subject to the LMI boundary conditions*

$$P_h > 0, \ P_h > H'P_0H \tag{7.33}$$

is feasible with respect to the matrix-valued function $P(t)$, and diagonal matrices $\Lambda \geq 0$, $\Theta \geq 0$, then the hybrid Lur'e system (7.7)–(7.10) is globally asymptotically stable and evolving from the initial condition $\psi(0^-) = \psi_0$ satisfies

$$\int_0^\infty z(t)'z(t)dt < \psi_0'\left(P_h + G_p'K^{1/2}\Theta K^{1/2}G_p\right)\psi_0 \tag{7.34}$$

for all $\phi \in \Phi$.

Proof From Lemmas 7.1 and 7.2, the global stability property for all $\phi \in \Phi$ follows immediately. We proceed by calculating the upper bound indicated in (7.34) that, from Theorem 7.1, equals $v_d(H\psi_0)$ with $v_d(\cdot)$ given in (7.25) with $S = P_0 > 0$. We have

$$\begin{aligned} v_d(H\psi_0) &= \psi_0' H' P_0 H \psi_0 + 2\sum_{i\in\mathbb{K}} \theta_i \int_0^{G_{pi}H\psi_0} \phi_i(\eta)d\eta \\ &< \psi_0' P_h \psi_0 + 2\sum_{i\in\mathbb{K}} \theta_i \int_0^{G_{pi}\psi_0} \kappa_i \eta d\eta \\ &= \psi_0' \left(P_h + \sum_{i\in\mathbb{K}} \theta_i \kappa_i G_{pi}' G_{pi} \right) \psi_0 \end{aligned} \tag{7.35}$$

which can be rewritten as the right hand side of (7.34), by noticing that K and Θ are non-negative diagonal matrices, completing the proof. □

From this result, it is worth noticing that the convex programming problem

$$\inf_{P(\cdot),\Lambda\in\mathbb{D},\Theta\in\mathbb{D}} \left\{ \psi_0' \left(P_h + G_p' K^{1/2} \Theta K^{1/2} G_p \right) \psi_0 : (7.28), (7.33) \right\} \tag{7.36}$$

provides (if any) the conditions to ensure asymptotic stability and a guaranteed level of performance of the hybrid Lur'e system, for all $\phi \in \Phi$. The fact that this analysis property that generalizes the celebrated Popov stability criterion to hybrid Lur'e systems can be characterized by means of a convex programming problem is a major result. It is important to point out that the constraints $\Lambda \in \mathbb{D}$ and $\Theta \in \mathbb{D}$, where $\mathbb{D}$ is the set of all $N \times N$ diagonal strictly positive (and not non-negative) matrices, have been included with no loss of generality, only to make it possible to solve (7.36) by any available LMI solver.

We now focus on the control design of the sampled-data system (7.1)–(7.5). To this end, we have to consider the hybrid Lur'e system with the specific matrices given in (7.11) and (7.12). First of all, it is immediate to verify that the condition $G_p H = G_p$ is fulfilled. Hence, we can go further by noticing that the main difficulty we have to overcome is to include the state feedback control gain L in the set of variables and to keep the problem (7.36) jointly convex. The next lemma provides useful information about the structure of the state feedback gain we want to design.

Lemma 7.3 *Consider $P_h > 0$ and $P_0 > 0$ partitioned as indicated in (4.9) and $I_x = [I\ 0]'$. The optimal choice of the gain matrix $L = [L_x\ L_u]$, such that the constraint $P_h > H' P_0 H$ becomes the less restrictive possible, is $L_x^* = -\hat{X}_0^{-1} V_0'$ and $L_u^* = 0$. Moreover, the constraint reduces to $P_h > I_x (I_x' P_0^{-1} I_x)^{-1} I_x'$.*

Proof First, rewrite the matrix $H = I_{xx} + I_u L$, where $I_{xx} = \text{diag}(I, 0)$ and $I_u = [0\ I]'$, see (4.136), and verify that the following factorization:

$$H'P_0H = (I_{xx} + I_uL)'P_0(I_{xx} + I_uL)$$
$$= I'_{xx}\left(P_0 - P_0I_u(I'_uP_0I_u)^{-1}I'_uP_0\right)I_{xx} + (L - L^*)'(I'_uP_0I_u)(L - L^*) \tag{7.37}$$

holds with $L^* = -(I'_uP_0I_u)^{-1}I'_uP_0I_{xx}$. Hence, taking into account the partitioning (4.9) and the factorization $I_{xx} = I_xI'_x$, this inequality enables us to say that

$$H'P_0H \geq I_x(I'_xP_0^{-1}I_x)^{-1}I'_x \tag{7.38}$$

for all L and the equality occurs if and only if $L = L^*$. Finally, the previous formula provides the optimal gain of the form $L^* = [L_x^* \; L_u^*]$ with $L_x^* = -\hat{X}_0^{-1}V'_0$ and $L_u^* = 0$. The proof is concluded. □

It is interesting to see that the degree of liberty promoted by the existence of L_u is useless as far as the sampled-data control design problem under consideration is concerned. On the other hand, it appears that we have no chance to succeed in our task to keep convexity untouchable. The main reason is that in order to linearize the boundary condition (7.38) we need to work with $Q(t) = P(t)^{-1}, \forall t \in [0, h]$ but, doing this, the convexity of the DLMI (7.28) is destroyed. A way to circumvent this difficulty is to impose to the multiplies Λ and Θ the constraint $\Theta = \gamma\Lambda$, where $\gamma \geq 0$ is a given scalar that preserves $\Theta \geq 0$ diagonal whenever $\Pi = \Lambda^{-1} \in \mathbb{D}$.

Theorem 7.3 *Let $h > 0$ and $\gamma \geq 0$ be given. If the DLMI*

$$\begin{bmatrix} -\dot{Q} + QF' + FQ & J_q\Pi - Q(KG_p + \gamma G_pF)' & QG' \\ \bullet & -\Pi(I + \gamma G_pJ_q)' - (I + \gamma G_pJ_q)\Pi & 0 \\ \bullet & \bullet & -I \end{bmatrix} < 0 \tag{7.39}$$

for all $t \in [0, h)$ subject to the LMI boundary conditions

$$Q_0 > 0, \; I'_x(Q_0 - Q_h)I_x > 0 \tag{7.40}$$

is feasible with respect to the matrix-valued function $Q(t)$ and matrix $\Pi \in \mathbb{D}$, then the closed-loop Lur'e system (7.1)–(7.5), governed by the sampled-data control with state feedback gains $L_x = -\hat{X}_0^{-1}V'_0$ and $L_u = 0$, constructed from the partitions of matrix $P_0 = Q_0^{-1}$, is globally asymptotically stable. Moreover, evolving from the initial condition $\psi(0^-) = \psi_0$, it satisfies

$$\int_0^\infty z(t)'z(t)dt < \psi'_0\left(Q_h^{-1} + \gamma G'_pK^{1/2}\Pi^{-1}K^{1/2}G_p\right)\psi_0 \tag{7.41}$$

for all $\phi \in \Phi$.

Proof It follows from the results of Theorem 7.2 together with Lemma 7.3. Actually, since $Q_0 > 0$ implies that $P(t) = Q(t)^{-1} > 0$ for all $t \in [0, h]$ and

$\Pi \in \mathbb{D}$ together with $\gamma \geq 0$ implies that $\Lambda = \Pi^{-1} \in \mathbb{D}$ and $\Theta = \gamma\Pi^{-1} \geq 0$, multiplying the DLMI (7.39) to the left by $\mathrm{diag}(Q^{-1}, \Pi^{-1}, I)$ and to the right by its transpose, we obtain the DLMI (7.28). It was established that, due to (7.38), for the given state feedback gains L_x and L_u, the boundary conditions (7.40) are equivalent to (7.33). Finally, (7.41) is nothing else than (7.34) evaluated at $\Theta = \gamma\Lambda$, proving thus the theorem proposed. □

It is important to assess the scope of this result. In order to keep convexity, we have added the constraint $\Theta = \gamma\Lambda$ with $\gamma \geq 0$ given. In particular, for $\gamma = 0$, the effect is to impose $\Theta = 0$, which means that the functions $V_c(\psi, t)$ and $v_d(\psi)$ become purely quadratic because the portions depending on the nonlinear function $\phi \in \Phi$ are eliminated.

Let us introduce the matrix variable $W > 0$ with appropriate dimensions, satisfying the LMI

$$\begin{bmatrix} W & I & G_p'(\gamma K)^{1/2} \\ \bullet & Q_h & 0 \\ \bullet & \bullet & \Pi \end{bmatrix} > 0 \tag{7.42}$$

which, as usual, is used to linearize the objective function. Hence, based on the result of Theorem 7.3, we may adopt the following strategy for the numerical determination of the sampled-data state feedback control. For a given $\gamma \geq 0$, the convex programming problem

$$\varrho_{pop}^2(\gamma) = \inf_{Q(\cdot),\Pi\in\mathbb{D}} \{\psi_0' W \psi_0 : (7.39)\text{–}(7.40), (7.42)\} \tag{7.43}$$

is solved. The minimum guaranteed cost associated with $\gamma^* \geq 0$ is then calculated by line search minimization of $\varrho_{pop}^2(\gamma)$, and the optimal matrix $P_0 = Q_0^{-1} > 0$, whenever partitioned provides the optimal gain L^*. We are now in position to solve some numerical examples with the main purpose of evaluating the conservativeness of the theoretical results reported so far.

Example 7.1 This is an example of sampled-data control design of a 4th order series integrator ($n = 4$), whose matrices of the state space realization (7.1)–(7.4) are given by

$$A = \begin{bmatrix} 0_{n-1} & I_{n-1} \\ 0 & 0_{n-1}' \end{bmatrix}, \; B = \begin{bmatrix} 0_{n-1} \\ 1 \end{bmatrix}, \; C_z = \begin{bmatrix} I_n \\ 0_n' \end{bmatrix}, \; D_z = \begin{bmatrix} 0_n \\ 1 \end{bmatrix}$$

and $C_p = 1_n'$, $E = B/10$, where I_n, 0_n, and 1_n denote the identity $n \times n$ matrix, the null $n \times 1$ and the one $n \times 1$ vectors, respectively. We consider the set Φ, composed of all continuous scalar-valued functions ($N = 1$) belonging to the sector characterized by $\kappa = 20$. For simulation purposes, we have chosen $\phi(p) = 10p(1 + \cos(7\pi p)) \in \Phi$, initial condition $x_0 = 1_n$ yielding $\psi_0 = [x_0' \; 0]'$, and sampling period $h = 1.0$. With $n_\phi = 32$, we have solved problem (7.43) three times:

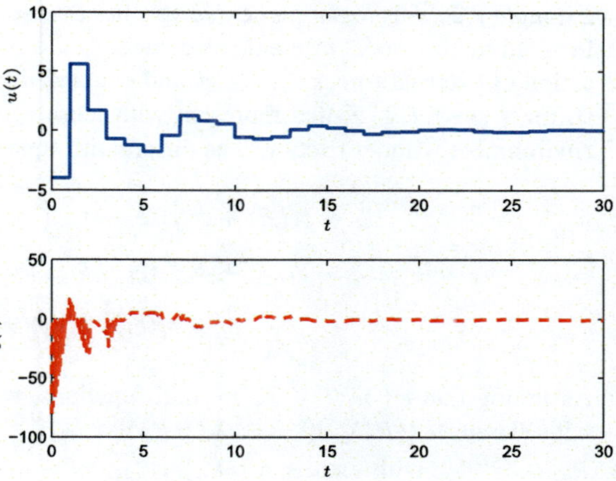

Fig. 7.1 Closed-loop system behavior

(a) For $\gamma = 0$, the problem is unfeasible, meaning that we are unable to determine a stabilizing sampled-data state feedback control. In this context, it is not possible to determine a sampled-data state feedback control associated with a pure quadratic Lyapunov function.

(b) In order to reproduce a result available in the literature, see the Bibliography notes at the end of this chapter, we have considered $\gamma = 3.00$ and the additional constraint $\Pi = 1.00$ that yields $\Theta = 3.00$. The optimal gain found is

$$L_x = \begin{bmatrix} -0.1847 & -0.8633 & -1.3239 & -1.3232 \end{bmatrix}$$

with the associated guaranteed cost $\varrho^2_{pop}(\gamma) = 2.3772e03$.

(c) Considering $\gamma = 3.00$, the optimal multiplied is $\Pi = 1.8452$, and the minimum guaranteed cost reduces to $\varrho^2_{pop}(\gamma) = 1.4770e03$. With a line search procedure, we have calculated a near-optimal guaranteed cost $\varrho^2_{pop}(\gamma^*) = 0.9518e03$ for $\gamma^* \approx 8.0$, $\Pi = 5.1190$, and the associated state feedback gain

$$L_x = \begin{bmatrix} -0.1291 & -0.8475 & -1.4999 & -1.4484 \end{bmatrix}$$

Comparing the guaranteed cost, the importance of calculating a near-optimal solution γ^* is clear. It is expected that this positive effect increases as N increases.

Figure 7.1 shows a time simulation of the closed-loop sampled-data system corresponding to the control design reported in item c). On the top, the sampled-data control signal is shown. On the bottom of the same figure, it appears the signal produced by the nonlinear perturbation. The control is actually effective to cope with this severe nonlinear perturbation. □

Example 7.2 This example explores the case of multiple nonlinearities. It is inspired by the model of the displacement $\xi_i, i \in \mathbb{K}$ of N unitary masses, under the action of external forces $u_i, i \in \mathbb{K}$ and coupled together by springs with nonlinear stiffness $\phi_i(\cdot), i \in \mathbb{K}$ that represent with more accuracy the reality, moving in an environment without friction. The differential equations can be written as

$$\ddot{\xi}_i + \phi_i(\xi_i) + \phi_{i+1}(\xi_i - \xi_{i+1}) = u_i, \; i = 1$$
$$\ddot{\xi}_i + \phi_i(\xi_i - \xi_{i-1}) + \phi_{i+1}(\xi_i - \xi_{i+1}) = u_i, \; 2 \leq i \leq N-1$$
$$\ddot{\xi}_i + \phi_i(\xi_i - \xi_{i-1}) = u_i, \; i = N$$

Assuming that $\phi(\cdot), \forall i \in \mathbb{K}$ are odd functions, then with matrices $R_p \in \mathbb{R}^{N\times N}$ with elements $\{(R_p)_{i,i}\}_{i=1}^N = 1$, $\{(R_p)_{i,(i-1)}\}_{i=2}^N = -1$ and zero otherwise and $R_q \in \mathbb{R}^{N\times N}$ with elements $\{(R_q)_{i,i}\}_{i=1}^N = -1$, $\{(R_q)_{i,(i+1)}\}_{i=1}^{N-1} = 1$ and zero otherwise, the model can be written as $\ddot{\xi} + R_q q = u$, $p = R_p \xi$, $q = -\phi(p)$, where (ξ, p, q) are vectors with components $(\xi_i, p_i, q_i), i \in \mathbb{K}$. Defining the state variable $x = [\xi' \; \dot{\xi}']' \in \mathbb{R}^{2N}$, this dynamic system can be written as (7.1)–(7.3) with the indicated matrices given by

$$A = \begin{bmatrix} 0 & I \\ 0 & 0 \end{bmatrix}, \; B = \begin{bmatrix} 0 \\ I \end{bmatrix}, \; E = \begin{bmatrix} 0 \\ -R_q \end{bmatrix}, \; C_p = \begin{bmatrix} R_p & 0 \end{bmatrix}$$

and the ones that define the cost

$$C_z = \begin{bmatrix} I & 0 \\ 0 & 0 \end{bmatrix}, \; D_z = \begin{bmatrix} 0 \\ 1 \end{bmatrix}$$

which indicates how one wants to bring all masses to their rest positions. We have considered $\kappa_i = 1, i \in \mathbb{K}$, sampling period $h = 0.5$, and a system composed of three ($N = 3$) masses that evolve from the rest at positions $\xi_0 = [0 \; 1 \; 0]'$ at $t = 0$. Problem (7.43) has been solved using piecewise linear approximation with $n_\phi = 32$ and $\gamma = 0$ yielding the state feedback gain $L_x \in \mathbb{R}^{N\times 2N}$ and the multipliers $\Lambda = \Pi^{-1} = \text{diag}(6.2776, 3.5553, 3.2167)$, $\Theta = \gamma\Pi^{-1} = 0$ associated with the minimum guaranteed cost $\varrho_{pop}^2 = 4.5887$.

Figure 7.2 provides two time simulations corresponding to the closed-loop system (solid lines) and open-loop system (dashed lines). For simulation, we have considered identical odd nonlinear functions

$$\phi_i(p_i) = \text{sign}(p_i)|p_i|e^{-|p_i|/3}, i \in \mathbb{K}$$

which clearly satisfies $\phi \in \Phi$. The displacements of the second mass $z_2(t) = \xi_2(t)$ appear on the top, while the control effort made on it, namely $u_2(t)$, appears on the bottom. Since the masses are friction free, in open loop the second mass oscillates

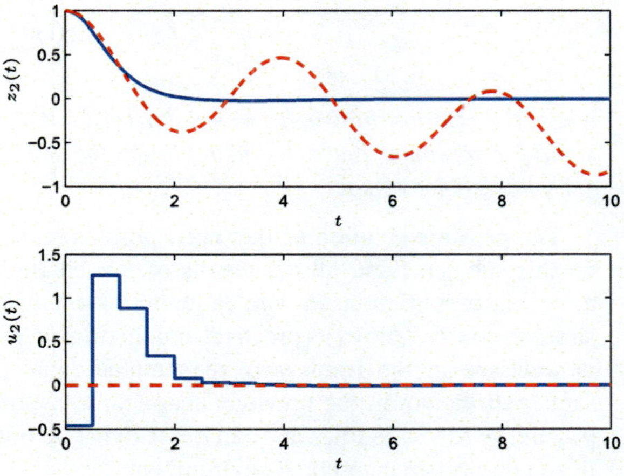

Fig. 7.2 Open- and closed-loop system behavior

according to the trajectory shown on the top of Fig. 7.2. However, the oscillation stops whenever the sampled-data state feedback control is connected to the plant. □

These examples put in evidence the usefulness of the results reported in Theorem 7.3 and summarized in the convex programming problem (7.43). Albeit the state feedback control design conditions are only sufficient, they have been solved for open-loop unstable systems, one of them, with multiple nonlinearities. Observe that in the second example, the open-loop system has a relatively large oscillation amplitude, which is quickly and severely attenuated by the action of the designed sampled-data state feedback control. Numerically speaking, we have treated dynamical systems of 4th and 6th orders with no major difficulty. In this sense, the possibility to adopt n_ϕ relatively small in the piecewise linear approximation procedure is a key and decisive issue.

There are many aspects that merit to be developed such as, for instance, filters and full order dynamic output controllers. However, to accomplish these tasks, a paramount open problem should be precisely and successfully addressed.

Open Problem 7.1 *For the open-loop sampled-data control system*

$$\dot{x}(t) = Ax(t) + Bu(t) + Eq_c(t) \tag{7.44}$$

$$p_c(t) = C_{pc}x(t) \tag{7.45}$$

$$p_d[k] = C_{pd}x[k] \tag{7.46}$$

$$z(t) = C_z x(t) + D_z u(t) \tag{7.47}$$

$$u(t) = q_d[k],\ t \in [t_k, t_{k+1}),\ \forall k \in \mathbb{N} \tag{7.48}$$

determine a convex guaranteed cost ϱ^2_{pop} satisfying

$$\sup_{\phi_c \in \Phi_c, \phi_d \in \Phi_d} \int_0^\infty z(t)' z(t) dt \le \varrho_{pop}^2 \tag{7.49}$$

where the relationships $q_c(t) = -\phi_c(p_c(t)), t \in \mathbb{R}_+$, $q_d[k] = -\phi_d(p_d[k]), k \in \mathbb{N}$ *stand for nonlinear functions belonging to the sets* Φ_c *and* Φ_d *that are appropriate component-wise sectors.*

The possible solution of this open problem can be classified as a tipping point on this subject. First, all the results of this chapter are readily generalized, since nonlinear perturbations in both continuous- and discrete-time domains are allowed, simultaneously. Due to its practical and theoretical importance, this accomplishment is welcome in the framework of sampled-data systems. Second, following the same path drawn in the previous chapters, the solution of this problem opens the possibility to tackle filter and full order dynamic output feedback controller design in an appropriate theoretical environment.

7.3 Bibliography Notes

In the context of analysis and control, Lur'e continuous-time-invariant systems have a paramount importance in control theory that was originated more than half a century ago with the celebrated stability study proposed by V. M. Popov and the remarkable contributions of V. A. Yakubovich and R. E. Kalman, leading to the Kalman–Yakubovich–Popov Lemma, see [35, 45] and [52] as sources of results and information including an expressive number of cited references. The key issue established by the KYP Lemma is the equivalence between the existence of a positive definite matrix and a certain frequency domain inequality, yielding the notion of real positiveness and passivity. The stability criterion that follows from the Lyapunov function candidate that emerges from Popov, Yakubovich and Kalman studies can, alternatively, be expressed through a linear matrix inequality, see [10] and the related references therein.

For this reason, it is natural to generalize these ideas to cope with sampled-data systems, analysis, and control design. Particularly, as far as stability analysis is concerned, the reader is asked to see the interesting and important papers [29] and [30]. Our main purpose in this chapter was to focus and establish sampled-data control synthesis conditions, expressed through DLMIs and LMIs, that can be handled by any LMI solver. This task has been accomplished for the special case of state feedback sampled-data control that minimizes the controlled output energy upper bound. Moreover, compared to the recent published work [19], several improvements on the theory of DLMI applied to sampled-data Lur'e systems have been reported and discussed, see the Example 7.1. It is worth mentioning the possibility to cope with multivariable nonlinearities ($N > 1$) and the complete answer to the question stated in that reference concerning the design of a control law with memory. It has been proven that, in the framework of state feedback control, memory as proposed is useless as far as the minimization of the associated guaranteed cost is concerned.

Finally, it is important to mention the effort to reduce the computational burden needed to solve numerically the control synthesis conditions. To this end, it was possible to reuse successfully the ideas already adopted in Chaps. 4–6 for state feedback control synthesis. To support this claim, the reader is invited to see the illustrative examples presented at the end of this chapter. They are representative in the sense that they are open-loop unstable, and the second one is constituted by $N = 3$ multivariable sector nonlinearities and is of relatively large (6th) order.

Chapter 8
Model Predictive Control

8.1 Introduction

Model Predictive Control (MPC) is widely recognized as an important control strategy in both practical and theoretical frameworks. Indeed, as it is informed in the Bibliography notes, it has been successfully applied to many systems of practical importance. One of the main reasons for that success is the fact that MPC is able to cope with state and control variable constraints. Hence, our main purpose is to generalize the previous state feedback design procedures in order to incorporate the mentioned constraints keeping unchanged the convex nature of the problems we need to solve. Of course, questions involving closed-loop system feasibility, stability, and robustness that naturally arise are tackled and adequately answered. The computational burden involved, for solving the control design conditions for systems of moderate order, is not prohibitive, making their calculation and online implementation possible in many instances.

8.2 Model Predictive Control

Let us first discuss and put in clear evidence the main features of the classical MPC strategy. Consider a discrete-time invariant system whose time evolution is defined by the state space equation

$$x(k+1) = g(x(k), u(k)) \tag{8.1}$$

where $x(\cdot) : \mathbb{N} \to \mathbb{R}^{n_x}$ and $u(\cdot) : \mathbb{N} \to \mathbb{R}^{n_u}$ are the state and control variables, respectively. The MPC strategy applied to (8.1) requires the optimal solution of the nonlinear control design problem

J. C. Geromel, *Differential Linear Matrix Inequalities*,
https://doi.org/10.1007/978-3-031-29754-0_8

$$\min_{(\eta(\cdot),\upsilon(\cdot))\in\mathbb{X}} \sum_{i=0}^{M-1} c(\eta(i),\upsilon(i)) + f(\eta(M)) \tag{8.2}$$
$$\text{s.t. } \eta(i+1) = g(\eta(i),\upsilon(i))$$

which evolves from an arbitrary but feasible initial condition $\eta(0) = \eta_0$. The set $\mathbb{X} \subset \mathbb{R}^{n_x} \times \mathbb{R}^{n_u}$ is given and, in general, imposes upper and lower bounds to the state and control variables and $(c(\cdot), f(\cdot))$ are the functions that define the cost of interest. Clearly, (8.2) is a finite horizon cost that takes into account M steps ahead. Formally, its optimal solution, whenever exists, can be expressed as

$$\upsilon^*(i) = \varphi_i(\eta_0),\ i = 0,\cdots,M-1 \tag{8.3}$$

Hence, thanks to the fact that the system under consideration is time-invariant, the MPC strategy states that the control to be implemented has the form

$$u(k) = \varphi_0(x(k)),\ \forall k \geq 0 \tag{8.4}$$

where it is very important to notice that the function $\varphi_0(\cdot) : \mathbb{R}^{n_x} \to \mathbb{R}^{n_u}$ is never calculated, instead, only the value of $\varphi_0(x(k))$ assessed for the current measured state variable at time $k \in \mathbb{N}$ needs to be determined. The state feedback control (8.4) admits the following interpretations and needs:

- The closed-loop control strategy $\varphi_0(\cdot)$ in (8.4) is virtually impossible to calculate. However, the value of $\varphi_0(x(k))$, evaluated only for a given state measured at time $k \in \mathbb{N}$, is much simpler. It suffices to solve (8.2) at time k subject to the specific initial condition $\eta_0 = x(k)$.
- The numerical effort needed to solve (8.2) depends on the functions c, f, and g and the set $\mathbb{X}$. Convexity requires $g(\cdot)$ to be linear and $\mathbb{X}$, $c(\cdot)$, $f(\cdot)$ jointly convex.
- Closed-loop stability is a serious requirement that is imposed by choosing M large enough and adequate terminal cost $f(\cdot)$. Another possibility is to add the constraint $x(M) = 0$.
- Robustness against parameter uncertainty is an issue that must be well characterized and imposed in the context of sampled-data MPC.

All points listed before are addressed afterward in the context of sampled-data MPC. The idea is to follow the same path and to use the algebraic manipulations already developed to deal with $\mathcal{H}_2$ and $\mathcal{H}_\infty$ optimal sampled-data state feedback control provided in Chaps. 4 and 5, respectively. The successful inclusion of state and control constraints is a tipping point on sampled-data control design.

Remark 8.1 Considering $\eta \in \mathbb{R}^{Mn_x}$ the vector with entries equal to the state variables $\eta(i+1)$ and $\upsilon \in \mathbb{R}^{Mn_u}$ the vector with entries equal to the control variables $\upsilon(i)$ for all $i = 0,\cdots,M$, the problem (8.2) we have to solve can be rewritten in the generic form

$$\min_{(\eta,\upsilon)\in\mathcal{Y}} \{F(\eta, \upsilon, \eta_0) : G(\eta, \upsilon, \eta_0) = 0\}$$

for some vector-valued functions $F(\cdot)$, $G(\cdot)$ and a set $\mathcal{Y}$ derived from c, f, g, and $\mathbb{X}$. Notice that the initial state has not been included in η to make explicit the dependence of η_0. As a result, its optimal solution (if any) provides $\upsilon^* = \varphi_0(\eta_0)$. Hence, if we are able to solve that problem in the elapsed time defined by successive time instants $k \mapsto k+1$, then the state feedback control (8.4) can be implemented online.

There is an aspect that must be brought to the discussion. It is related to the fact that M finite implies that the problem to be solved has finite dimension. Hence, in principle, it can be solved by some nonlinear programming method. This would not be the case if the original problem (8.2) had been formulated in continuous-time which, to keep the dimension finite and moderate, must be approximated by some discretization procedure often based on Euler's method. If problem (8.2) has linear constraints and quadratic objective function, then some efficient quadratic programming algorithm is the most suitable for the determination of $u(k)$ based on the measured state $x(k)$, for all $k \in \mathbb{N}$. □

Remark 8.2 In order to put in correct perspective the stability of the closed-loop system whenever the state feedback control (8.4) is used, let us consider the simplest case when (8.2) is the linear quadratic problem

$$\begin{aligned} \min_{\eta(\cdot),\upsilon(\cdot)} \; & \sum_{i=0}^{M-1} \big(\eta(i)'Q\eta(i) + \upsilon(i)'R\upsilon(i)\big) + \eta(M)'Q_f\eta(M) \\ \text{s.t.} \;\; & \eta(i+1) = A\eta(i) + B\upsilon(i) \end{aligned}$$

where the state and control variables are unconstrained and matrices Q and Q_f and R is positive definite with appropriate dimensions. Its optimal solution exists and provides the linear function $\varphi_i(\eta_0) = K_i\eta_0$, calculated backward by

$$\begin{aligned} P_i &= (A + BK_i)'P_{i+1}(A + BK_i) + K_i'RK_i + Q, \; P_M = Q_f \\ K_i &= -(R + B'P_{i+1}B)^{-1}B'P_{i+1}A \end{aligned}$$

for $i = M-1, \cdots, 0$. With matrices $Q = I$, $R = 1$ and

$$A = \begin{bmatrix} 0 & 1 \\ 10 & 0 \end{bmatrix}, \; B = \begin{bmatrix} 0 \\ 1 \end{bmatrix}$$

which represent an unstable discrete-time open-loop system, we have performed the following calculations to evaluate whether or not the closed-loop system $x(k+1) = Ax(k) + Bu(k)$ and $u(k) = K_0x(k)$ is asymptotically stable:

(a) With $Q_f = 0$ and $1 \leq M \leq 3$, the closed-loop system is unstable. For $M \geq 4$, it is stable and for $M = 9$ the gain K_0 has converged to the stationary solution (infinite horizon).
(b) With $Q_f = 10I$ and $M \geq 1$, the closed-loop system is always stable, and for $M \geq 8$ the gain K_0 has converged to the stationary solution, identical to the previous one.

This illustrates the fact that there are no theoretical reasons for the linear control $u(k) = K_0 x(k)$ to be stabilizing. Indeed, it may not stabilize the closed-loop system, unless the optimization horizon M or the final state weight matrix Q_f is properly defined.

The linearity of the function $\varphi_0(\eta_0)$ strongly depends on the fact that problem (8.2) has linear constraints, quadratic objective function, and $\mathbb{X} = \mathbb{R}^{n_x} \times \mathbb{R}^{n_u}$. However, for linear quadratic problems with constrained state and control variables, that is, $(\eta(\cdot), \upsilon(\cdot))) \in \mathbb{X} \subset \mathbb{R}^{n_x} \times \mathbb{R}^{n_u}$, the function $\varphi_0(\eta_0)$, in general, becomes nonlinear. As a consequence, for a given η_0, there is no difficulty to calculate $\varphi_0(\eta_0)$, but this task may be prohibitive whenever one wants to determine it for all $\eta_0 \in \mathbb{R}^{n_x}$ or even for a subset of its domain. □

In the next section we show how to include convex state and control variable constraints in the sampled-data state feedback control design while optimizing the $\mathcal{H}_2$ performance index. The main ideas introduced in Chap. 4 are once again used here, but the results, as one can expect, are only sufficient for characterizing stability and optimality.

8.3 Nonlinear Sampled-Data Control

Let us consider a sampled-data control system with the following state space realization:

$$\dot{x}(t) = Ax(t) + Bu(t) + Ew_c(t), \; x(0) = 0 \tag{8.5}$$

$$s[k] = C_s x[k] + D_s u[k] \tag{8.6}$$

$$z(t) = C_z x(t) + D_z u(t) \tag{8.7}$$

$$u(t) = \upsilon[k] + E_u w_d[k], \; t \in [t_k, t_{k+1}), \; \forall k \in \mathbb{N} \tag{8.8}$$

where as usual $x(\cdot) : \mathbb{R}_+ \rightarrow \mathbb{R}^{n_x}$, $u(\cdot) : \mathbb{R}_+ \rightarrow \mathbb{R}^{n_u}$, $w_c(\cdot) : \mathbb{R}_+ \rightarrow \mathbb{R}^{r_c}$, and $z(\cdot) : \mathbb{R}_+ \rightarrow \mathbb{R}^{n_z}$ are the state, the control, the exogenous perturbation, and the controlled output of the continuous-time process, respectively. This model has two important particularities that are relevant for the control synthesis results presented afterward, namely:

- The controlled output $s[\cdot] : \mathbb{N} \rightarrow \mathbb{R}^{n_s}$ is used to impose the constraint $s[k] \in \mathbb{S}, \forall k \in \mathbb{N}$, where $\mathbb{S}$ is a convex set with nonempty interior such that $0 \in \mathbb{S}$.

Matrices C_s and D_s, of appropriate dimensions, are freely chosen by the designer in order to build up the state and control constraints of interest.

- The sampled-data control signal $u(t)$ given in (8.8) is corrupted by the discrete-time exogenous perturbation $w_d[\cdot] : \mathbb{N} \to \mathbb{R}^{r_d}$ and

$$v[k] = \varphi(x[k]) \tag{8.9}$$

is the state feedback control, with $\varphi[\cdot]$ possibly nonlinear, to be determined.

It is important to stress that the controlled output $s(t) = C_s x(t) + D_s u(t)$ is constrained to belong to $\mathbb{S}$ at sampling times only. This means that $s[k] = s(t_k) \in \mathbb{S}$ for all $k \in \mathbb{N}$, but it may occur that $s(t) \notin \mathbb{S}$ for some $t \neq t_k \in \mathbb{R}_+$. Moreover, at first glance, the necessity to satisfy this constraint prevents the function $\varphi(\cdot)$ to be linear.

The performance index we want to optimize takes into account plant parameter uncertainty and it is defined as being a guaranteed cost associated with the worst case $\mathcal{H}_2$ performance. It is denoted ϱ_{2rob}^2 and satisfies

$$\sup_{(A,B)\in\mathcal{A}} \sum_{i=1}^{r_c+r_d} ||z_i||_2^2 \leq \varrho_{2rob}^2 \tag{8.10}$$

where, according to Definition 4.1, $z = z_i$, $1 \leq i \leq r_c + r_d$, are the controlled outputs corresponding to continuous and discrete-time impulses injected in each channel of the exogenous perturbations w_c and w_d, respectively. Concerning the set $\mathcal{A}$, it naturally depends on the model of parameter uncertainty we want to tackle, namely *norm bounded* or *convex bounded* uncertainty. We will be concerned with parameter uncertainty in due course.

For the moment, let us mention that our strategy to solve the state feedback control design problem is to split it into two steps. Initially, the cost to be optimized (8.10) is well characterized and determined assuming that $\mathbb{S} \equiv \mathbb{R}^{n_s}$. Afterward, the constraint $s[k] \in \mathbb{S}$, $\forall k \in \mathbb{N}$, is taken into account in the optimization process. Observe that we have to generalize the results provided in Chap. 4 to cope with a nonlinear state feedback control of the form (8.9). To this end, let us write the nonlinear hybrid system that emerges from the sampled-data control system (8.5)–(8.8) governed by the state feedback control (8.9), that is,

$$\dot{\psi}(t) = F\psi(t) + J_c w_c(t) \tag{8.11}$$

$$z(t) = G\psi(t) \tag{8.12}$$

$$\psi(t_k) = H_\varphi(\psi(t_k^-)) + J_d w_d[k] \tag{8.13}$$

where $t \in [t_k, t_{k+1})$ and $k \in \mathbb{N}$ defines the interval of time between successive samplings $\{t_k = kh\}_{k\in\mathbb{N}}$ with $h > 0$ and $\psi = [x' \; u']' \in \mathbb{R}^{n_x+n_u}$ is its state variable. The involved matrices, of appropriate dimensions, are repeated here for convenience

$$F = \begin{bmatrix} A & B \\ 0 & 0 \end{bmatrix}, \; G = \begin{bmatrix} C_z & D_z \end{bmatrix}, \; J_c = \begin{bmatrix} E \\ 0 \end{bmatrix}, \; J_d = \begin{bmatrix} 0 \\ E_u \end{bmatrix} \tag{8.14}$$

and the nonlinear vector-valued function

$$H_\varphi(\psi) = \begin{bmatrix} x \\ \varphi(x) \end{bmatrix} \tag{8.15}$$

where the control action is concentrated. Notice that, as it is already known, the control acts only at the jump instants by means of the nonlinear function (8.15). If the initial condition of the original system is $x(0) = x_0 \in \mathbb{R}^{n_x}$ and it is free of external perturbations, then the initial condition of the hybrid system is $\psi(0) = H_\varphi(\psi(0^-))$, where $\psi(0^-) = \psi_0 = [x_0' \; 0]'$. The next theorem provides a central result of this chapter.

Theorem 8.1 *Let $h > 0$, $\varepsilon > 0$, the initial condition $\psi(0) \neq 0$, and the function $\varphi(x) : \mathbb{R}^{n_x} \to \mathbb{R}^{n_u}$ be given. Besides, assume that w_c and w_d are null perturbations, that is, $w_c = 0$ and $w_d = 0$. If there exist a positive function $v_d(\psi)$ and a function $V_c(\psi, t)$ satisfying*

$$\left\{ \frac{\partial V_c'}{\partial \psi} F\psi + \psi' G' G \psi \right\} + \frac{\partial V_c}{\partial t} \leq -\varepsilon \psi' \psi, \; t \in [0, h) \tag{8.16}$$

subject to the boundary conditions

$$v_d(\psi) \geq V_c(\psi, 0), \; V_c(\psi, h) \geq v_d(H_\varphi(\psi)), \; \forall \psi \in \mathbb{R}^{n_x + n_u} \tag{8.17}$$

then, the hybrid nonlinear system (8.11)–(8.13) is globally asymptotically stable and satisfies

$$\int_0^\infty z(t)' z(t) dt < v_d(\psi(0)) \tag{8.18}$$

Proof Invoking the time-invariant nature of the system under consideration, simple integration of inequality (8.16) in the time interval $[0, h)$ yields

$$V_c(\psi(t_k), t_k) \geq \int_{t_k}^{t_{k+1}} \big(z(t)' z(t) + \varepsilon \psi(t)' \psi(t) \big) dt + V_c(\psi(t_{k+1}^-), t_{k+1}) \tag{8.19}$$

for all $k \in \mathbb{N}$. In addition, with the boundary conditions (8.17), we immediately have $v_d(\psi(t_k)) \geq V_c(\psi(t_k), t_k)$ and $V_c(\psi(t_{k+1}^-), t_{k+1}) \geq v_d(H_\varphi(\psi(t_{k+1}^-))) = v_d(\psi(t_{k+1}))$ where the last equality is due to the jump equation (8.13). Hence, we rewrite (8.19) as

$$v_d(\psi(t_k)) \geq \int_{t_k}^{t_{k+1}} \big(z(t)'z(t) + \varepsilon\psi(t)'\psi(t)\big)dt + v_d(\psi(t_{k+1})) \tag{8.20}$$

valid for all $k \in \mathbb{N}$. Taking into account that $\varepsilon > 0$ and $v_d(\cdot)$ is a positive function, then $v_d(\psi(t_{k+1})) < v_d(\psi(t_k)), \forall k \in \mathbb{N}$, which implies that the discrete-time sequence $\{\psi(t_k)\}_{k\in\mathbb{N}}$ converges asymptotically to zero. Hence, the telescoping sum of (8.20) provides

$$\begin{aligned} \int_0^\infty \big(z(t)'z(t) + \varepsilon\psi(t)'\psi(t)\big)dt &\leq \sum_{k\in\mathbb{N}} \Big(v_d(\psi(t_k)) - v_d(\psi(t_{k+1}))\Big) \\ &= v_d(\psi(0)) \end{aligned} \tag{8.21}$$

from which two conclusions can be drawn. First, since $\varepsilon > 0$, then (8.21) yields

$$\int_0^\infty \psi(t)'\psi(t)dt \leq \varepsilon^{-1}v_d(\psi(0)) < \infty \tag{8.22}$$

and consequently $\psi(t) \to 0$ as $t \to \infty$, proving global asymptotic stability. Second, for the same reason, (8.21) implies that (8.18) holds as well, concluding thus the proof.

This is a genuine generalization of the results reported in Chap. 4 to cope with nonlinear control. However, it is important to put in evidence the sufficient nature of the result in which case we can further enforce the quadratic forms $v_d(\psi) = \psi'S\psi$ with $S > 0$ and $V_c(\psi, t) = \psi'P(t)\psi$, where $P(t)$ is a matrix-valued function, both to be determined. Of course, the nonlinearity of the state feedback control law makes it clear that, in general, these choices introduce some conservativeness whose impact needs verification through numerical examples. Before proceeding, we need to characterize the two classes of parameter uncertainty we are interested in.

Remark 8.3 The parameter uncertainty in matrices $(A, B) \in \mathcal{A}$ reflects only on matrix F of the hybrid system. For norm bounded uncertainty, we can say that $F \in \mathcal{F}_{nb}$, where

$$\mathcal{F}_{nb} = \{F = F_0 + J_0\Delta G_0 : |\Delta| \leq \gamma^{-1}\}$$

is a convex set defined by the uncertainty level (or magnitude) $\gamma > 0$ and matrices

$$J_0 = \begin{bmatrix} E_0 \\ 0 \end{bmatrix}, \; G_0 = \begin{bmatrix} C_0 & D_0 \end{bmatrix}$$

of appropriate dimensions. Since $\Delta = 0$ is feasible, then the nominal matrix F_0 belongs to $\mathcal{F}_{nb}$. On the other hand, for convex bound uncertainty, we have $F \in \mathcal{F}_{cb}$ where

$$\mathcal{F}_{cb} = \left\{ F = \sum_{i \in \mathbb{K}} \lambda_i F_i : \lambda \in \Lambda \right\}$$

with $\Lambda \subset \mathbb{R}^N$ being the unity simplex. Hence, $\mathcal{F}_{cb}$ is a convex set yielded by the convex hull of the extreme matrices $\{F_i\}_{i \in \mathbb{K}}$ and, consequently, a nominal matrix F_0 is not explicitly identified. □

These ideas consolidate the next corollaries of Theorem 8.1. Indeed, the two classes of parameter uncertainty described in the last remark are considered. An amazing aspect is that only the DLMIs are different, while the boundary conditions remain exactly the same. As it is clear in the sequel, this is of particular importance as far as control design is concerned.

Corollary 8.1 *Let $h > 0$ and the set $\mathcal{F}_{nb}$ be given. Consider that the function $\varphi(x) : \mathbb{R}^{n_x} \to \mathbb{R}^{n_u}$ is given and there exist $S = P_0 > 0$, $P_h > 0$, and a matrix-valued function $P(t)$ satisfying the DLMI*

$$\begin{bmatrix} \dot{P}(t) + F_0' P(t) + P(t) F_0 + G'G & P(t) J_0 & G_0' \\ \bullet & -\gamma^2 I & 0 \\ \bullet & \bullet & -I \end{bmatrix} < 0, \; t \in [0, h) \tag{8.23}$$

subject to the boundary condition

$$\psi' P_h \psi \geq H_\varphi(\psi)' P_0 H_\varphi(\psi), \; \forall \psi \in \mathbb{R}^{n_x + n_u} \tag{8.24}$$

and then, the hybrid nonlinear system (8.11)–(8.13) is globally asymptotically stable and satisfies the performance constraint (8.10) with

$$\varrho_{2rob}^2 = \mathbf{tr}(J_c' P_h J_c) + \mathbf{tr}(J_d' P_0 J_d) \tag{8.25}$$

Proof It is a direct application of Theorem 8.1. As we know, due to $P_h > 0$, any feasible solution to the DLMI (8.23) is such that $P(t) > 0, \forall t \in [0, h]$, meaning in particular that $P_0 > 0$. Keeping in mind that the upper bound

$$\begin{aligned} (J_0 \Delta G_0)' P(t) + P(t)(J_0 \Delta G_0) &\leq P(t) J_0 \Delta \Delta' J_0' P(t) + G_0' G_0 \\ &\leq \gamma^{-2} P(t) J_0 J_0' P(t) + G_0' G_0 \end{aligned} \tag{8.26}$$

holds for all $|\Delta| \leq \gamma^{-1}$, performing the Schur Complement with respect to the second and third rows and columns of the DLMI (8.23), the conclusion is

$$\dot{P}(t) + F' P(t) + P(t) F + G'G < 0, \; \forall F \in \mathcal{F}_{nb} \tag{8.27}$$

which, imposing $V_c(\psi, t) = \psi' P(t) \psi$, the differential inequality (8.16) is satisfied for $\varepsilon > 0$ sufficiently small. In addition, choosing $v_d(\psi) = \psi' P_0 \psi$, which is a positive definite function because $P_0 > 0$, then (8.24) implies that the boundary conditions (8.17) hold as well. It remains to evaluate the total cost involved. A continuous-time impulse at $t = 0^-$ in the i-th channel of w_c is equivalent to the initial condition $\psi(0^-) = J_{ci}$, where J_{ci} is the i-th column of matrix J_c that yields $\psi(0) = H_\varphi(J_{ci})$, and thus with (8.18), we obtain

$$\begin{aligned} \int_0^\infty z_i(t)' z_i(t) dt &< v_d(\psi(0)) \\ &= H_\varphi(J_{ci})' P_0 H_\varphi(J_{ci}) \\ &\leq J_{ci}' P_h J_{ci} \end{aligned} \tag{8.28}$$

where the last inequality is due to the boundary condition (8.24). On the other hand, a discrete-time impulse at $k = 0$ in the i-th channel of w_d is equivalent to the initial condition $\psi(0) = J_{di}$, where J_{di} is the i-th column of matrix J_d which together with (8.18) gives rise to

$$\begin{aligned} \int_0^\infty z_i(t)' z_i(t) dt &< v_d(\psi(0)) \\ &= J_{di}' P_0 J_{di} \end{aligned} \tag{8.29}$$

Summing up the effects of all impulses and remembering that inequalities (8.28) and (8.29) are valid for all $F \in \mathcal{F}_{nb}$, it is seen that (8.10) is valid for ϱ_{2rob}^2 given in (8.25), proving thus the claim.

This corollary supports an interesting and useful interpretation. Indeed, the conditions of Corollary 8.1, whenever satisfied, imply that the nominal hybrid nonlinear system (with $F = F_0$) exhibits the $\mathcal{H}_\infty$ performance level bounded above, that is, $\varrho_\infty < \gamma$. This is the key issue that allows the determination of the corresponding guaranteed $\mathcal{H}_2$ performance level ϱ_{2rob}. A similar but slightly different situation occurs whenever convex bounded uncertainty is concerned.

Corollary 8.2 *Let $h > 0$ and the set $\mathcal{F}_{cb}$ be given. Consider that the function $\varphi(x) : \mathbb{R}^{n_x} \to \mathbb{R}^{n_u}$ is given and there exist $S = P_0 > 0$, $P_h > 0$, and a matrix-valued function $P(t)$ satisfying the uncoupled DLMIs*

$$\dot{P}(t) + F_i' P(t) + P(t) F_i + G'G < 0, \; t \in [0, h) \tag{8.30}$$

for $i \in \mathbb{K}$, subject to the boundary condition (8.24), and then, the hybrid nonlinear system (8.11)–(8.13) is globally asymptotically stable and satisfies the performance constraint (8.10) with guaranteed performance ϱ_{2rob}^2 given in (8.25).

Proof Multiplying the DLMI (8.30) by $\lambda_i \geq 0$ for each $i \in \mathbb{K}$, the sum of all terms provides the DLMI

$$\dot{P}(t) + F'P(t) + P(t)F + G'G < 0, \ \forall F \in \mathcal{F}_{cb} \tag{8.31}$$

from which, the rest of the proof follows close to the one of Corollary 8.1, being thus omitted.

These corollaries put in clear evidence a twofold appearance of the same singular issue. That is, the state feedback control rule $\varphi(\cdot)$ appears only on the boundary condition (8.24). This fact has, of course, a major importance to the solution of the following optimal control problem with which one tries to determine the best control action, as far as the guaranteed $\mathcal{H}_2$ cost is concerned. Actually, the most favorable boundary constraint yielded by the choice of the function $\varphi(\cdot)$ is

$$\begin{aligned}\psi' P_h \psi &\geq \min_{\varphi} H_{\varphi}(\psi)' P_0 H_{\varphi}(\psi) \\ &= \min_{\varphi} \begin{bmatrix} x \\ \varphi \end{bmatrix}' P_0 \begin{bmatrix} x \\ \varphi \end{bmatrix}\end{aligned} \tag{8.32}$$

which naturally provides the optimal decision that is readily expressed as

$$\varphi^*(x) = \arg\min_{\varphi} \begin{bmatrix} x \\ \varphi \end{bmatrix}' P_0 \begin{bmatrix} x \\ \varphi \end{bmatrix} \tag{8.33}$$

At this point, we have to mention two remarkable and different situations. The first and obvious one concerns problem (8.32)–(8.33) with no constraint on $\varphi(\cdot)$, which when solved must provide the results reported in Chap. 4 obtained with $\varphi(\cdot)$ linear. The second, more important in the present context, is the possibility to constrain $\varphi(\cdot)$ such that it produces a feasible trajectory $s[k] \in \mathbb{S}, \forall k \in \mathbb{N}$. Both possibilities are now treated separately.

8.3.1 Unconstrained Control

We start by splitting matrix $P_0 > 0$ into four blocks, as already done before, and repeated here for convenience

$$P_0 = \begin{bmatrix} X_0 & V_0 \\ V_0' & \hat{X}_0 \end{bmatrix} \tag{8.34}$$

where the block matrices are of the indicated dimensions, $X_0 \in \mathbb{R}^{n_x \times n_x}$, $V_0 \in \mathbb{R}^{n_x \times n_u}$, and $\hat{X}_0 \in \mathbb{R}^{n_u \times n_u}$. The problem to be solved, namely (8.32), is strictly convex whose solution is fully characterized by zeroing the gradient of the quadratic objective function that yields

$$\varphi^*(x) = -\hat{X}_0^{-1} V_0' x \tag{8.35}$$

which is a linear function, familiar to us, firstly appeared in Theorem 4.4. Plugging this solution into (8.32), we obtain

$$\begin{aligned} \psi' P_h \psi &\geq x' \left(X_0 - V_0 \hat{X}_0^{-1} V_0' \right) x \\ &= \psi' I_x (I_x' P_0^{-1} I_x)^{-1} I_x' \psi \end{aligned} \tag{8.36}$$

where we have used the equality $x = I_x' \psi$ with $I_x' = [I \ 0]$. In order to impose that the strict version of this inequality holds for all $0 \neq \psi \in \mathbb{R}^{n_x+n_u}$, its Schur Complement yields the LMI conditions

$$Q_0 > 0, \ I_x'(Q_0 - Q_h) I_x > 0 \tag{8.37}$$

expressed through the initial $Q_0 > 0$ and final $Q_h > 0$ matrix values of $Q(t) = P(t)^{-1} > 0, \forall t \in [0, h]$, which are the boundary conditions that we have already obtained many times before. Rewriting the DLMI (8.23) in the equivalent form

$$\begin{bmatrix} -\dot{Q}(t) + Q(t)F_0' + F_0 Q(t) + \gamma^{-2} J_0 J_0' & Q(t)G_0' & Q(t)G' \\ \bullet & -I & 0 \\ \bullet & \bullet & -I \end{bmatrix} < 0, \ t \in [0, h) \tag{8.38}$$

the conditions stated in Corollary 8.1 are fulfilled by the state feedback control law (8.9) with $\varphi = \varphi^*(x)$ being the linear function (8.35) extracted from the partitions (8.34) of matrix $P_0 = Q_0^{-1} > 0$, which is the optimal solution to the convex programming problem

$$\varrho_{2rob}^2 = \inf_{Q(\cdot)} \left\{ \mathbf{tr}(J_c' Q_h^{-1} J_c) + \mathbf{tr}(J_d' Q_0^{-1} J_d) : (8.37)\text{–}(8.38) \right\} \tag{8.39}$$

This problem deserves some comments. First, the constraint $Q_0 > 0$ imposes that $Q(t) > 0$ in the whole time interval $[0, h]$ as it is required for the existence of the inverse. Second, setting $J_0 = 0$ and $G_0 = 0$, the uncertainty is removed, and the result of Theorem 4.4 applied to the nominal system with $F = F_0$ is exactly recovered.

Adopting similar reasoning to the set of uncoupled DLMIs (8.30), it produces the equivalent ones

$$\begin{bmatrix} -\dot{Q}(t) + Q(t)F_i' + F_i Q(t) & Q(t)G' \\ \bullet & -I \end{bmatrix} < 0, \ t \in [0, h) \tag{8.40}$$

for all $i \in \mathbb{K}$ and, as in the previous case, the conditions of Corollary 8.2 are satisfied whenever the state feedback control law (8.9), with $\varphi = \varphi^*(x)$ being the linear function (8.35), is extracted from the partitions of matrix $P_0 = Q_0^{-1} > 0$, which is the optimal solution to the convex programming problem

$$\varrho_{2rob}^2 = \inf_{Q(\cdot)} \left\{ \mathbf{tr}(J_c' Q_h^{-1} J_c) + \mathbf{tr}(J_d' Q_0^{-1} J_d) : (8.37),\ (8.40) \right\} \tag{8.41}$$

Corollary 4.2 is once again reproduced by the algebraic manipulations adopted in this section. Of course, these facts were expected because identical design conditions were faced. In the next section, we generalize those results in order to take into account the convex constraint on the controlled output of the plant $s(k) \in \mathbb{S}, \forall k \in \mathbb{N}$, which is of great interest in the framework of MPC.

8.3.2 Constrained Control

We now tackle a problem that is at the core of the MPC design. It consists of the determination of a state feedback control law $u[k] = v[k] = \varphi(x[k])$ such that the hybrid nonlinear system (8.11)–(8.13), free of external continuous and discrete-time perturbations, is globally asymptotically stable and $s[k] \in \mathbb{S}, \forall k \in \mathbb{N}$. To this end, we need to impose that

$$\begin{bmatrix} C_s & D_s \end{bmatrix} H_\varphi(\psi[k]) \in \mathbb{S},\ \forall k \in \mathbb{N} \tag{8.42}$$

The set $\mathbb{S}$ is supposed to be convex with nonempty interior and $0 \in \mathbb{S}$. Hence, there exists a symmetric matrix $0 \leq R_s \in \mathbb{R}^{n_s \times n_s}$ such that the generic ellipsoid $\mathbb{E}_s = \{s \in \mathbb{R}^{n_s} : s' R_s s \leq 1\} \subseteq \mathbb{S}$. The determination of such matrix is discussed in the Bibliography notes of this chapter and is explicitly determined in due course for the important case of lower and upper bounds to the controlled output $s[k], k \in \mathbb{N}$. Defining the symmetric non-negative definite matrix

$$\begin{bmatrix} C_s' \\ D_s' \end{bmatrix} R_s \begin{bmatrix} C_s' \\ D_s' \end{bmatrix}' = R_\psi \tag{8.43}$$

and the convex set $\mathbb{E}_\psi = \{\psi \in \mathbb{R}^{n_x + n_u} : \psi' R_\psi \psi \leq 1\}$, it is immediate to verify that, by construction, the implication

$$H_\varphi(\psi[k]) \in \mathbb{E}_\psi \implies s[k] \in \mathbb{E}_s \subseteq \mathbb{S},\ \forall k \in \mathbb{N} \tag{8.44}$$

holds. It is clear that we have all interest to determine $R_s \geq 0$, in order to have $\mathbb{E}_s$ of maximum volume, such that the inclusion $\mathbb{E}_s \subseteq \mathbb{S}$ is as tight as possible. Based on the algebraic manipulations presented so far, a way to enforce (8.42) is to replace inequality (8.32) by

$$
\begin{aligned}
\psi' P_h \psi &\geq \min_{\varphi}\{H_\varphi(\psi)' P_0 H_\varphi(\psi) : H_\varphi(\psi) \in \mathbb{E}_\psi\} \\
&= \min_{\varphi} \left\{ \begin{bmatrix} x \\ \varphi \end{bmatrix}' P_0 \begin{bmatrix} x \\ \varphi \end{bmatrix} : \begin{bmatrix} x \\ \varphi \end{bmatrix} \in \mathbb{E}_\psi \right\}
\end{aligned} \tag{8.45}
$$

which immediately provides the optimal decision expressed as

$$
\varphi_*(x) = \arg\min_{\varphi} \left\{ \begin{bmatrix} x \\ \varphi \end{bmatrix}' P_0 \begin{bmatrix} x \\ \varphi \end{bmatrix} : \begin{bmatrix} x \\ \varphi \end{bmatrix} \in \mathbb{E}_\psi \right\} \tag{8.46}
$$

Observe that due to (8.44), the constraint of this problem whenever satisfied provides a feasible state feedback control but at the expense of some degree of conservatism that has been inevitably added by the previous calculations. In addition, notice that (8.45) is convex and falls into the class of quadratic programming problems with a quadratic objective function subject to a quadratic constraint. In principle, at least from the numerical viewpoint, it can be solved with no difficulty. Even so, we follow a simpler alternative based on the next lemma.

Lemma 8.1 *Consider that $P_0 > \theta^{-1} R_\psi \geq 0$ and $\theta > 0$ such that*

$$
x' \left(I_x' P_0^{-1} I_x \right)^{-1} x \leq \theta^{-1} \tag{8.47}
$$

The convex programming problem (8.45) admits a solution, and it satisfies

$$
\min_{\varphi}\{H_\varphi(\psi)' P_0 H_\varphi(\psi) : H_\varphi(\psi) \in \mathbb{E}_\psi\} \leq x' \left(I_x' P_0^{-1} I_x \right)^{-1} x \tag{8.48}
$$

and its objective function equals the right hand side of (8.48) for $\varphi = \varphi^*(x)$ given in (8.35).

Proof From the assumption that $\theta P_0 > R_\psi \geq 0$ and $\theta > 0$, it follows that

$$
\mathbb{E}_0 = \{\psi \in \mathbb{R}^{n_x+n_u} : \psi'(\theta P_0)\psi \leq 1\} \subseteq \mathbb{E}_\psi \tag{8.49}
$$

and $\mathbb{E}_0 \neq \emptyset$ because $\varphi = \varphi^*(x)$ given in (8.35) yields $H_{\varphi^*}(\psi)'(\theta P_0) H_{\varphi^*}(\psi) = \theta x'(I_x' P_0^{-1} I_x)^{-1} x$, which together with (8.47) shows that $H_{\varphi^*}(\psi) \in \mathbb{E}_0 \subseteq \mathbb{E}_\psi$, leading to

$$
\begin{aligned}
\min_{\varphi}\{H_\varphi(\psi)' P_0 H_\varphi(\psi) : H_\varphi(\psi) \in \mathbb{E}_\psi\} &\leq H_{\varphi^*}'(\psi) P_0 H_{\varphi^*}(\psi) \\
&= x' \left(I_x' P_0^{-1} I_x \right)^{-1} x
\end{aligned} \tag{8.50}
$$

which is the claim. The proof is complete.

This result is interesting and useful because, under the assumption that inequality (8.47) holds, then plugging $\varphi = \varphi^*(x)$ into the constraint (8.24), it becomes (8.36) being expressed once again by the LMIs (8.37). Clearly, $\varphi^*(x)$ is not, in general, equal to the optimal function $\varphi_*(x)$ given in (8.46). Indeed, $\varphi^*(x)$ is simply a feasible suboptimal solution to the constrained control design problem (8.45). It has the same formula, but the matrix $P_0 > 0$ needs to satisfy an additional constraint for the result of Lemma 8.1 to hold. Moreover, it requires the knowledge of the matrix $R_\psi \geq 0$, which is readily determined from (8.43) whenever $R_s \geq 0$ is known. The reader is requested to see the Bibliography notes for more information on this important issue. However, for the often used special case

$$\mathbb{S} = \{s \in \mathbb{R}^{n_s} : |s_i| \leq \bar{s}_i, i = 1, \cdots, n_s\} \tag{8.51}$$

with $\bar{s}_i > 0, i = 1, \cdots, n_s$, the ellipsoidal set $\mathbb{E}_s$ with maximal volume inside $\mathbb{S}$ is the one defined by the positive diagonal matrix

$$R_s = \text{diag}\left(\frac{1}{\bar{s}_1^2}, \cdots, \frac{1}{\bar{s}_{n_s}^2}\right) \tag{8.52}$$

which is very simple to calculate and produces, in general, precise results mainly for n_s of moderate size. We are now in position to prove one of the most important results of this chapter stated in the next theorem. Since the proof for both uncertainty classes is identical, they are presented together.

Theorem 8.2 *Let $h > 0$, matrix $R_\psi = D_\psi' D_\psi \geq 0$ of compatible dimensions, and the initial condition $\psi_0 = [x_0' \ 0]'$ be given. If the matrix-valued function $Q(t)$ satisfies the DLMI (8.38) (DLMIs (8.40)), subject to the boundary conditions expressed by the LMIs (8.37) and*

$$\theta I - D_\psi Q_0 D_\psi' > 0 \tag{8.53}$$

$$I_x' Q_0 I_x - \theta x_0 x_0' > 0 \tag{8.54}$$

then the following statements are true:

(a) *For $x_0 = 0$, the closed-loop sampled-data control system (8.5)–(8.8) governed by the state feedback control $v[k] = \varphi^*(x[k])$ is globally asymptotically stable for all $F \in \mathcal{F}_{nb}$ ($F \in \mathcal{F}_{cb}$) and satisfies the guaranteed performance with*

$$\varrho_{2rob}^2 = \mathbf{tr}(J_c' Q_h^{-1} J_c) + \mathbf{tr}(J_d' Q_0^{-1} J_d) \tag{8.55}$$

(b) *For $x_0 \in \mathbb{R}^{n_x}$ arbitrary and null w_c and w_d perturbations, the closed-loop sampled-data control system (8.5)–(8.8) governed by the state feedback control*

$u[k] = \varphi^*(x[k])$ is globally asymptotically stable for all $F \in \mathcal{F}_{nb}$ ($F \in \mathcal{F}_{cb}$) and the controlled output is feasible, that is, $s[k] \in \mathbb{S}, \forall k \in \mathbb{N}$.

Proof The proof of item (a) is immediate and follows directly from Corollary 8.1 (Corollary 8.2), because any feasible solution to the DLMI (8.38) (DLMIs (8.40)), subject to the boundary condition (8.37) provides the state feedback control (8.35), which is stabilizing and ensures the guaranteed performance (8.55) for all $F \in \mathcal{F}_{nb}$ ($F \in \mathcal{F}_{cb}$).

To prove item (b), we first notice that the LMI (8.53) is equivalent to $P_0 = Q_0^{-1} > \theta^{-1} D_\psi' D_\psi = \theta^{-1} R_\psi$, and (8.54) leads to

$$\begin{aligned}\theta^{-1} &> x_0' \left(I_x' P_0^{-1} I_x \right)^{-1} x_0 \\ &= H_{\varphi^*}'(\psi_0) P_0 H_{\varphi^*}(\psi_0) \\ &= \psi(0)' P_0 \psi(0)\end{aligned} \tag{8.56}$$

where the second equality follows from the jump equation (8.13) keeping in mind that $w_d = 0$. Furthermore, we know from the proof of Theorem 8.1 that $v_d(\psi(t_{k+1})) < v_d(\psi(t_k))$ for all $k \in \mathbb{N}$ provided that $v_d(\psi) = \psi' P_0 \psi$. As a consequence, $\psi(t_k)' P_0 \psi(t_k) < \psi(0)' P_0 \psi(0)$ which, together with (8.56), provides the inequality $\psi[k]' P_0 \psi[k] < \theta^{-1}$ valid for all $k \in \mathbb{N}$. The jump equation $\psi(t_k) = H_{\varphi^*}(\psi(t_k^-))$ imposes continuity to the state variable $x(t_k) = x(t_k^-)$. Therefore, it yields

$$\psi[k] = \begin{bmatrix} x(t_k) \\ \varphi^*(x(t_k)) \end{bmatrix} = H_{\varphi^*}(\psi[k]) \tag{8.57}$$

which, together with the last inequality, multiplied both sides by $\theta > 0$ and $P_0 > \theta^{-1} R_\psi$, allow the conclusion that $H_{\varphi^*}(\psi[k]) \in \mathbb{E}_0 \subseteq \mathbb{E}_\psi$ meaning that, see (8.44), the controlled output of the closed-loop system (8.7)–(8.8) governed by the sampled-data control $u[k] = \varphi^*(x[k])$ is feasible, that is, $s[k] \in \mathbb{S}, \forall k \in \mathbb{N}$.

Finally, we also obtain the inequality

$$\begin{aligned}\theta^{-1} &> \psi[k]' P_0 \psi[k] \\ &= H_{\varphi^*}'(\psi[k]) P_0 H_{\varphi^*}(\psi[k]) \\ &= x[k]' \left(I_x' P_0^{-1} I_x \right)^{-1} x[k]\end{aligned} \tag{8.58}$$

making it clear that the inequality (8.47) of Lemma 8.1 is verified for $x = x[k], \forall k \in \mathbb{N}$, completing the proof.

Several comments are in order. First, the minimum guaranteed $\mathcal{H}_2$ cost associated with the norm bounded uncertainty follows from the optimal solution of the convex programming problem

$$\inf_{Q(\cdot),\theta} \left\{ \mathbf{tr}(J_c' Q_h^{-1} J_c) + \mathbf{tr}(J_d' Q_0^{-1} J_d) : \text{(8.37)–(8.38)}, \ \text{(8.53)–(8.54)} \right\} \tag{8.59}$$

which has an additional scalar variable θ to be determined. The constraint $Q_0 > 0$ appearing in (8.37) makes the feasible solutions to the DLMI (8.38) positive definite in the time interval $[0, h]$, ensuring the existence of the matrix inverse $P(t) = Q(t)^{-1} > 0, \forall t \in [0, h]$. Moreover, the LMI (8.53) with $Q_0 > 0$ implies that $\theta > 0$ as desired. Second, if in problem (8.59) the DLMI (8.38) is replaced by the DLMIs (8.40), then the same properties remain valid but now for convex bounded uncertainty. Third, let us evaluate the conservativeness of the design conditions we have obtained, by verifying what happens in two extremes situations, namely:

- If $D_\psi = 0$, the constraint $s[k] \in \mathbb{S}, \forall k \in \mathbb{N}$, is removed. The constraints (8.53)–(8.54) are satisfied for $\theta > 0$ sufficiently small, which makes problems (8.59) and (8.39) identical.
- If $|x_0|$ is small, then $\theta > 0$ large enough satisfies the constraints (8.53)–(8.54) and once again the unconstrained sampled-data control is obtained. This is comprehensive since for $|x_0|$ small the trajectories tend to remain in the interior of $\mathbb{S}$.

Of course, the final validation of the proposed sampled-data control design must be made by solving a series of representative (in terms of order) academical examples and practical applications of interest. These aspects are tackled in the illustrative examples presented in the sequel.

Remark 8.4 The result of Theorem 8.2 must be understood and interpreted carefully. The objective function of problem (8.59) has been obtained by converting the effect of impulses into initial conditions whose correspondent trajectories were not subject to the constraint $s[k] \in \mathbb{S}, \forall k \in \mathbb{N}$. The second part of Theorem 8.2 guarantees that, whenever the closed-loop system is free of external perturbations, then $s[k] \in \mathbb{S}, \forall k \in \mathbb{N}$ provided that the initial condition $x(0) = x_0$ is the one appearing in the constraint (8.54). Only for this trajectory, feasibility is granted.

There is another possibility which consists of evaluating the guaranteed cost

$$\sup_{(A,B)\in\mathcal{A}} \int_0^\infty z(t)'z(t)dt \le \varrho_{2rob}^2$$

where $z(t)$ is the controlled output corresponding to the initial condition $x(0) = x_0$ that yields $\psi_0 = [x_0' \ 0]'$. From the proof of Theorem 8.1 and Corollary 8.1, we have

$$\begin{aligned} \varrho_{2rob}^2 &= H_{\varphi^*}(\psi_0)' P_0 H_{\varphi^*}(\psi_0) \\ &= x_0'(I_x' P_0^{-1} I_x)^{-1} x_0 \\ &= x_0'(I_x' Q_0 I_x)^{-1} x_0 \end{aligned}$$

Hence, if, instead of (8.59), one solves the convex programming problem

$$\inf_{Q(\cdot),\theta} \left\{ x_0'(I_x' Q_0 I_x)^{-1} x_0 : \ (8.37)-(8.38), \ \ (8.53)-(8.54) \right\}$$

then the minimized guaranteed cost and the LMIs (8.53)–(8.54) that ensure $s[k] \in \mathbb{S}, \forall k \in \mathbb{N}$ both depend on the same initial condition. This remark brings to light some useful information for MPC design. □

Example 8.1 The third order system to be controlled with $h = 1.5$ is free of external perturbations, which implies that $E = 0$, $E_u = 0$, and it has the state space realization (8.5)–(8.8) with matrices

$$A = \begin{bmatrix} 0 & 1 & 0 \\ 0 & 0 & 1 \\ 9 & -5 & -9 \end{bmatrix}, \ B = \begin{bmatrix} 0 \\ 0 \\ 2 \end{bmatrix}, \ C_z = \begin{bmatrix} 1\,0\,0 \\ 0\,1\,0 \\ 0\,0\,1 \\ 0\,0\,0 \end{bmatrix}, \ D_z = \begin{bmatrix} 0 \\ 0 \\ 0 \\ 1 \end{bmatrix}$$

It is open-loop unstable and we suppose that it evolves from the initial condition $x(0) = x_0 = [0\ 1\ 1]'$. The controlled output is defined by matrices

$$C_s = \begin{bmatrix} 1\,0\,0 \\ 0\,0\,0 \end{bmatrix}, \ D_s = \begin{bmatrix} 0 \\ 1 \end{bmatrix}$$

because we want to impose the bounds constraint $s[k] \in \mathbb{S}$, defined in (8.51), to the first state variable and the control variable, independently. We have solved the control design problem discussed in Remark 8.4, for the case of convex bounded uncertainty with only one vertex, by the piecewise linear approximation procedure with $n_\phi = 32$, considering three different situations, namely:

(a) With $\bar{s}_1 = \bar{s}_2 = 100$, large enough, the constraint $s[k] \in \mathbb{S}$ has been dropped. We have obtained the optimal control

$$\varphi^*(x) = \begin{bmatrix} -5.8658 & -4.6739 & -0.4798 \end{bmatrix} x$$

and the corresponding cost $\varrho_{2rob}^2 = 49.8990$. Figure 8.1 shows, in solid lines (blue), that the bounds $|x_1[k]| \leq 0.4$ and $|u[k]| \leq 4$ are both violated.

(b) With $\bar{s}_1 = 0.4$, $\bar{s}_2 = 100$, and the constraint $s[k] \in \mathbb{S}$, we have determined

$$\varphi^*(x) = \begin{bmatrix} -7.9852 & -6.0969 & -0.6211 \end{bmatrix} x$$

and the corresponding cost $\varrho_{2rob}^2 = 129.3986$. Figure 8.1 shows, in dashed lines (red), that the bound $|x_1[k]| \leq 0.4$ is satisfied, but the bound $|u[k]| \leq 4$ is violated.

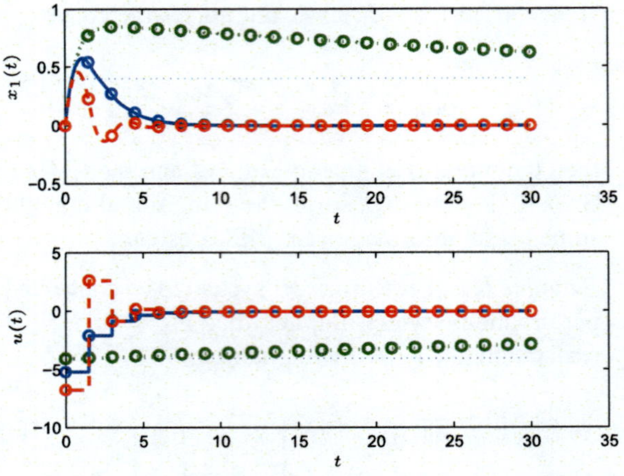

Fig. 8.1 Closed-loop system trajectories of Example 8.1

(c) With $\bar{s}_1 = 100$, $\bar{s}_2 = 4$ and the constraint $s[k] \in \mathbb{S}$, we have calculated

$$\varphi^*(x) = \begin{bmatrix} -4.5393 & -3.6272 & -0.3724 \end{bmatrix} x$$

and the corresponding cost $\varrho_{2rob}^2 = 721.1336$. Figure 8.1 shows, in dotted lines (green), that the bound $|x_1[k]| \leq 0.4$ is violated, but the bound $|u[k]| \leq 4$ is satisfied.

This example illustrates two interesting occurrences that are important in practice. In case b), the state $x_1[k]$ never violates the bound, but $x_1(t)$ does at some instant of time $t \in (0, 1.5)$. In case c), the hard bound on the control signal makes the time response of the closed-loop system very slow. □

We have to put in evidence the fact that the constrained control synthesis has been successfully accomplished at the expense of some conservatism which could be reduced if we are able to solve the next formulated open problem.

Open Problem 8.1 *Replace inequality (8.45) by*

$$\psi' P_h \psi \geq \min_{\varphi} \left\{ \begin{bmatrix} x \\ \varphi \end{bmatrix}' P_0 \begin{bmatrix} x \\ \varphi \end{bmatrix} : C_s x + D_s \varphi \in \mathbb{S} \right\} \tag{8.60}$$

and show that Theorem 8.2, or an adapted version of it, remains true. This is of interest even if it could be solved only for the special convex set $\mathbb{S}$ *given in (8.51).*

The solution of this problem (if any) may have an enormous impact on the development of a new sampled-data MPC strategy, closer to reality, which clearly is a net gain as far as practical applications are concerned.

8.4 Sampled-Data MPC

We are now in position to apply the late control design technique to the implementation of the MPC strategy. The next developments are done for systems subject to norm bounded uncertainty, but as the reader can verify, the case of convex bound uncertainty can be dealt with similarly, with no difficulty.

As usual, we denote $x_{k|k} = x[k]$ the actual state variable measured and made available at time $t_k, k \in \mathbb{N}$. The k-th MPC update is calculated from the optimal solution to the convex programming problem, stemmed from (8.59)

$$\inf_{Q(\cdot),\theta} \left\{ \mathbf{tr}(J_c' Q_h^{-1} J_c) + \mathbf{tr}(J_d' Q_0^{-1} J_d) : (8.37)\text{–}(8.38),\ (8.62)\text{–}(8.64) \right\} \tag{8.61}$$

with the constraints (8.53)–(8.54) being replaced by the following ones:

$$\theta I - D_\psi Q_0 D_\psi' > 0 \tag{8.62}$$

$$I_x' Q_0 I_x - \theta x_{k|k} x_{k|k}' > 0 \tag{8.63}$$

$$\begin{bmatrix} x_{k|k}'(I_x' Q_{0(k-1)} I_x)^{-1} x_{k|k} & x_{k|k}' \\ \bullet & I_x' Q_0 I_x \end{bmatrix} \geq 0 \tag{8.64}$$

where matrix $Q_{0(k-1)} > 0$ has been calculated in the precedent iteration. For each $k \geq 0$, the optimal solution of (8.61) produces the variables of interest $(Q_{0(k)}, Q_{h(k)}, \theta_{(k)})$. For $k = 0$, matrix $Q_{0(-1)} > 0$ is taken with arbitrarily small norm, meaning that $(Q_{0(0)}, Q_{h(0)}, \theta_{(0)})$ is the optimal solution to the original problem (8.59). The next theorem provides the major result of this section and retains one of the most important and useful features of MPC.

Theorem 8.3 *If problem (8.61) is feasible at the initial time $t_0 = 0$, then it remains feasible at any subsequent update at time t_k, $\forall k \geq 1$ and, consequently, the minimum cost is a non-increasing sequence. In the affirmative case, for $x_{0|0} = x_0$ and null w_c and w_d perturbations, the closed-loop sampled-data control system (8.5)–(8.8) governed by the state feedback MPC $u[k] = u_{k|k} = \varphi_{(k)}^*(x_{k|k})$ is globally asymptotically stable for all $F \in \mathcal{F}_{nb}$ ($F \in \mathcal{F}_{cb}$) and the controlled output is feasible, that is, $s[k] \in \mathbb{S}, \forall k \in \mathbb{N}$.*

Proof By induction, assume that problem (8.61) is feasible at the initial time $\{t_k\}_{k=0}$. We have to show that its optimal solution at time $t_k = kh$ represented by the variables of interest $(Q_{0(k)}, Q_{h(k)}, \theta_{(k)})$ is feasible for the same problem at time $t_{k+1} = (k+1)h$. To this end, we have to verify the feasibility of the constraints (8.63) and (8.64), the only ones among all that depend on the current measured state $x[k] = x_{k|k}$.

Consider $k \geq 0$. The constraint (8.63) yields

$$\theta_{(k)}^{-1} > x[k]'(I_x' Q_{0(k)} I_x)^{-1} x[k]$$

$$
\begin{aligned}
&= H_{\varphi^*_{(k)}}(\psi[k])' P_{0(k)} H_{\varphi^*_{(k)}}(\psi[k]) \\
&= \psi[k]' P_{0(k)} \psi[k]
\end{aligned} \tag{8.65}
$$

where the last equality is due to (8.57). Now, invoking once again the fact that $v_d(\psi(t_{k+1})) < v_d(\psi(t_k))$ provided that $v_d(\psi) = \psi' P_{0(k)} \psi$, we have

$$
\begin{aligned}
\theta_{(k)}^{-1} &> \psi[k+1]' P_{0(k)} \psi[k+1] \\
&\geq \min_{\varphi} \begin{bmatrix} x[k+1] \\ \varphi \end{bmatrix}' P_{0(k)} \begin{bmatrix} x[k+1] \\ \varphi \end{bmatrix} \\
&= x[k+1]' (I_x' Q_{0(k)} I_x)^{-1} x[k+1]
\end{aligned} \tag{8.66}
$$

Applying the Schur Complement, it is rewritten in the equivalent form

$$
I_x' Q_{0(k)} I_x - \theta_{(k)} x_{k+1|k+1} x_{k+1|k+1}' > 0 \tag{8.67}
$$

which shows that $(Q_{0(k)}, \theta_{(k)})$ satisfies the constraint (8.63) with the state measurement $x_{k+1|k+1} = x[k+1]$. The same can be seen to hold for constraint (8.64), by evaluating it at $k+1$ with $Q_0 = Q_{0(k)}$.

Let us now move our attention to the stability property of the closed-loop system. Consider that $F \in \mathcal{F}_{nb}$ ($F \in \mathcal{F}_{cb}$) is arbitrary but given and define the following time-varying Lyapunov function candidate:

$$
\vartheta(x, k) = x' (I_x' Q_{0(k)} I_x)^{-1} x \tag{8.68}
$$

Evaluating the constraint (8.64) at $k+1$ with the optimal solution $Q_0 = Q_{0(k+1)}$, from the Schur Complement, it readily follows that

$$
\begin{aligned}
\vartheta(x[k+1], k+1) &= x[k+1]' (I_x' Q_{0(k+1)} I_x)^{-1} x[k+1] \\
&\leq x[k+1]' (I_x' Q_{0(k)} I_x)^{-1} x[k+1]
\end{aligned} \tag{8.69}
$$

and invoking once again the inequalities in Eqs. (8.65) and (8.66), we have

$$
\begin{aligned}
\vartheta(x[k+1], k+1) &< x[k]' (I_x' Q_{0(k)} I_x)^{-1} x[k] \\
&= \vartheta(x[k], k)
\end{aligned} \tag{8.70}
$$

which proves that the closed-loop sampled-data control system (8.5)–(8.8) governed by the state feedback MPC $u[k] = u_{k|k} = \varphi^*_{(k)}(x_{k|k})$ is globally asymptotically stable for all $F \in \mathcal{F}_{nb}$ ($F \in \mathcal{F}_{cb}$). Finally, the feasibility of the controlled output $s[k] \in \mathbb{S}$, $\forall k \in \mathbb{N}$, follows immediately from part b) of Theorem 8.2.

This theorem has been proven in the general context of $\mathcal{H}_2$ performance optimization. We have proposed a Lyapunov function candidate that needed the supplementary constraint (8.64) to be included, in order to be useful in guaranteeing the asymptotic behavior of the MPC . Of course, this is the price to be paid if one wants to implement the proposed MPC strategy successfully. It will be shown by means of some illustrative examples that the MPC strategy may improve considerably, in terms of the cost under consideration, the performance of the closed-loop system.

Following Remark 8.4, let us consider, instead of (8.61), the convex programming problem

$$\inf_{Q(\cdot),\theta} \left\{ x'_{k|k}(I'_x Q_0 I_x)^{-1} x_{k|k} : (8.37)-(8.38),\ (8.62)-(8.63) \right\} \tag{8.71}$$

which provides the state feedback sampled-data control that minimizes the guaranteed cost $\|z\|_2^2$, where z is the controlled output associated with the initial condition $x(0) = x_0$. Its importance stems from the fact that it is identical to the linear quadratic regulator design formulated in the context of MPC. See the Bibliography notes of this chapter for a brief discussion on this matter.

Corollary 8.3 *Theorem 8.3 remains valid for problem (8.71).*

Proof Observe that constraint (8.64) has been dropped from problem (8.71), and the same algebraic manipulations adopted in the proof of Theorem 8.3 are still valid to show that the optimal solution $(Q_{0(k)}, Q_{h(k)}, \theta_{(k)})$ is feasible at $k+1$. Consequently, its optimal solution at $k+1$ provides

$$x[k+1]'(I'_x Q_{0(k+1)} I_x)^{-1} x[k+1] \le x[k+1]'(I'_x Q_{0(k)} I_x)^{-1} x[k+1] \tag{8.72}$$

which implies that constraint (8.64) is indeed superfluous. With the Lyapunov function candidate (8.68), inequalities (8.69) and (8.70) remain true, completing the proof.

Problem (8.71) has all flavors that contribute decisively for the success of the MPC strategy in both theoretical and practical viewpoints. A pure quadratic cost is minimized (output energy bound reduction), global asymptotic stability is preserved under parameter uncertainty (norm and convex bounded), and the output constraint $s[k] \in \mathbb{S}, \forall k \in \mathbb{N}$, is imposed. All just mentioned characteristics are due to the sampled-data state feedback control $u[k] = u_{k|k} = \varphi^*_{(k)}(x_{k|k})$ synthesized from the optimal solution of a convex programming problem with constraints expressed by a DLMI and LMIs.

Remark 8.5 It is possible to update the current measured state in the MPC strategy at some multiple, say $m \ge 1$, of the sampling period h. Doing this, the update occurs at time instants $\{t_n = n(mh)\}_{n\in\mathbb{N}}$. For this implementation, it suffices to replace the indices $n \to k$ on constraints (8.63), (8.64) and on the objective function of problem (8.71).

Theorem 8.3 and Corollary 8.3 remain true, because the quadratic Lyapunov function $v_d(\psi) = \psi' P_{0(n)}\psi$ is such that $v_d(\psi(t_{\ell+1})) < v_d(\psi(t_\ell))$ for all $\ell \in \{0, 1, \cdots, m-1\}$ where $t_\ell = n(mh) + \ell h$. Therefore, this inequality applied successively yields

$$\psi(t_{n+1})' P_{0(n)} \psi(t_{n+1}) < \psi(t_n)' P_{0(n)} \psi(t_n)$$

leading to $\vartheta(x[n+1], n+1) < \vartheta(x[n], n)$ for all $n \in \mathbb{N}$ and $\vartheta(x, n) = x'(I_x' Q_{0(n)} I_x)^{-1} x$. Inside the time interval defined by two consecutive updates, the MPC strategy keeps the state feedback control law invariant and equal to $u[k] = u_{k|k} = \varphi_{(n)}^*(x_{k|k})$. According to Theorem 8.3, the MPC provided by the solution of problem (8.61) produces two non-increasing sequences, namely

$$\left\{\mathbf{tr}(J_c' Q_{h(n)}^{-1} J_c) + \mathbf{tr}(J_d' Q_{0(n)}^{-1} J_d)\right\}_{n\in\mathbb{N}}, \quad \left\{x_{n|n}'(I_x' Q_{0(n)} I_x)^{-1} x_{n|n}\right\}_{n\in\mathbb{N}}$$

whereas, according to Corollary 8.3, for the MPC provided by the solution of problem (8.71) only the second sequence has this property, since it coincides with its objective function. □

There are several aspects involving the theoretical results presented so far deserving illustration and detailed discussion. This is done through the illustrative examples solved in the sequel. To that end, we give particular and exclusive attention to the numerical solution and implementation of the MPC provided by the optimal solution of the convex programming problem (8.71).

Example 8.2 This example treats a continuous-time double integrator supposed to be free of external perturbations ($E = 0$ and $E_u = 0$), to be controlled with $h = 1.0$. It has the state space realization (8.5)–(8.8) with matrices

$$A = \begin{bmatrix} 0 & 1 \\ 0 & 0 \end{bmatrix}, \quad B = \begin{bmatrix} 0 \\ 1 \end{bmatrix}, \quad C_z = \begin{bmatrix} 1 & 0 \\ 0 & 0 \end{bmatrix}, \quad D_z = \begin{bmatrix} 0 \\ 1/\sqrt{10} \end{bmatrix}$$

and evolves from the initial condition $x(0) = x_0 = [10\ -5]'$. The controlled output is defined by matrices

$$C_s = \begin{bmatrix} 0 & 0 \end{bmatrix}, \quad D_s = \begin{bmatrix} 1 \end{bmatrix}$$

because we want to impose the bound constraint $|u[k]| \leq 1, \forall k \in \mathbb{N}$, defined in (8.51), to the control variable, exclusively. We have solved the control design problem (8.71), for the case of convex bounded uncertainty with only one vertex, by the piecewise linear approximation procedure with $n_\phi = 32$. Following Remark 8.5, the MPC strategy has been updated with a time interval equal to mh. Table 8.1 provides the true value of the cost for certain values of m. Since the time simulation was taken equal to $T_{simul} = 40.0$ and $h = 1.0$, then $m = 40$ means that the MPC strategy has been updated only once, whereas $m = 1$ means that it has been

Table 8.1 True $\mathcal{H}_2$ cost produced by MPC strategy

m	40	20	8	4	1
Cost	271.2359	269.4239	242.1640	161.0034	123.0780

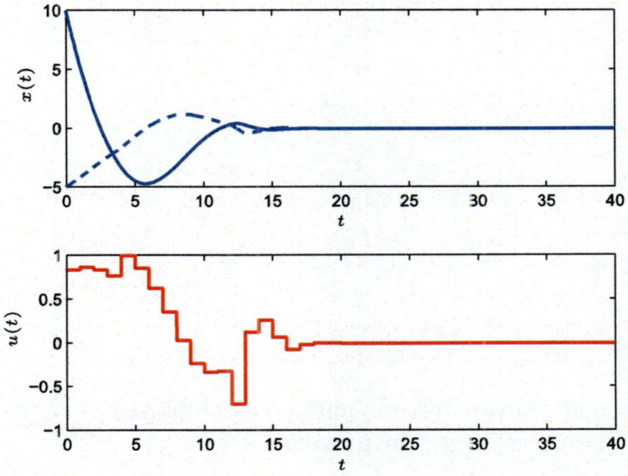

Fig. 8.2 Closed-loop system trajectories of Example 8.2

updated every sampling time. In this case, the expressive cost reduction shows the effectiveness of the MPC strategy. Notice however that, in general, the true cost sequence is non-increasing, but this is not always the case because, due to the sufficient nature of the results, only a guaranteed cost has been minimized. As indicated in Remark 8.5, they constitute a non-increasing sequence. Figure 8.2 shows on the top the state trajectories of the closed-loop system, while the control signal is shown on the bottom of the same figure for $m = 4$. Observe that $u[k]$ near $k = 5$ hits the prescribed bound. In this case, the design conditions are not conservative. □

Example 8.3 The system to be dealt with is composed of two unitary masses kept together by means of a spring with unknown spring constant $0.5 \le \kappa \le 10.0$. A force is applied to the first mass in order to move the second one to a desired steady-state position. After shifting the origin of the inertial frame of reference, the continuous-time system supposed to be free of external perturbations ($E = 0$ and $E_u = 0$) is controlled with $h = 0.1$. It has the state space realization (8.5)–(8.8) with matrices

$$A = \begin{bmatrix} 0 & 0 & 1 & 0 \\ 0 & 0 & 0 & 1 \\ -\kappa & \kappa & 0 & 0 \\ \kappa & -\kappa & 0 & 0 \end{bmatrix}, \; B = \begin{bmatrix} 0 \\ 0 \\ 1 \\ 0 \end{bmatrix}, \; C_z' = \begin{bmatrix} 0 & 0 \\ 1 & 0 \\ 0 & 0 \\ 0 & 0 \end{bmatrix}, \; D_z = \begin{bmatrix} 0 \\ 1 \end{bmatrix}$$

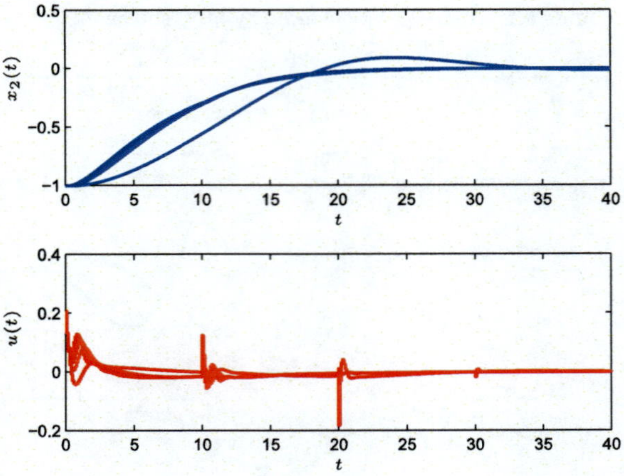

Fig. 8.3 Closed-loop system trajectories of Example 8.3

and evolves from the initial condition $x(0) = x_0 = [-1\ \ -1\ 0\ 0]'$. The controlled output is defined by matrices

$$C_s = \begin{bmatrix} 0\,0\,0\,0 \end{bmatrix},\ D_s = \begin{bmatrix} 1 \end{bmatrix}$$

because we want to impose the bound constraint $|u[k]| \leq 1, \forall k \in \mathbb{N}$, defined in (8.51), to the control variable, exclusively.

We have solved the control design problem (8.71), for the case of convex bounded uncertainty with $N = 2$ vertices, by the piecewise linear approximation procedure with $n_\phi = 32$. Following Remark 8.5, the MPC strategy has been updated with a time interval equal to mh. Since the time simulation was taken equal to $T_{simul} = 40.0$ and $h = 0.1$, then $m = 100$ means that the MPC strategy has been updated only four times during simulation. Figure 8.3 shows on the top the position $x_2(t)$ of the second mass, while the control signal is shown on the bottom of the same figure for $m = 100$. It has been drawn by plotting the trajectories of the closed-loop system for many feasible values of the spring constant, that is, $\kappa \in \{0.500, 2.515, 5, 250, 7.625, 10.0\}$. Observe that, in this case, $u[k]$ never hits the prescribed bound. □

These illustrative examples borrowed from the literature, see the Bibliography notes of this chapter for details, put in evidence the importance of MPC strategy from the practical viewpoint. Actually, the flexibility to decide when the measured state update is taken into account is important to make the calculation of the controller and its online implementation possible. The improvement in terms of the cost function reduction appears in Table 8.1 and is a positive assessment of the MPC strategy. The main difficulty is the computational burden involved in the determination of the optimal solution of problem (8.71) mainly if the sampling period needs to be small. The possibility to choose m large enough circumvents this drawback but, of course, at the expense of increasing the cost function. Finding the good trade-off between these two alternatives is an art.

8.5 Bibliography Notes

The literature provides countless books and papers dealing with Model Predictive Control (MPC) involving all relevant aspects related to stability, performance optimization, numeric implementation, and others. See for instance the survey papers [42–44] and the references therein. However, the same cannot be said with respect to sampled-data MPC. Indeed, one of the few exceptions is reference [2] that generalizes the MPC framework to hybrid systems. We believe that the results reported in this book and those of the aforementioned reference are complementary, in the sense that here we provide new and more specific stability and optimality conditions, which do not depend on control persistence, and we are able to cope with parameter uncertainty.

Perhaps the main reason for the success of MPC in practice is its ability to handle state and control constraints, see [42, 47] and [50] for details. However, the price to be paid is that, in general, continuous-time systems and optimal control problems with infinity optimization horizon are difficult to treat, since in order to cope with the mentioned hard constraints efficiently, one needs to work in finite dimensional spaces. In this area, the closed-loop system stability, mainly when the plant is subject to parameter uncertainty, is a paradigm to be attained. To this end, appropriate cost penalty or final state constraints are applied and discussed in references [7] and [8]. The case of infinity optimization horizon has been first addressed in [47] and [50], where interesting and useful theoretical properties have been determined. In this context, the important issue related to robustness with respect to parameter uncertainty has been addressed and solved in [38] through conditions expressed by linear matrix inequalities (LMIs), see also [44]. The results reported in this book are based on the recent reference [22] and, in our opinion, the proposed sampled-data MPC strategy goes beyond the aforementioned quoted references, since we are able to handle continuous-time systems without any kind of approximation.

In the context of MPC, the maximization involving constrained ellipsoids given in [10] is useful, mainly if one wants to go further toward a possible solution to the proposed open-problem. The two last illustrative examples solved and discussed in this chapter have been borrowed from references [7] and [38], respectively, because they are easy to understand and comparisons can be made in order to raise and evaluate the effects of sampling on sampled-data MPC design.

Chapter 9
Numerical Experiments

9.1 Introduction

This chapter aims to apply and present with detail some examples with practical appeal that can be tackled by the sampled-data control design procedures introduced in the previous chapters of this book. In fact, the designed control will not be practically implemented but numerically simulated, in order to evaluate stability, robustness, performance and the impact that the sampling period magnitude $h > 0$ has on those important attributes of the closed-loop system. Several control strategies will be assessed and compared carefully, with the main objective to identify eventual weaknesses. All variables and data are related to the International System of Units (ISU).

9.2 Inverted Pendulum

The inverted pendulum is depicted in Fig. 9.1. It is composed of a small cart with mass M_p that moves in a horizontal plane, under the action of an external horizontal force with intensity u. The pendulum has mass m_p and length ℓ_p. The movement is restricted to the plane, and the cart and the pendulum are supposed to move in a viscous frictionless environment. In real life this assumption is not true, but it characterizes the worst case as far as a control design is concerned. Indeed, viscous friction always acts in the sense of attenuating displacements. Force f is the traction acting on the pendulum rod and g is the gravitational acceleration.

Denoting by $\eta(t)$ the position of the cart center of mass at time $t \in \mathbb{R}_+$, the dynamic equations with respect to the inertial reference system (the orthogonal axes represented in dashed lines) follow from the direct application of Newton's second law. The horizontal movement of the cart, assuming that it is always in contact with the ground, is governed by the differential equation

J. C. Geromel, *Differential Linear Matrix Inequalities*,
https://doi.org/10.1007/978-3-031-29754-0_9

Fig. 9.1 Inverted pendulum

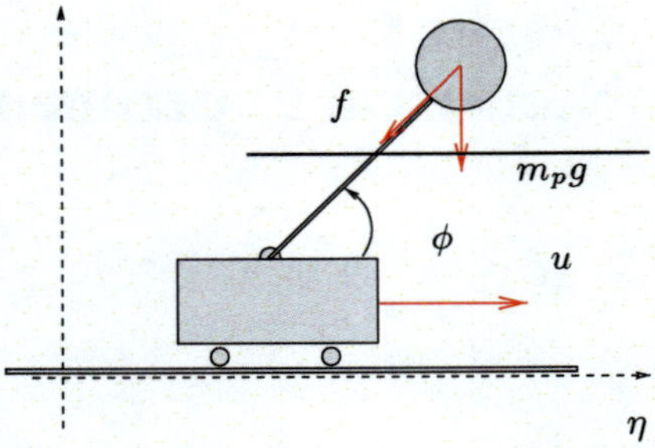

$$M_p \frac{d^2}{dt^2}(\eta) = f\cos(\phi) + u \tag{9.1}$$

Since the horizontal position of the pendulum is $\eta + \ell_p\cos(\phi)$, its movement on this axes is given by

$$m_p \frac{d^2}{dt^2}(\eta + \ell_p\cos(\phi)) + f\cos(\phi) = 0 \tag{9.2}$$

whereas since the vertical position of the pendulum is $\ell_p\sin(\phi)$, its movement on this direction is defined by

$$m_p \frac{d^2}{dt^2}(\ell_p\sin(\phi)) + f\sin(\phi) + m_p g = 0 \tag{9.3}$$

To conclude our model, let us assume that the rod never breaks, which means that the traction f does not need to be determined. Eliminating it from the previous equations is a simple, but tedious task. Indeed, after some algebraic manipulations, Eqs. (9.1) and (9.2) yield

$$(M_p + m_p)\ddot{\eta} - m_p\ell_p\sin(\phi)\ddot{\phi} - m_p\ell_p\cos(\phi)\dot{\phi}^2 = u \tag{9.4}$$

and multiplying (9.2) by $-\sin(\phi)$, (9.3) by $\cos(\phi)$ and summing up the results, we obtain

$$\ell_p\ddot{\phi} - \sin(\phi)\ddot{\eta} + g\cos(\phi) = 0 \tag{9.5}$$

which, both together, constitute the model for the movement of the cart and the pendulum under the action of the external force with intensity u. Clearly, our concern is to synthesize the control action such that the closed-loop system presents a desired behavior, as, for instance, it always brings the pendulum to the upright position by moving the cart adequately. To this end, we need to work with a linear system, derived by the first order approximation valid in a neighborhood of $(\eta, \delta) = (0, 0)$ where $\delta = \phi - \pi/2$. It is given by

$$(M_p + m_p)\ddot{\eta} - m_p \ell_p \ddot{\delta} = u \tag{9.6}$$

$$\ell_p \ddot{\delta} - \ddot{\eta} - g\delta = 0 \tag{9.7}$$

We are now in position to express the model of the inverted pendulum in the form adopted at the beginning of Chap. 3, that is

$$\dot{x}(t) = Ax(t) + Bu(t) + Ew_c(t) \tag{9.8}$$

$$y[k] = C_y x[k] + E_y w_d[k] \tag{9.9}$$

$$z(t) = C_z x(t) + D_z u(t) \tag{9.10}$$

where $x = [\eta\ \delta\ \dot{\eta}\ \dot{\delta}]' \in \mathbb{R}^4$ is the state variable, which together with (9.6)–(9.7), allow us to determine

$$A = \begin{bmatrix} 0 & 0 & 1 & 0 \\ 0 & 0 & 0 & 1 \\ 0 & (m_p/M_p)g & 0 & 0 \\ 0 & (1 + m_p/M_p)(g/\ell_p) & 0 & 0 \end{bmatrix}, \quad B = \begin{bmatrix} 0 \\ 0 \\ 1/M_p \\ 1/(M_p \ell_p) \end{bmatrix} \tag{9.11}$$

As usual, matrices C_z and D_z with appropriate dimensions are used to synthesize the performance index to be optimized. In the present case, we have adopted

$$C_z = \begin{bmatrix} 1 & 0 & 0 & 0 \\ 0 & \ell_p & 0 & 0 \\ 0 & 0 & 0 & 0 \end{bmatrix}, \quad D_z = \begin{bmatrix} 0 \\ 0 \\ d_z \end{bmatrix} \tag{9.12}$$

in which case the controlled output

$$\int_0^\infty z(t)'z(t)dt = \int_0^\infty \left(\eta(t)^2 + \ell_p^2 \delta(t)^2 + d_z^2 u(t)^2\right) dt \tag{9.13}$$

penalizes the actual distance from the cart and the pendulum positions, namely $(\eta(t), \ell_p\delta(t))$, to the desired final ones $(0, 0)$ and the force intensity $u(t)$. The parameter $d_z > 0$ is free and must be adjusted by the designer, in order to establish, by trial and error, a good trade-off between these quantities. The remaining matrices and variables need the following observations to be well defined:

- **Control channel perturbation:** It is appropriate to assume that the control action is corrupted by a continuous-time exogenous perturbation w_c with gain $\alpha_c \in (0, 1)$, depending on the precision of the actuator devices. This imposes the choice $E = \alpha_c B$.
- **Measurement channel perturbation:** It is also assumed that the measurement action is corrupted by a discrete-time exogenous perturbation w_d with gain $\alpha_d \in (0, 1)$, depending on the precision of the gauge devices. If we want to measure

the horizontal position of the pendulum $\eta + \ell_p\cos(\phi) \approx \eta - \ell_p\delta$, then $C_y = [1\ -\ell_p\ 0\ 0]$ and $E_y = \alpha_d$.

Finally, to complete the model to be dealt with we need to explicit the control law to be synthesized. It obeys Eq. (3.4) with $E_u = \alpha_c$ or (3.5) with $E_u = \alpha_d$, depending on its continuous or sampled-data nature. Additionally, parameter uncertainty can be taken into account. For instance, one may suppose that the cart mass is unknown, but belongs to a known interval, that is, $1/M_p \in [1/M_{c1}, 1/M_{c2}]$, which means that this scalar uncertainty can be viewed as norm or convex bounded types, indistinctly. All these aspects will be raised and discussed again in due course, more precisely, when each control design problem to be solved is properly stated.

Remark 9.1 In all calculations and simulations, we have considered the following numerical values $M_p = 1.30$ [kg], $m_p = 0.23$ [kg], $\ell_p = 0.33$ [m] for the cart-pendulum, $g = 9.81$ [m/s^2] for the gravitational acceleration, and $\alpha_c = \alpha_d = 0.20$ for the exogenous perturbation gains. The uncertainty in the cart mass is such that $1/M_p \in [0.5, 1.0]$. We have chosen the sampling period equal to $h = 100$ [ms] and $d_z = 0.5$. □

Our goal is to compare the performance of different control strategies, in order to put in evidence potentialities and eventual weaknesses. Hence, it is natural and appropriate to adopt, in all cases, a unique performance index to be optimized. For this reason, in this section, only the $\mathcal{H}_2$ performance index is tackled but by incorporating the design of robust sampled-data control to face convex bounded uncertainty. All control design problems are solved by the piecewise linear approximation procedure with $n_\phi = 32$.

9.2.1 State Feedback Control

The first experiment was accomplished with the nominal system defined by the cart mass $M = 1.30$ [kg] and $E_y = 0$. The solution of the convex programming problem (4.64) of Theorem 4.4, provided the optimal sampled-data state feedback control law $u(t) = u[k] = Lx[k] + E_u w_d[k]$, $t \in [t_k, t_{k+1})$, $\forall k \in \mathbb{N}$, where

$$L = \begin{bmatrix} 0.9376 & -27.6368 & 1.5909 & -4.7429 \end{bmatrix} \tag{9.14}$$

and whose associated minimum $\mathcal{H}_2$ performance index is $\varrho_2^2 = 0.1517$. The robust control design has also been addressed for the convex bounded uncertainty $1/M_p \in [0.5, 1.0]$, which yields $\{F_i\}_{i=1,2}$, keeping the matrix E unchanged and depending only on the nominal value of the mass M. Problem (4.69) stated in Corollary 4.2 gives the matrix gain

$$L = \begin{bmatrix} 0.7444 & -33.1665 & 1.4934 & -5.7213 \end{bmatrix} \tag{9.15}$$

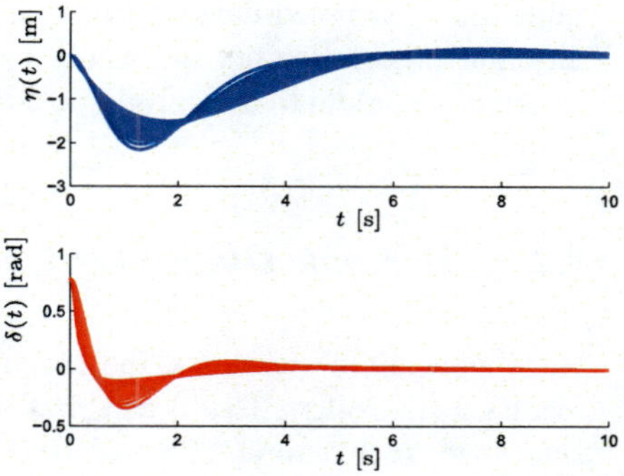

Fig. 9.2 Inverted pendulum closed-loop trajectories - state feedback robust control

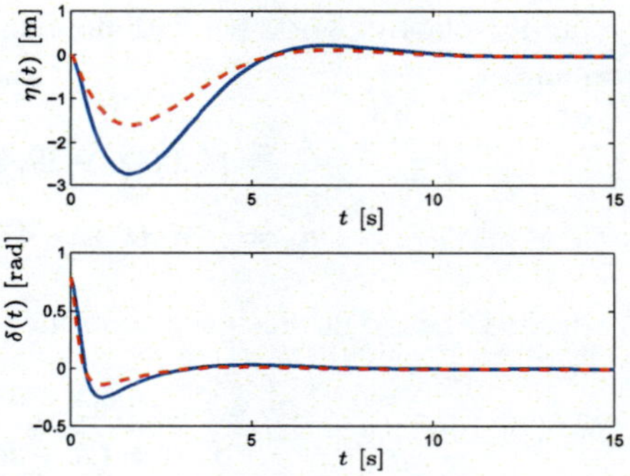

Fig. 9.3 Inverted pendulum closed-loop trajectories - linear and nonlinear models

associated with the minimum guaranteed $\mathcal{H}_2$ performance index $\varrho_{2rob}^2 = 0.4878$. Comparing both indices we see an expressive loss of performance due to uncertainty, whose amplitude is about 50% around the nominal value of the cart mass.

Figure 9.2 shows the time simulation of the closed-loop system. The movement starts at $t = 0$ from the rest, with the cart placed at $\eta(0) = 0$ and the pendulum at $\delta(0) = \pi/4$ [rad]. The trajectories on the top represent the displacement of the cart and the ones on the bottom show the angular displacement of the pendulum. Observe that, initially, the cart moves away from the original position and then comes back to it. By its turn, the pendulum moves toward the upright position but, due to inertia, passes it and then comes back as desired. Different trajectories are due to different values of the cart mass, with $1/M_p$ belonging to the uncertainty set.

Let us now move our attention to a situation that often occurs in practice and is related to the necessity of validating the control law, by controlling the true nonlinear model and not the liner approximation used to design it.

We have simulated this situation with the robust state feedback control gain (9.14), placed in the feedback loop of the nonlinear system, defined by Eqs. (9.4)–(9.5) starting from the same initial condition considered before. Figure 9.3 shows the displacement of the cart (on the top) and the angular displacement of the pendulum (on the bottom). In dashed lines, the responses of the linear model are plotted, while in solid lines are those of the nonlinear model, all of them controlled by the same robust sampled-data control. The initial condition $\delta(0) = \pi/4$ [rad] produces different trajectories but the whole picture is that the main goal of bringing the inverted pendulum to the upright position has been properly accomplished.

9.2.2 Dynamic Output Feedback Control

Now, we want to investigate the design of sampled-data full order dynamic output feedback controllers. To this end, first, we want to design the optimal controller with state space realization (4.106)–(4.108). We have applied the result of Theorem 4.7, more precisely, we have solved the convex programming problem (4.121) considering the plant (9.8)–(9.10) with nominal parameters. We have obtained the optimal gains

$$\hat{C}_c = \begin{bmatrix} 2.0000 & -39.4812 & 3.1589 & -6.8138 \end{bmatrix} \tag{9.16}$$

$$\hat{B}_d' = \begin{bmatrix} -0.6001 & -4.2942 & -1.8231 & -25.4784 \end{bmatrix} \tag{9.17}$$

needed to build the designed controller of the aforementioned class, with state space realization

$$\dot{\hat{x}}(t) = \left(A + B\hat{C}_c\right)\hat{x}(t), \ \hat{x}(0) = 0 \tag{9.18}$$

$$\hat{x}(t_k) = \left(I - \hat{B}_d C_y\right)\hat{x}(t_k^-) + \hat{B}_d y[k] \tag{9.19}$$

$$v(t) = \hat{C}_c\hat{x}(t) \tag{9.20}$$

which imposes to the closed-loop system the minimum $\mathcal{H}_2$ performance cost $\varrho_2^2 = 515.7093$. It is important to remember that the control signal is corrupted by the continuous-time perturbation, implying that $u(t) = v(t) + E_u w_c(t)$ and $E_y \neq 0$. This controller acts, indistinctly, during all $t \in \mathbb{R}_+$, including inter-sampling and sampling times $\{t_k\}_{k\in\mathbb{N}}$.

Roughly speaking, since they are not strictly comparable due to the fact that the output feedback controller acts during the whole time interval and not only at the sampling instants of time as the state feedback controller does, the $\mathcal{H}_2$ performance has been deteriorated significantly. This is a consequence of the lack of information available to the controller, because only the horizontal displacement of the pendulum and not the velocities of the cart and the pendulum, is measured. Figure 9.4 confirms this claim. Starting from the same initial condition the cart moves, and by consequence, the pendulum swings much more prior to reach the desired equilibrium point. However, we must stress that the stabilization time is similar and asymptotic stability of the closed-loop system is preserved.

Second, for comparison purposes, we have determined the pure discrete-time optimal full order dynamic output feedback controller with state space realization

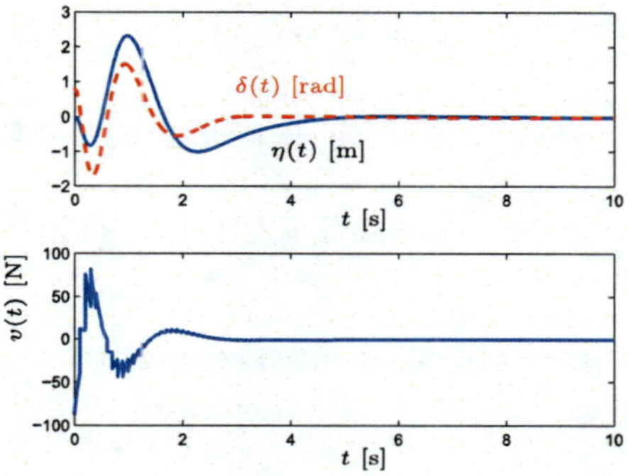

Fig. 9.4 Inverted pendulum closed-loop trajectories - dynamic output feedback control

(3.60)–(3.61). The solution of convex programming problem (3.73), stated in Theorem 3.1, yields its transfer function

$$\mathcal{C}_h(\zeta) = \frac{209.3\zeta^4 - 533.7\zeta^3 + 439.6\zeta^2 - 115.1\zeta - 0.1312}{\zeta^4 - 1.378\zeta^3 + 1.322\zeta^2 - 0.4085\zeta + 0.05689} \tag{9.21}$$

and the associated $\mathcal{H}_2$ performance index $\varrho_2^2 = 335.5946$. Notice that, in this case, the control signal $u[k] = v[k] + E_u w_d[k]$ is corrupted by the discrete-time exogenous perturbation and $E_y \neq 0$. This is an important characteristic of this optimal design, being able to cope with exogenous perturbations in the measurement channel $E_y \neq 0$.

Of course, as we did before, these two controllers of different classes can be further validated against parameter uncertainty and nonlinear modeling. Norm bounded uncertainty can be addressed by the $\mathcal{H}_\infty$ theory studied in Chap. 5 and applied in many instances of this book, in particular, in Chap. 8. The effect of the true nonlinear model can be evaluated by simulation, as we have just done before in this section. The interested reader can pursue working on this task.

9.2.3 Model Predictive Control

Let us now apply the MPC strategy to the inverted pendulum free of continuous and discrete-time exogenous perturbations. Each update at time instants $\{t_n = n(mh)\}_{n\in\mathbb{N}}$, where m is an integer greater or equal to one, produces the MPC action given by

$$u(t) = u[k] = \varphi_{(n)}^*(x_{k|k}), \ \forall t \in [t_k, t_{k+1}), \ k \in \mathbb{N} \tag{9.22}$$

where the function $\varphi_{(n)}^*(x)$, constructed with the matrix blocks of $P_{0(n)} = Q_{0(n)}^{-1}$ as indicated in (8.35), remains unchanged between successive measurement data update and at t_n follows from

$$Q_{0(n)} = \arg \inf_{Q(\cdot),\theta} \left\{ x'_{n|n}(I'_x Q_0 I_x)^{-1} x_{n|n} : (9.24)\text{–}(9.27) \right\} \tag{9.23}$$

where to improve readability the DLMIs

$$\begin{bmatrix} -\dot{Q}(t) + Q(t)F'_i + F_i Q(t) & Q(t)G' \\ \bullet & -I \end{bmatrix} < 0, \ t \in [0, h) \tag{9.24}$$

for $i \in \mathbb{K}$ and the LMI constraints

$$Q_0 > 0, \ I'_x(Q_0 - Q_h)I_x > 0 \tag{9.25}$$

$$\theta I - D_\psi Q_0 D'_\psi > 0 \tag{9.26}$$

$$I'_x Q_0 I_x - \theta x_{n|n} x'_{n|n} > 0 \tag{9.27}$$

are once again made explicit here. The MPC strategy represented by this convex programming problem has been developed in Chap. 8 where its properties and behavior have been established. Our purpose is to apply it to the inverted pendulum free of parameter uncertainty (nominal system with $N = 1$) and to the same model subject to convex bounded uncertainty.

In a first experiment, we have imposed a constraint to the control, by setting $C_s = [0\ 0\ 0\ 0]$ and $D_s = [1]$. We have simulated a time horizon of 10 [s], whose closed-loop system trajectories $\delta(t)$ and $u(t)$ are shown in Fig. 9.5. They should be understood as follows:

- Measurement update $m = 100$. Since $h = 100$ [ms] this means that the MPC function (9.22) was determined only once. It has been calculated with $\bar{s} = 150$ [N] (very large) which yielded the optimal unconstrained controller (solid lines without oscillation). With $\bar{s} = 15$ [N], it is seen that the optimal unconstrained controller is not feasible. Hence, imposing this constraint, we have determined the MPC function (9.22), which provided the oscillatory trajectories represented in solid lines.

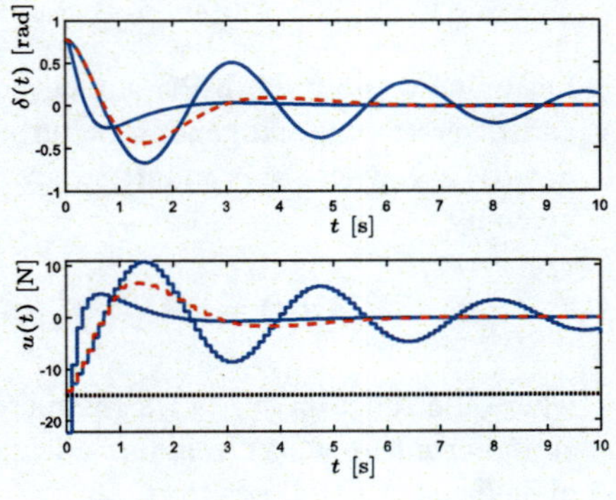

Fig. 9.5 Inverted pendulum closed-loop trajectories - nominal system model predictive control

Table 9.1 True $\mathcal{H}_2$ cost produced by the MPC strategy

m	100	50	25	10	5
Cost	88.9846	80.9361	62.3490	50.4069	54.8586

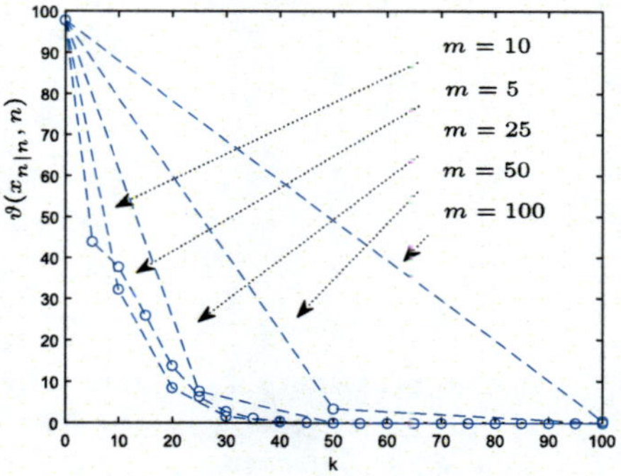

Fig. 9.6 Objective function of the MPC

- Measurement update $m = 5$. The MPC function (9.22), under the same control constraint, was determined 20 times to cope with new measurements. The trajectories, drawn in dashed lines, are feasible but not oscillatory. This behavior is due to the successful action of the MPC strategy.

Moreover, Table 9.1 contains the true values of the total cost, namely

$$||z||_2^2 = \sum_{n \in \mathbb{N}} \left\{ \int_{t_n}^{t_{n+1}} z(t)'z(t)dt : x(t_n) = x_{n|n} \right\} \tag{9.28}$$

which depends on the function $\varphi^*_{(n)}(\cdot)$ calculated at the time instants t_n whenever a new measurement is available and valid in the time interval $[t_n, t_{n+1})$. As we have already mentioned, in general, this sequence is non-increasing but this behavior is not mandatory. Let us investigate this point with more details:

- For a certain system, if you do not take into account the constraint $s[k] \in \mathbb{S}, \forall k \in \mathbb{N}$ by setting matrix $R_s > 0$ small enough, see (8.51)–(8.52), then the MPC strategy is useless, because $\varphi^*_{(n)}(\cdot)$ is always the same linear function, identical to the optimal one.
- For a certain system, if you take into account the constraint $s[k] \in \mathbb{S}$, then the MPC strategy is an effective way to calculate the control action corresponding to the measured state $x_{n|n}$, that minimizes the guaranteed cost (not necessarily the true one) and preserves stability. Hence, it may occur what can be observed in the last two entries of Table 9.1.
- Notice that, as predicted by Corollary 8.3, and as it is shown in Fig. 9.6, the sequence composed by

$$\vartheta(x_{n|n}, n) = x'_{n|n}(I'_x Q_{0(n)} I_x)^{-1} x_{n|n} \tag{9.29}$$

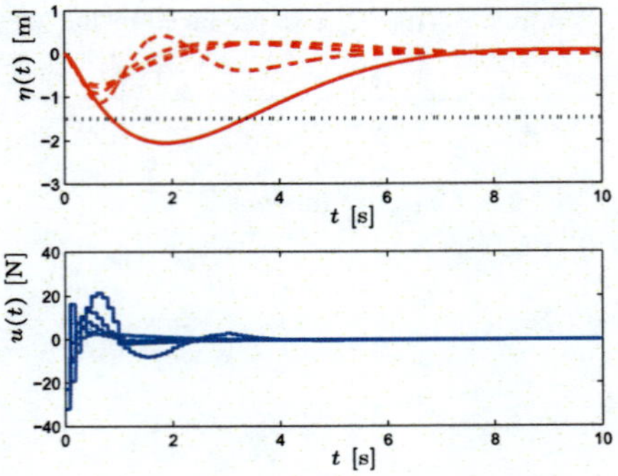

Fig. 9.7 Inverted pendulum closed-loop trajectories - uncertain system model predictive control

is non-increasing for all values of $m \in \{100, 50, 25, 10, 5\}$. This was expected, since $\vartheta(x, n)$ is a Lyapunov function associated with the closed-loop system controlled by the MPC strategy.

For uncertain systems, the control function follows from the minimization of a guaranteed cost, implying that the first item is not verified, in general. Finally, it is worth noticing that in all cases, that is, for any $m \geq 1$, the true value of the norm in (9.28) satisfies $||z||_2^2 \leq \vartheta(x_0, 0)$, being, as the MPC of the inverted pendulum indicates, much smaller than that upper bound.

For completeness, we have controlled the inverted pendulum, assuming that the mass of the cart is uncertain and belongs to the convex bounded set $1/M_p \in [0.5, 1.0]$ and $C_s = [1\ 0\ 0\ 0]$, $D_s = [0]$, which means that we want to limit the movement of the cart.

Figure 9.7 shows the simulation performed with the MPC strategy. On the top, it is shown in solid line (red) the car trajectory with $M_p = 1.33$ [kg] with unconstrained control. Notice that, in this case, the displacement of the car is bigger than 1.5 [m]. We have included the state constraint $|x_1[k]| \leq 1.5, \forall k \in \mathbb{N}$ and now all trajectories of the closed-loop system become feasible. On the bottom of Fig. 9.7, the corresponding robust MPC actions are shown. Once again, it is clear that the MPC strategy is efficient to deal with a 4th order system with practical appeal.

9.3 An Economic System

This a simplified and highly aggregated dynamical system obtained from the multiplier-accelerator model of Samuelson. In general lines, it provides the time evolution of *national income*, depending on the *governmental expenditure*, yielding the *consumption expenditure* and the *private investment*. It is assumed that the economy can operate in three different modes, namely *normal*, *boom*, and *slump*, respectively. The reader is requested to see the Bibliography notes of this chapter for more details, including the references we are based on.

The relevant aspect of this model is the fact that the transition among the $N = 3$ modes $\theta_i(t) \in \mathbb{K}$ is described by a Markov chain with transition rate matrix $\Lambda \in \mathbb{R}^{N\times N}$. It has been verified that the waiting time in each mode was exponentially distributed. Following the material included in Chap. 6, the model under consideration has the form

$$\dot{x}(t) = A_{\theta(t)}x(t) + B_{\theta(t)}u(t) + Ew_c(t) \tag{9.30}$$

$$y[k] = C_{y\theta[k]}x[k] + E_{y\theta[k]}w_d[k] \tag{9.31}$$

$$z(t) = C_z x(t) + D_z u(t) \tag{9.32}$$

where the control structure, state or dynamic output feedback, and the exogenous perturbation gains are specified in each case, in order to tackle several aspects involving stability and robustness. In order to reproduce the results already available in the literature, we adopt the following perturbation structure:

- **Control channel perturbation:** It is considered that the control action is corrupted by a continuous-time exogenous perturbation w_c with independent unitary gains. This imposes the choice $E = I$.
- **Measurement channel perturbation:** It is assumed that the measurement action is corrupted by a discrete-time exogenous perturbation w_d with gain $\alpha_d \in (0, 1)$. The third state variable is measured, leading to $C_y = [0\ 0\ 1]$ and $E_y = \alpha_d$.

Moreover, it is assumed that $E_u = 0$ and all parameters are precisely known, meaning that the study of robustness is not among our goals. In this framework, the involved matrix data are given in the next remark.

Remark 9.2 The numerical data are those provided in the references quoted in the Bibliography notes of this chapter. They are

$$A_1 = \begin{bmatrix} 0 & 0 & 0 \\ 0 & -0.545 & 0.626 \\ 0 & -1.570 & 1.465 \end{bmatrix}, \quad B_1 = \begin{bmatrix} 0 \\ -0.283 \\ 0.333 \end{bmatrix}$$

$$A_2 = \begin{bmatrix} 0 & 0 & 0 \\ 0 & -0.106 & 0.087 \\ 0 & -3.810 & 3.861 \end{bmatrix}, \quad B_2 = \begin{bmatrix} 0 \\ 0 \\ 0.087 \end{bmatrix}$$

$$A_3 = \begin{bmatrix} 1.80 & -0.3925 & 4.52 \\ 3.14 & 0.100 & -0.28 \\ -19.06 & -0.148 & 1.56 \end{bmatrix}, \quad B_3 = \begin{bmatrix} -0.064 \\ 0.195 \\ -0.080 \end{bmatrix}$$

We consider $\alpha_d = 0.20$, the initial condition $x_0 = [0\ 0\ 1]'$ and equal probabilities $\pi_{01} = \pi_{02} = \pi_{03} = 1/3$. In some instances, we also consider $\pi_{0i} = 1$, and $\pi_{0j} = 0$ for all $i \neq j \in \mathbb{K}$, in order to select a deterministic value for the initial mode $\theta_0 = i \in \mathbb{K}$. Finally, the transition rate matrix is given by

$$\Lambda = \begin{bmatrix} -0.53 & 0.32 & 0.21 \\ 0.50 & -0.88 & 0.38 \\ 0.40 & 0.13 & -0.53 \end{bmatrix}$$

We have adopted the sampling period $h = 1/12$ [year], meaning that the state vector is measured only once each month. □

In the sequel, we solve and discuss several control design conditions provided in Chap. 6. They consider, as usual, that the Markov mode is measured but, for completeness, the possibility of implementing a sampled-data control with no access to this information is briefly discussed. As before, our goals are restricted to the $\mathcal{H}_2$ performance index optimization. The control design conditions are solved by the piecewise linear approximation procedure with a smaller number of time subintervals $n_\phi = 16$, when compared to that usually adopted.

9.3.1 State Feedback Control

First, let us focus on the design of a state feedback controller of the usual form $u(t) = u[k] = L_{\theta[k]}x[k]$, for $t \in [t_k, t_{k+1})$, $\forall k \in \mathbb{N}$, which needs the measurement of the Markov mode $\theta[k] \in \mathbb{K}$, $k \in \mathbb{N}$ to be implemented. The following matrices of compatible dimensions are those necessary to define the controlled output

$$C_z = \begin{bmatrix} I \\ 0 \end{bmatrix}, \; D_z = \begin{bmatrix} 0 \\ 1 \end{bmatrix} \tag{9.33}$$

and recall that $E_u = 0$ indicates that the system is not subject to discrete-time exogenous perturbation. With the convex programming problem (6.41) provided in Theorem 6.3, the following experiments have been made:

- Setting $h = 10^{-4}$ very small, we have obtained

$$\begin{bmatrix} L_1 \\ L_2 \\ L_3 \end{bmatrix} = \begin{bmatrix} 2.0350 & 14.5097 & -23.5826 \\ 1.0170 & 73.0769 & -78.7410 \\ 93.6472 & -11.4913 & 11.6906 \end{bmatrix} \tag{9.34}$$

  with the associated $\mathcal{H}_2$ performance index $\varrho_2^2 = 2.3567 \times 10^3$, which is very close to the solution available in the literature, provided by the optimal continuous-time controller.
- Setting $h = 1/12$ [year], we have obtained

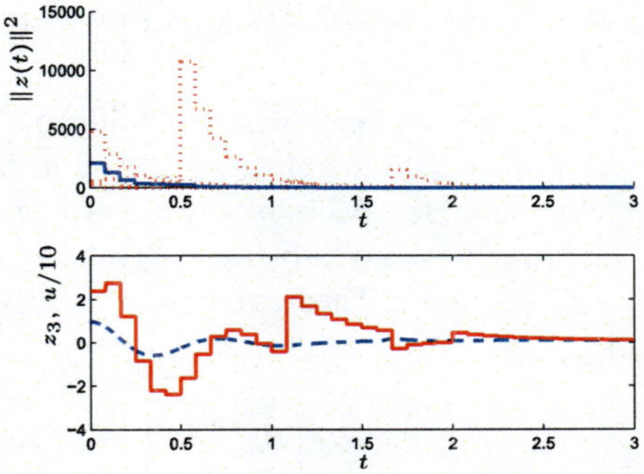

Fig. 9.8 Controller performance and a trajectory sample - state feedback control

$$\begin{bmatrix} L_1 \\ L_2 \\ L_3 \end{bmatrix} = \begin{bmatrix} -0.0420 & 11.0640 & -17.2424 \\ -2.3074 & 63.7051 & -69.1828 \\ 58.2449 & -10.0479 & 23.9421 \end{bmatrix} \tag{9.35}$$

with the associated $\mathcal{H}_2$ performance index $\varrho_2^2 = 4.0033 \times 10^3$. This is close, but not identical to the solution available in the literature due to the use of n_ϕ not big enough, keeping the computational burden reasonable.

- We have tried to solve a modified version of Theorem 6.2 with the additional constraints $W_i = W$ and $K_i = K$ for all $i \in \mathbb{K}$, in order to calculate the state feedback gains $L_i = L = KW^{-1}$, $i \in \mathbb{K}$ independent of the Markov modes. Unfortunately, doing this, problem (6.38) became unfeasible.

These calculations put in evidence that there is no theoretical difficulty to design the optimal state feedback controller, since it follows from the solution of a convex programming problem. With some conservatism, it is also possible to cope with parameter uncertainty (also in the transition rate matrix) and independence of the Markov modes. Some difficulty appears because the computational effort involved may be expressive. In this sense, it is important to have the possibility to work with n_ϕ small as the second experiment illustrates.

With the state feedback control of the second case, we have implemented a Monte Carlo simulation with 100 runs and the closed-loop system always evolving from the initial condition $x_0 = [0\ 0\ 1]'$ and $\theta_0 \in \mathbb{K}$ with equal probability. On the top of Fig. 9.8 it is shown, against time, the value of $||z(t)||^2$ in dotted lines and the value of its mean in solid line, over all runs. Observe that the existence of events with large norms, caused by the control effort, are not sufficient to move the mean away from zero as time passes, but it defines the shape of the trajectories. On the bottom of the same figure, a sample of the controlled output trajectories $z_3(t)$ in dashed line, and $u(t)/10$ in solid line are given. The magnitude of the control signal has been divided by a factor of 10, in order to draw it in the same figure with the same scale. Since each Markov mode matrix $A_i, i \in \mathbb{K}$ is unstable, the effectiveness of the sampled-data state feedback control is apparent.

9.3.2 Dynamic Output Feedback Control

We now investigate the possibility of implementing a dynamic output feedback controller with the clear advantage to measure online only one component of the state variables. Moreover, we consider a more realistic experiment, by including measurement perturbations. We have solved the convex programming problem (6.77) stated in Theorem 6.6 that yielded the constant gains

$$\left[\hat{B}_{d1}\ \hat{B}_{d2}\ \hat{B}_{d3} \right] = \begin{bmatrix} -0.0118 & 0.0084 & -0.4731 \\ -0.8368 & -0.8243 & 0.0174 \\ 0.9712 & 0.9772 & 0.9959 \end{bmatrix} \tag{9.36}$$

$$\begin{bmatrix} \hat{C}_{c1} \\ \hat{C}_{c2} \\ \hat{C}_{c3} \end{bmatrix} = \begin{bmatrix} 2.0362 & 14.5178 & -23.5921 \\ 1.0198 & 73.0958 & -78.7596 \\ 93.6885 & -11.4911 & 11.6749 \end{bmatrix} \tag{9.37}$$

and the associated optimal $\mathcal{H}_2$ performance index $\varrho_2^2 = 4.0980 \times 10^3$. The good surprise comes to light if we compare this performance with that corresponding to the state feedback controller (9.35). They are very close! This fact can be interpreted by noticing that the lack of information provided by the output feedback controller (only the third state variable is measured) is, in this case, compensated by the action of this controller in all $t \in \mathbb{R}_+$ and not only at the sampling instants t_k, $k \in \mathbb{N}$ as the state feedback controller does. The optimal controller has been implemented by the following state space realization

$$\dot{\hat{x}}(t) = (A_{\theta(t)} + B_{\theta(t)}\hat{C}_{c\theta(t)})\hat{x}(t), \ \hat{x}(0) = 0 \tag{9.38}$$

$$\hat{x}(t_k) = (I - \hat{B}_{d\theta[k]}C_{y\theta[k]})\hat{x}(t_k^-) + \hat{B}_{d\theta[k]}y[k] \tag{9.39}$$

$$u(t) = \hat{C}_{c\theta(t)}\hat{x}(t) \tag{9.40}$$

in order to run 100 Monte Carlo simulations with the closed-loop system evolving from the initial condition $x_0 = [0\ 0\ 1]'$. On the top of Fig. 9.9, it can be seen the square norm $||z(t)||^2$ in dotted lines and its mean in solid line. Once again, the occurrence of large control effort is clearly observed and confirmed with the trajectories drawn on the bottom of the same figure. Indeed, one can see the trajectories $z_3(t)$ in dashed line and the normalized control $u(t)/10$. It is apparent that the designed controller works very well and is simple to be implemented given its time-invariant nature.

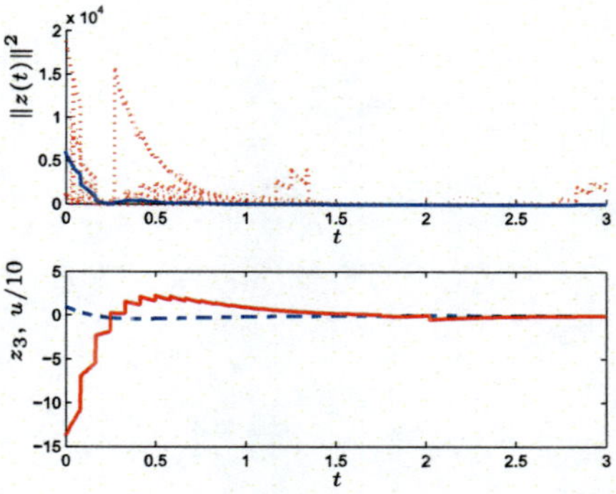

Fig. 9.9 Controller performance and a trajectory sample - dynamic output feedback control

9.4 Bibliography Notes

There are many references where the reader can find the model of the inverted pendulum with concentrated or distributed mass and viscous friction. However, for completeness, we have included the model determination from Newton's law which, as the reader can see, is a simple task, mainly because the mass of the pendulum is concentrated and the couple cart-pendulum moves in a frictionless environment. The control design experiments involving sampled-data state and dynamic output feedback controllers are not available in the literature. The same is true for the MPC strategy, where some interesting theoretical points have been raised and clarified.

The economic system based on the multiplier-accelerator model of Samuelson, within the context of continuous-time Markov jump linear systems, has its origin in the reference [9], where the reader can find an interesting and useful discussion about this model, including the way the system data has been obtained. Concerning control design, the references [9], and [15] treat the case of state feedback only. The second one solves the associated optimal $\mathcal{H}_2$ performance design by Riccati equation solvers. The case of sampled-data state feedback control design has been tackled in [23] where the solution of [15] has been determined by setting the sampling period very small. The same validation has been adopted in this chapter. This numerical experiment ends with the design of a sampled-data dynamic output feedback time-invariant controller, which works very well and is simple to implement in practice. A similar solution is not available in the literature.

These numerical experiments put in evidence the need of more research effort toward the development of new algorithms more amenable and efficient to solve DLMIs. For the moment, the possibility to apply the linear piecewise procedure with a small number of subintervals is welcome in the sense of computational effort reduction.

References

1. L.I. Allerhand, U. Shaked, Robust control of linear systems via switching. IEEE Trans. Autom. Control **58**, 506–512 (2013)
2. B. Altın, P. Ojaghi, R.G. Sanfelice, A model predictive control framework for hybrid dynamical system, *Proceedings of the 6th IFAC Conference on Nonlinear Model Predictive Control* (2018), pp. 128–133
3. M. Ariola, G. De Tommasi, A. Mele, G. Tartaglione, On the numerical solution of differential linear matrix inequalities. J. Optim. Theory Appl. **185**, 540–553 (2020)
4. B. Bamieh, J.B. Pearson, The H_2 problem for sampled-data systems. Syst. Control Lett. **19**, 1–12 (1992)
5. B.R. Barmish, Necessary and sufficient conditions for quadratic stabilizability of an uncertain system. J. Optim. Theory Appl. **46**, 399–408 (1985)
6. R. Bellman, *Introduction to Matrix Analysis*, 2nd edn. (McGraw-Hill, New York, 1970)
7. A. Bemporad, M. Morari, V. Dua, E.N. Pistikopoulos, The explicit linear quadratic regulator for constrained systems. Automatica **38**, 3–20 (2002)
8. A. Bemporad, F. Borrelli, M. Morari, Model predictive control based on linear programming – the explicit solution. IEEE Trans. Autom. Control **47**, 1974–1985 (2002)
9. W.P. Blair, D.D. Sworder, Continuous-time regulation of a class of econometric models. IEEE Trans. Syst. Man Cybern. **5**, 341–346 (1975)
10. S. Boyd, L. El Ghaoui, E. Feron, V. Balakrishnan, *Linear Matrix Inequalities in Systems and Control Theory* (SIAM Studies in Applied Mathematics, Philadelphia, 1994)
11. C. Briat, Convex conditions for robust stability analysis and stabilization of linear aperiodic impulsive and sampled-data systems under dwell-time constraints. Automatica **49**, 3449–3457 (2013)
12. T. Chen, A simple derivation of the H_2-optimal sampled-data controllers. Syst. Control Lett. **20**, 49–56 (1993)
13. T. Chen, B.A. Francis, *Optimal Sampled-Data Control Systems* (Springer, London, 1995)
14. P. Colaneri, J.C. Geromel, A. Locatelli, *Control Theory and Design: An $\mathcal{H}_2/\mathcal{H}_\infty$ Viewpoint* (Academic Press, London, 1997)
15. O.L.V. Costa, M.D. Fragoso, M.G. Todorov, *Continuous-Time Markov Jump Linear Systems* (Springer, Berlin, 2012)
16. G.S. Deaecto, P. Bolzern, L. Galbusera, J.C. Geromel, $\mathcal{H}_2$ and $\mathcal{H}_\infty$ control of time-varying delay switched linear systems with application to sampled-data control. Nonlinear Anal. Hybrid Syst **22**, 43–54 (2016)
17. D.P. de Farias, J.C. Geromel, J.B.R. do Val, O.L.V. Costa, Output feedback control of Markov jump linear systems in continuous-time. IEEE Trans. Autom. Control **45**, 944–949 (2000)

J. C. Geromel, *Differential Linear Matrix Inequalities*,
https://doi.org/10.1007/978-3-031-29754-0

18. G.W. Gabriel, Optimal Sampled-Data State Feedback Control Applied to Markov Jump Linear Systems, PhD thesis. University of Campinas (2016)
19. G.W. Gabriel, J.C. Geromel, Sampled-data control of Lur'e systems. Nonlinear Anal. Hybrid Syst **40** (2021). https://doi.org/10.1016/j.nahs.2020.100994
20. G.W. Gabriel, T.R. Gonçalves, J.C. Geromel, Optimal and robust sampled-data control of Markov jump linear systems: a differential LMI approach. IEEE Trans. Autom. Control **63**, 3054–3060 (2018)
21. J.C. Geromel, Optimal linear filtering under parameter uncertainty. IEEE Trans. Signal Process. **47**, 168–175 (1999)
22. J.C. Geromel, Sampled-data model predictive control. IEEE Trans. Autom. Control **67**, 2466–2472 (2021)
23. J.C Geromel, G.W. Gabriel, Optimal $\mathcal{R}_2$ state feedback sampled-data control design of Markov jump linear systems. Automatica **54**, 182–188 (2015)
24. J.C. Geromel, P.L.D. Peres, J. Bernussou, On a convex parameter space method for linear control design of uncertain systems. SIAM J. Control Optim. **29**, 381–402 (1991)
25. J.C. Geromel, M.C. de Oliveira, J. Bernussou, Robust filtering of discrete-time linear systems with parameter dependent Lyapunov functions. SIAM J. Control Optim. **41**, 700–711 (2002)
26. J.C. Geromel, M. Souza, On an LMI approach to optimal sampled-data state feedback control design. Int. J. Control **88**, 2369–2379 (2015)
27. J.C. Geromel, P. Colaneri, P. Bolzern, Differential linear matrix inequality in optimal sampled-data control. Automatica **100**, 289–298 (2019)
28. T.R. Gonçalves, G.W. Gabriel, J.C. Geromel, Differential linear matrix inequalities optimization. IEEE Control Syst. Lett. **3**, 380–385 (2019)
29. T. Hagiwara, G. Kuroda, M. Araki, Popov-type criterion for stability of nonlinear sampled-data systems. Automatica **34**, 671–682 (1998)
30. T. Hagiwara, Nyquist stability criterion and positive-realness of sampled-data systems. Syst. Control Lett. **45**, 283–291 (2002)
31. T. Holicki, A Complete Analysis and Design Framework for Linear Impulsive and Related Hybrid Systems, PhD thesis. Department of Mathematics, University of Stuttgart (2021)
32. R.A. Horn, C.R. Johnson, *Matrix Analysis* (Cambridge University Press, Cambridge, 1985)
33. A. Ichicawa, H. Katayama, *Linear Time Varying Systems and Sampled-Data Systems* (Springer, London, 2001)
34. P. Kabamba, Control of linear systems using generalized sampled-data hold functions. IEEE Trans. Autom. Control **32**, 772–783 (1987)
35. H.K. Khalil, *Nonlinear Systems*, 3rd edn. (Prentice Hall, Upper Saddle River, 2002)
36. P.P. Khargonekar, N. Sivashankar, H_2 optimal control for sampled-data systems. Syst. Control Lett. **17**, 425–436 (1991)
37. J.H. Kim, T. Hagiwara, The generalized H_2 controller synthesis problem of sampled-data systems. Automatica **142** (2022). https://doi.org/10.1016/j.automatica.2022.110400
38. M.V. Kothare, V. Balakrishnan, M. Morari, Robust constrained model predictive control using linear matrix inequalities. Automatica **32**, 1361–1369 (1996)
39. B.P. Lathi, *Signal Processing & Linear Systems* (Berkeley-Cambridge Press, Carmichael, 1998)
40. A. Leon-Garcia, *Probability, Statistics, and Random Processes for Electrical Engineering*, 3rd edn. (Pearson Prentice Hall, Upper Saddle River, 2007)
41. A.H. Levis, R.A. Schluete, M. Athans, Behaviour of optimal linear sampled-data regulators. Int. J. Control **13**, 343–361 (1971)
42. D.Q. Mayne, J.B. Rawlings, C.V. Rao, P.O.M. Scokaert, Constrained model predictive control: stability and optimality. Automatica **36**, 789–814 (2000)
43. M. Morari, J.H. Lee, Model predictive control: past, present and future. Comput. Chem. Eng. **23**, 667–682 (1999)
44. X. Ping, J. Hu, T. Lin, B. Ding, P. Wang, Z. Li, A survey of output feedback robust MPC for linear parameter varying systems. IEEE/CAA J. Autom. Sin. **9**, 1717–1751 (2022)
45. A. Rantzer, On the Kalman-Yakubovich-Popov Lemma. Syst. Control Lett. **28**, 7–10 (1996)

46. C.W. Scherer, Mixed $\mathcal{H}_2/\mathcal{H}_\infty$ control, in *Trends in Control – A European Perspective*, ed. by A. Isidori (Springer, London, 1995)
47. P.O.M. Scokaert, J.B. Rawlings, Constrained linear quadratic regulation. IEEE Trans. Autom. Control **43**, 1163–1169 (1998)
48. I. Smears, Hamilton-Jacobi-Bellman equations analysis and numerical analysis. Durham University (2022). Available online: http://fourier.dur.ac.uk/Ug/projects/highlights/PR4/Smears_HJB_report.pdf
49. W. Sun, K.M. Nagpal, P.P. Khargonekar, $\mathcal{H}_\infty$ control and filtering for sampled-data systems. IEEE Trans. Autom. Control **38**, 1162–1175 (1993)
50. M. Sznaier, M.J. Damborg, Suboptimal control of linear systems with state and control inequality constraints, in *Proceedings of the 26th IEEE Conference on Decision and Control* (1987), pp. 761–762
51. M. Unser, Sampling – 50 years after Shannon. Proc. IEEE **88**, 569–587 (2000)
52. M. Vidyasagar, *Nonlinear Systems Analysis* (SIAM Studies in Applied Mathematics, Philadelphia, 2002)
53. W.M. Wonham, On a matrix Riccati equation of stochastic control. SIAM J. Control **6**, 681–697 (1968)

Index

J. C. Geromel, *Differential Linear Matrix Inequalities*,
https://doi.org/10.1007/978-3-031-29754-0